Properties of Impurity States in Superlattice Semiconductors

NATO ASI Series

Advanced Science Institutes Series

A series presenting the results of activities sponsored by the NATO Science Committee, which aims at the dissemination of advanced scientific and technological knowledge, with a view to strengthening links between scientific communities.

The series is published by an international board of publishers in conjunction with the NATO Scientific Affairs Division

A	**Life Sciences**	Plenum Publishing Corporation
B	**Physics**	New York and London
C	**Mathematical and Physical Sciences**	Kluwer Academic Publishers Dordrecht, Boston, and London
D	**Behavioral and Social Sciences**	
E	**Applied Sciences**	
F	**Computer and Systems Sciences**	Springer-Verlag
G	**Ecological Sciences**	Berlin, Heidelberg, New York, London,
H	**Cell Biology**	Paris, and Tokyo

Recent Volumes in this Series

Volume 177—Instabilities and Chaos in Quantum Optics II
edited by N. B. Abraham, F. T. Arecchi, and L. A. Lugiato

Volume 178—High-Brightness Accelerators
edited by Anthony K. Hyder, M. Franklin Rose, and Arthur H. Guenther

Volume 179—Interfaces, Quantum Wells, and Superlattices
edited by C. Richard Leavens and Roger Taylor

Volume 180—The Physics of Submicron Semiconductor Devices
edited by Harold L. Grubin, David K. Ferry, and C. Jacoboni

Volume 181—Fundamental Processes of Atomic Dynamics
edited by J. S. Briggs, H. Kleinpoppen, and H. O. Lutz

Volume 182—Physics, Fabrication, and Applications of Multilayered Structures
edited by P. Dhez and C. Weisbuch

Volume 183—Properties of Impurity States in Superlattice Semiconductors
edited by C. Y. Fong, Inder P. Batra, and S. Ciraci

Volume 184—Narrow-Band Phenomena—Influence of Electrons with Both Band and Localized Character
edited by J. C. Fuggle, G. A. Sawatzky, and J. W. Allen

Series B: Physics

Properties of Impurity States in Superlattice Semiconductors

Edited by

C. Y. Fong

University of California, Davis
Davis, California, USA

Inder P. Batra

IBM Almaden Research Center
San Jose, California, USA

and

S. Ciraci

Bikent University
Ankara, Turkey

Plenum Press
New York and London
Published in cooperation with NATO Scientific Affairs Division

Proceedings of a NATO Advanced Research Workshop on
the Properties of Impurity States in Superlattice Semiconductors,
held September 7–11, 1987,
at the University of Essex, Colchester, United Kingdom

Library of Congress Cataloging in Publication Data

NATO Advanced Research Workshop on the Properties of Impurity States in Superlattice Semiconductors (1987: University of Essex)
Properties of impurity states in superlattice semiconductors.

(NATO ASI series. Series B, Physics; v. 183)
"Proceedings of a NATO Advanced Research Workshop on the Properties of Impurity States in Superlattice Semiconductors, held September 7–11, 1987, at the University of Essex, Colchester, United Kingdom"—Verso t.p.
"Published in cooperation with NATO Scientific Affairs Division."
Includes bibliographical references and index.
1. Doped semiconductor superlattices—Congresses. I. Fong, C. Y. (Ching-yao) II. Batra, Inder P. III. Ciraci, S. IV. North Atlantic Treaty Organization. Scientific Affairs Division. V. Title. VI. Series.
QC611.8.S86N37 1987 530.4′1 88-25348
ISBN 0-306-43009-6

A Division of Plenum Publishing Corporation
233 Spring Street, New York, N.Y. 10013

Printed in the United States of America

SPECIAL PROGRAM ON CONDENSED SYSTEMS OF LOW DIMENSIONALITY

This book contains the proceedings of a NATO Advanced Study Institute held within the program of activities of the NATO Special Program on Condensed Systems of Low Dimensionality, running from 1983 to 1988 as part of the activities of the NATO Science Committee.

Other books previously published as a result of the activities of the Special Program are:

Volume 148 INTERCALATION IN LAYERED MATERIALS
edited by M. S. Dresselhaus

Volume 152 OPTICAL PROPERTIES OF NARROW-GAP LOW-DIMENSIONAL STRUCTURES
edited by C. M. Sotomayor Torres, J. C. Portal, J. C. Maan, and R. A. Stradling

Volume 163 THIN FILM GROWTH TECHNIQUES FOR LOW-DIMENSIONAL STRUCTURES
edited by R. F. C. Farrow, S. S. P. Parkin, P. J. Dobson, J. H. Neave, and A. S. Arrott

Volume 168 ORGANIC AND INORGANIC LOW-DIMENSIONAL CRYSTALLINE MATERIALS
edited by Pierre Delhaes and Marc Drillon

Volume 172 CHEMICAL PHYSICS OF INTERCALATION
edited by A. P. Legrand and S. Flandrois

Volume 182 PHYSICS, FABRICATION, AND APPLICATIONS OF MULTILAYERED STRUCTURES
edited by P. Dhez and C. Weisbuch

FOREWORD

A NATO workshop on "The Properties of Impurity States in Semiconductor Superlattices" was held at the University of Essex, Colchester, United Kingdom, from September 7 to 11, 1987.

Doped semiconductor superlattices not only provide a unique opportunity for studying low dimensional electronic behavior, they can also be custom-designed to exhibit many other fascinating electronic properties. The possibility of using these materials for new and novel devices has further induced many astonishing advances, especially in recent years.

The purpose of this workshop was to review both advances in the state of the art and recent results in various areas of semiconductor superlattice research, including: (i) growth and characterization techniques, (ii) deep and shallow impurity states, (iii) quantum well states, and (iv) two-dimensional conduction and other novel electronic properties.

This volume consists of all the papers presented at the workshop. Chapters 1–6 are concerned with growth and characterization techniques for superlattice semiconductors. The question of δ-layer is also discussed in this section. Chapters 7–15 contain a discussion of various aspects of the impurity states. Chapters 16–22 are devoted to quantum well states. Finally, two-dimensional conduction and other electronic properties are described in chapters 23–26.

We would like to thank the IBM Almaden Research Center for their partial financial support of this workshop. We would also like to express our gratitude to Professors Robert N. Shelton (Chairman) and John A. Jungerman, both of the Physics Department of the University of California at Davis, for their contributions to the support of this workshop. Also, special thanks are due to Nilda Muniz for her typing and administrative assistance.

C.Y. Fong

Inder P. Batra

S. Ciraci

CONTENTS

I. GROWTH TECHNIQUES AND CHARACTERIZATIONS OF SUPERLATTICE SEMICONDUCTORS

Chapter 1 Doping in Two Dimensions: The δ-layer 1
A. Zrenner and F. Koch

Chapter 2 Optical Measurements of Acceptor Concentration Profiles at GaAs/GaAlAs Quantum Well Interfaces 11
M.H. Meynadier

Chapter 3 Molecular Beam Epitaxy of $Ga_{0.99}Be_{0.01}As$ for Very High Speed Heterojunction Bipolar Transistors 19
J.L. Liévin, F. Alexandre, and C. Dubon-Chevallier

Chapter 4 Progress Report on Molecular Beam Epitaxy of III-V Semiconductors – from Fibonacci to Monolayer Superlattices 29
L. Tapfer, J. Nagle, and K. Ploog

Chapter 5 Interface Characterization of GaInAs-InP Superlattices Grown by Low Pressure Metalorganic Chemical Vapor Deposition 43
Manijeh Razeghi, Phillippe Maurel, and Franck Omnes

Chapter 6 Structural and Chemical Characterization of Semiconductor Interfaces by High Resolution Transmission Electron Microscopy 63
A. Ourmazd

II. DEEP AND SHALLOW IMPURITY STATES

Chapter 7 Deep Level Behavior in Superlattice 77
Jacques C. Bourgoin and Michel Lannoo

Chapter 8 Role of the Si Donors in Quantum and Ultraquantum Transport Phenomena in GaAs-GaAlAs Heterojunctions 85
André Raymond

Chapter 9 Defects Characterization in GaAs-GaAlAs Superlattices 107
Dominique Vuillaume and Didier Stiévenard

Chapter 10 Studies of the DX Centre in Heavily Doped $n^{+}GaAs$ 121
L. Eaves, J.C. Portal, D.K. Maude, and T.J. Foster

Chapter 11 Shallow and Deep Impurity Investigations: the Important Step Towards a Microwave Field-Effect Transistor Working at Cryogenic Temperatures 135
W. Prost, W. Brockerhoff, M. Heuken, S. Kugler, K. Heime, W. Schlapp, and G. Weimann

Chapter 12 Electronic States in Heavily and Ordered Doped Superlattice Semiconductors 147
Inder P. Batra and C.Y. Fong

Chapter 13 Properties of Impurity States in n-i-p-i Superlattice Structures 159
Gottfried H. Döhler

Chapter 14 Deep Impurity Levels in Semiconductors, Semiconductor Alloys, and Superlattices 175
John D. Dow, Shang Yuan Ren, and Jun Shen

Chapter 15 "Pinning" of Transition-Metal Impurity Levels 189
J. Tersoff

III. QUANTUM WELL STATES

Chapter 16 Theory of Impurity States in Superlattice Semiconductors 195
G.P. Srivastava

Chapter 17 Effective-Mass Theory of Electronic States in Heterostructures and Quantum Wells 219
U. Rössler, F. Malcher, and A. Ziegler

Chapter 18 Hot Electron Capture in GaAs MQW: NDR and Photo-Emission 229
N. Balkan and B.K. Ridley

Chapter 19 In-Plane Electronic Excitations in GaAs/GaAlAs Modulation Doped Quantum Wells 245
N. Mestres, G. Fasol, and K. Ploog

Chapter 20 Resonant Tunneling in Double Barrier Heterostructures 255
Mark A. Reed

Chapter 21 Extrinsic Photoluminescence in Unintentionally and Magnesium Doped GaInAs/GaAs Strained Quantum Wells 271
A.P. Roth, R. Masut, D. Morris, and C. Lacelle

Chapter 22 Magneto-Optics of Excitons in GaAs-(GaAl)As Quantum Wells 285
W. Ossau, B. Jäkel, E. Bangert, and G. Weimann

IV. TWO DIMENSIONAL AND OTHER ELECTRONIC PROPERTIES

Chapter 23 The Influence of Impurities on the Shubnikov-De Haas and Hall Resistance of Two-Dimensional Electron Gases in $GaAs/Al_xGa_{1-x}As$ Heterostructures Investigated by Back-Gating and Persistent Photoconductivity 297
J. Wolter, F.A.P. Blom, P. Koenraad, P.F. Fontein, and G. Weimann

Chapter 24 Cyclotron Resonance of Polarons in Two Dimensions 307
J.T. Devreese, Xiaoguang Wu, and F.M. Peeters

Chapter 25 Structure and Electronic Properties of Strained Si/Ge Semiconductor Superlattices 319
S. Ciraci and Inder P. Batra

Chapter 26 Theory of Raman Scattering from Plasmons Polaritons in $GaAs/Al_xGa_{1-x}As$ Superlattices 333
M. Babiker, N.C. Constantinou, and M.G. Cottam

Index 347

I. GROWTH TECHNIQUES AND CHARACTERIZATIONS OF SUPERLATTICE SEMICONDUCTORS

DOPING IN TWO DIMENSIONS: THE δ-LAYER

A. Zrenner and F. Koch

Physik-Department, Technische Universität München
D-8046 Garching, Federal Republic of Germany

ABSTRACT

Dopant atoms incorporated as a sheet in a single atomic layer during an interruption of molecular-beam-epitaxy growth are shown to provide a novel 2-D system of carriers. We consider the electronic properties of such a δ(z)-function doping layer of Si in GaAs.

INTRODUCTION

In todays layer-by-layer growth schemes of semiconductors, dopant impurities don't just happen - they are caused, engineered to suit the purpose. Donors and acceptors can be positioned with atomic layer precision during the growth of the crystal. When the areal density N_D of a donor such as Si in GaAs exceeds $\sim 10^{11}$ cm^{-2}, the system forms a degenerate conductory layer of electrons in 2-dimensional subband states. We review here various physical properties of the 2-D electrons in the δ-doping layer as they have been discussed in the literature.

SAMPLE PREPARATION - THE MAKING OF A δ-LAYER

Sheet and planar doping has been attempted in various epitaxial growth modes, but the most promising effort to achieve the ultimate atomic layer confinement of Si donors in GaAs has been the technique of interrupted MBE-growth as introduced by K. Ploog /1/. The specified, desired number of Si atoms is deposited while growth is temporarily discontinued. In the procedure the growth temperature is maintained in the typical 500 - 600° C range. Growth of undoped material is continued thereafter to achieve the desired total thickness.

For MOCVD growth the strategy has been to make use of very low growth rates achieved in low-pressure reactors and to use a high concentration of the doping gas. We have recently worked with S-layers in InP /2/. An earlier sample with Se embedded in MOCVD GaAs has been employed in Ref./3/.

To incorporate donors such as Sb in electron-beam MBE growth of Si is a more difficult task because of the low coefficient of incorporation at the growth temperature. In reference /4/ the strategy that has been pursued is to first cool the substrate to room temperature and then deposit the donor atoms. The layer is then covered by 20 - 40 Å of amorphous Si before it is recrystallized in a solid phase epitaxy step. The figure is exemplary of the confinement that has been achieved for Si (Fig. 1).

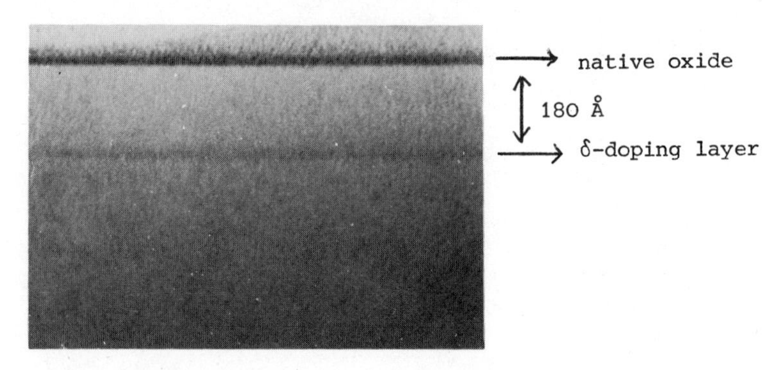

Fig. 1. TEM photograph of an Sb doping sheet incorporated in Si MBE growth (after H.P. Zeindl et al., Ref. /4/)

THE δ-LAYER IN THEORY

It is a straightforward matter to calculate the electronic subbands in the self-consistent potential of a uniform positive sheet of charge when the band structure follows a parabolic dispersion law. In Fig. 2 levels and wavefunctions, as they apply for Si in GaAs at a typical

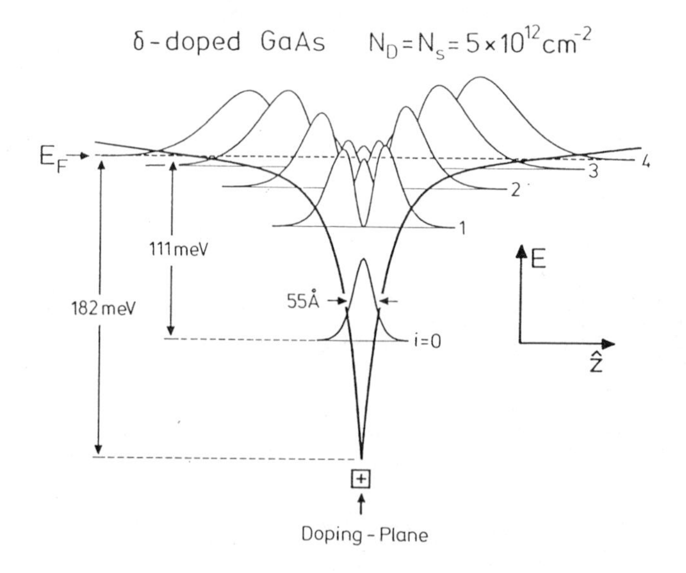

Fig. 2. Subbands and wavefunctions in the δ-layer calculated using a parabolic dispersion relation for the GaAs conduction band.

N_D of 5.0 x 10^{12} cm^{-2}, are shown. Note the increased confinement of charge in the i = 0 groundstate over that of the electron in a Bohr orbit (2 a^* ∿ 200 Å). Several subbands are found to be filled and E_F reaches 182 meV above the band edge.

We learn from this little exercise that a meaningful calculation of the subbands, for all but the very lowest doping levels ($N_D < 10^{12}$ cm^{-2}), demands the use of an appropriate nonparabolic dispersion relation. At the E_F in Fig. 2, electrons in the i = 0 level have nearly twice the m^* of the band edge states. Subband occupations and energies are changed. Such a calculation is described in Ref. /5/.

When the surface of the GaAs is close to the δ-layer and there is a Schottky-barrier potential, the potential and distribution of the charge can be strongly skewed. This situation is encountered in the tunneling spectroscopy experiments of Ref. /3/ for GaAs. It is also a characteristic feature of the δ-layer of Sb donors in Si as shown in Fig. 1 and the experiments in Ref. /4/.

Returning to GaAs, the theory must in principle take account of two additional features of the δ-layer. As a genuine "chemical" layer, the electrons sit right on top of the donor charges. One should expect a central-cell chemical-shift which is enhanced over that for the isolated donor state because of the stronger electron confinement. In addition, for sufficiently high N_D the Fermi energy E_F can reach to higher minima or to defect levels in the energy bands. Reference /6/ calls attention to this possibility.

Finally, it is clear from Fig. 2 that level spacings and relative occupations of the subbands will depend sensitively on the distribution and possible spreading of the positive ion charge. We first called attention to this fact in Ref. /7/ and showed that with the spreading the bottom of the potential well becomes parabolic. The lower levels become more equally spaced. Refs. /5/ and /8/ exploit this dependence in order to show that there is a finite spreading of the Si ions.

THE δ-LAYER - IN PRACTICE

The 2-D nature of the electronic states was unequivocally demonstrated in the magneto-oscillation experiments reported in Ref. /1/. Because there was approximate agreement of the measured occupation of the bands with that calculated using a parabolic dispersion law and an assumed $\delta(z)$-confinement of the ionic charge, it was optimistically claimed that true δ-doping had been achieved. How wrong! With improved accuracy of measurement and refinement of the calculation it is now certain that there is spreading of the ionic charge. The Si donors appear according to Ref. /8/ to be built into the growing GaAs at a volume density near the known solid solubility limit, which at the growth temperature is $\sim 7 \times 10^{18}$ cm^{-3}. Thus the δ-layer in practice is not really an atomic-plane-confined dopant. For layers with $N_D = 1 \times 10^{12}$ cm^{-2} or less the spreading is 2 lattice constants or less. With densities in the 10^{13} range, numbers like 130 Å of spreading must be expected.

Of those experiments that directly measure subband energies it is only tunneling spectroscopy that has been done successfully to date. In Ref. /3/ the tunneling current was used to identify subband-related features and thus to determine energies. Fig. 3 is an example from this work. The peaks as marked on the current derivative are the experimental results for the subband energies. The observed features show clearly that the density of states of the investigated electron system is structured, as expected for a 2-D system. Because of the broadness of the observed structures a quantitative analysis in terms of a self-consistent model calculation as shown in Ref. /3/ seems not to be too meaningful at present.

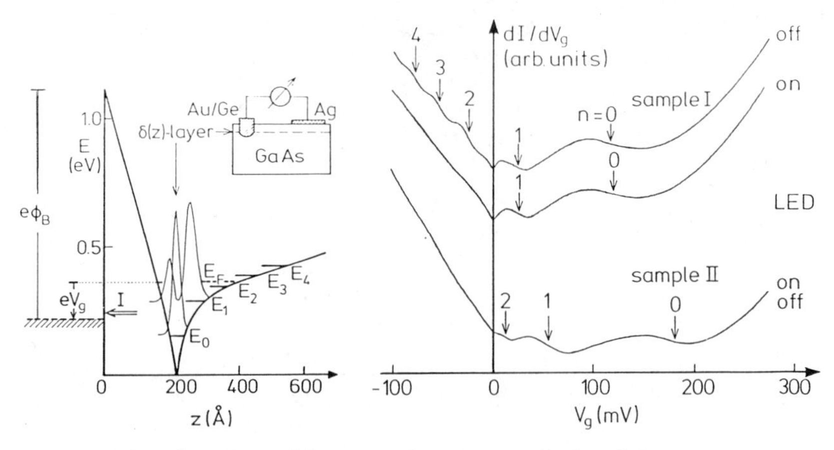

Fig. 3. Tunneling spectroscopy of the δ-layer subbands for $N_D = 8 \times 10^{12}$ cm^{-2} (after M. Zachau et al., Ref. /3/)

The most powerful experimental technique to examine the subbands has been the oscillatory magnetotransport. The Shubnikov-de Haas oscillations measure subband occupations N_s^i. In a perpendicular magnetic field the resistance of the layer is observed to vary periodically in 1/B. Each occupied subband gives a characteristic period that appears as a peak in the Fourier spectrum. Fig. 4 is an example of such work. The sample had a design doping of $N_D = 7 \times 10^{12}$ cm^{-2}. Counting the various periods, an $N_s = 6.7 \times 10^{12}$ of electronically active Si-ions per cm^2 are identified. Such an agreement between N_D and N_s is typical provided $N_D \lesssim 1.0 \times 10^{13}$ cm^{-2}. The figure shows each of the steps in the experiment. The measurement records at once the magnetoresistance ρ_{xx} and its field derivative $d\rho_{xx}/dB$. In a second step the data is displayed vs. 1/B and Fourier-transformed. The arrows identify the experimental periods along a 2-D density axis.

In the table we give the experimentally determined subband occupations and compare them with those calculated using a GaAs conduction band

Table 1. Comparison of experiment and model calculations for a layer with $N_s = 6.7 \times 10^{12}$ cm^{-2}.

Subband	:	0	1	2	3	4
N_s^i (experiment)	:	3.83	1.74	0.76	0.29	-
δ-model theory	:	4.17	1.57	0.69	0.23	0.05
dz = 78 Å	:	3.82	1.82	0.75	0.28	0.03

dispersion with nonparabolicity included. The δ-layer calculation applies for an atomically sharp arrangement of the positive charge. In a last step we show that really satisfactory agreement can be achieved when the Si-ions are assumed to be spread at uniform density over dz = 78 Å.

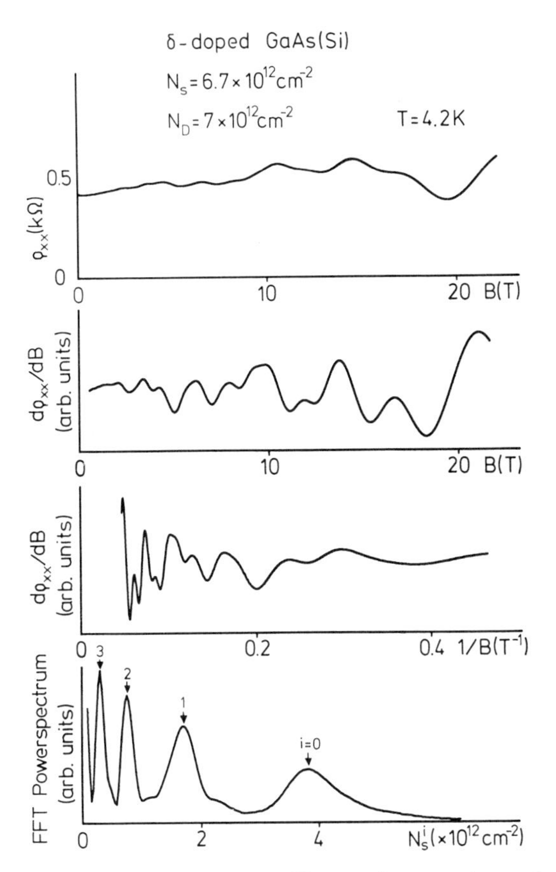

Fig. 4. Experimental recordings of magnetoresistance data and the identification of the oscillatory period.

While there is no claim that necessarily the density of ions is constant, the table shows that ion spreading in some form must be involved. The average density in the layer is 6.7 x 10^{12}/78 Å or equivalently 8.6 x 10^{18} cm^{-3}. A large number of δ-layer samples has been studied experimentally and analyzed in this way. The 3-D doping density has been found to vary between 5 - 9 x 10^{18} cm^{-2}. The conclusion that there is a finite spreading of the dopant has become inescapable for all samples with N_D in the several times 10^{12} cm^{-2} range.

An interesting other way to study the subband levels has been the diamagnetic Shubnikov-de Haas effect. With a magnetic field applied in the plane of δ-layer samples there is observed an oscillating conductivity that results when subbands are successively depopulated by the rising diamagnetic energy $(e^2/2m^*) \cdot B_{\parallel}^2 \langle z_i{}^2 \rangle$. Here $z_i{}^2$ is the extent of the wavefunction in the ith state. In the accompanying Fig. 5 below is shown an example from work in Ref. /9/.

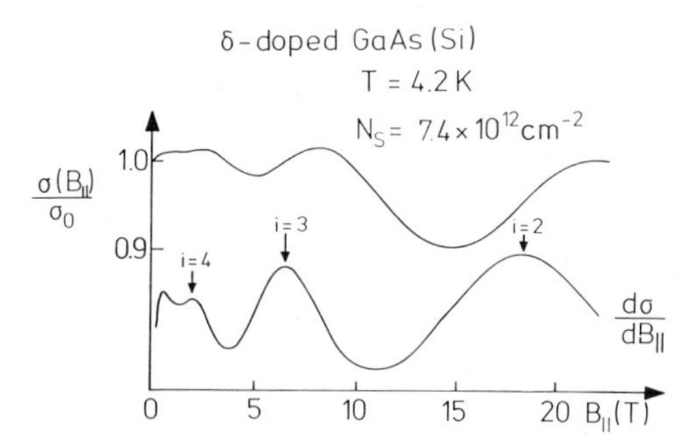

Fig. 5. Magnetoresistance oscillations in a B field parallel to the δ-layer plane. Arrows mark the field where a given subband is emptied.

The very high and low N_D limits of the δ-layer both pose exciting physical questions. At the upper end of the density scale it is the search for a limit of the electronically active N_s that has motivated our work. Using a series of samples with N_D up to 6.8 x 10^{13} cm^{-2} we have searched for a maximum N_S. In spite of the fact that there is spreading it appears that $N_S \sim 1.2 \times 10^{13}$ cm^{-2} is such an upper limit for samples grown using the interrupted growth procedure. When an N_D exceeding this number is deposited during the interruption, it appears that the excess will cluster and remain immobile in the δ-plane. Possibly Si atoms pair up in neighbouring Ga and As sites and thus are electronically inactive. Our hypothesis has been published in Ref. /8/. More detailed experiments remain to be done on this point.

At the lower end of the N_D scale it is the metal-insulator question that needs to be examined. At a density of $N_D \sim 1.35 \times 10^{11}$ cm^{-2} the δ-layer first shows conductivity at 4.2 K and we conclude that this is the critical density for the transition from the insulating to the conducting state. At somewhat higher densities the layers show a magnetically induced transition to the insulating state in a perpendicular field. Fig. 6 is an example of such work from Ref. /10/. At B ∿ 10 T the resistance rises quite

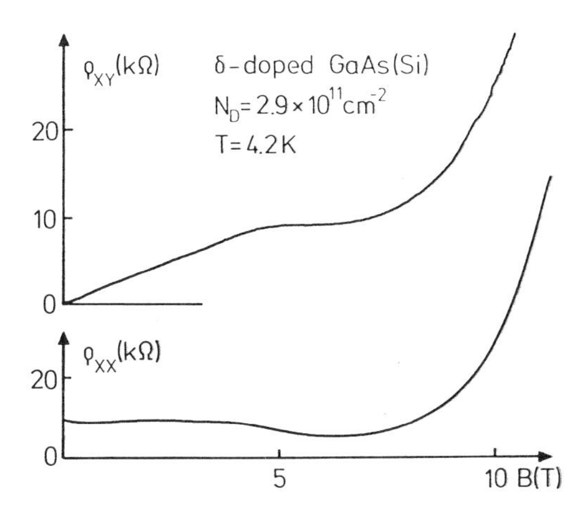

Fig. 6. Hall-resistance and magnetoresistance for a sample with N_D = 2.9×10^{11} cm^{-2}. At the upper end of the trace the resistance rises steeply to an insulating state.

steeply until all electrical contact to the sample is lost. Just prior to the rise there is observed a plateau in the Hall resistance and the magnetoresistance has a minimum. The effect is reminiscent of von Klitzing's Quantum Hall Effect. The exciting prospect of exploring both the quantum Hall effect and the metal-insulator transition side-by side in a δ-layer is opened up.

Because of the large E_F the δ-layer is also ideally suited to probe defect states resonant with the band-structure. The well-known DX center is a prominent example and it appears that its influence can be seen in the δ-layer magnetotransport experiments. In Fig. 7 we show an example from recent work /11/ where hydrostatic pressure has been applied. At a pressure of about 10 kbar (applied at 300 K) the periods abruptly begin to change and the Fourier amplitude of the i = 0 oscillation rises distinctly. In Ref. /11/ these effects are interpreted in terms of spilling some electrons into DX-traps located at the exact center of the layer potential.

Among the remaining 2-D layer experiments that have been done and can be found in the literature are the case of a δ-layer in the GaAs/AlGaAs barrier /10/, and the case of a δ-layer under hot carrier conditions /12/.

THE OUTLOOK FOR δ-LAYERS

The δ-layers have proven a physically interesting 2-D system, one that merits further attention with good prospects for exciting new results. What is particularly stimulating in this case is the versatility of this experimental system. Practically all the many features typical of 2-D have been observed and some extras, such as the chemical shift, are just waiting to be explored.

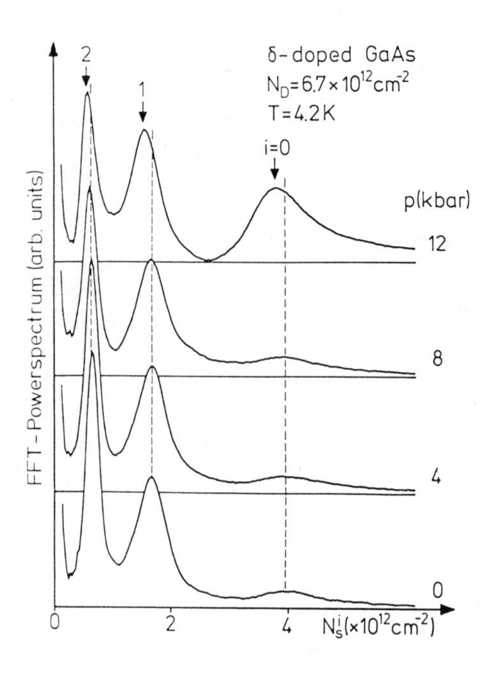

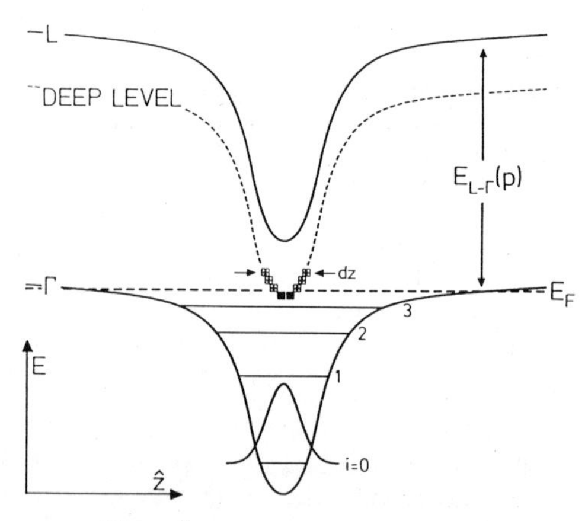

Fig. 7.
Changes in the Fourier transform of the low temperature magneto-resistance when pressure has been applied at 300 K prior to cooling. The explanation involves transfer of carriers to the DX-level (after A. Zrenner et al. in Ref. /11/).

Perhaps the most important feature of this layer system is the fact that it is an essential part of devices. In Ref. /13/ a MESFET base on the δ-layer has been explored. A recent review article mentions various other possibilities /14/. The most prominent example of impurity engineering in practice is the exact placing of the impurities for modulation doping. We believe that the understanding of impurity incorporation as it applies for δ-layers will be useful in designing the optimal strategy for remote doping.

REFERENCES

/1/ A. Zrenner, H. Reisinger, F. Koch, and K. Ploog: Proc. of the 17th International Conf. on the Physics of Semiconductors, San Francisco, 1984, edited by J.P. Chadi and W.A. Harrison (Springer Verlag, New York. 1985), p. 325

/2/ D. Grützmacher and P. Balk have prepared those samples; results will be published.

/3/ M. Zachau, F. Koch, K. Ploog, P. Roentgen, and H. Beneking, Solid State Commun. 59, 591 (1986)

/4/ H.P. Zeindl, T. Wegehaupt, I. Eisele, H. Oppolzer, H. Reisinger, G. Tempel, and F. Koch: Appl. Phys. Lett. 17, 1164 (1987)

/5/ A. Zrenner and F. Koch, Proc. of the EP2DS VII (Santa Fé, 1987)

/6/ A. Zrenner and F. Koch, Proc. of the 18th Int. Conf. on the Physics of Semiconductors, Stockholm 1986, ed. by O. Engström (World Scientific, Singapore 1987), p. 1523

/7/ F. Koch, A. Zrenner, and M. Zachau, Springer Series in Solid State Sciences 67, 175 (1986)

/8/ A. Zrenner, F. Koch, and K. Ploog, Proc. of the 14th Int. Symp. on Gallium Arsenide and Related Compounds (Crete 1987)

/9/ A. Zrenner, H. Reisinger, F. Koch, K. Ploog, and J.C. Maan, Phys. Rev. B 33, 5607 (1986)

/10/ A. Zrenner, F. Koch, and K. Ploog, Proc. of the Int. Conf. on the Application of High Magnetic Fields in Semiconductor Physics (Würzburg, 1986)

/11/ A. Zrenner et al., to be published in Semiconductor Science and Technology

/12/ P. Lugli, S. Goodnik and F. Koch, Superlattices and Microstructures 2, 335 (1986)

/13/ E.F. Schubert and K. Ploog, Jap. J. of Appl. Phys. 24, L608 (1985)

/14/ K. Ploog, Proc. of the 14th Int. Symp. on Gallium Arsenide and Related Compounds (Crete 1987)

OPTICAL MEASUREMENTS OF ACCEPTOR CONCENTRATION PROFILES AT GaAs/GaAlAs QUANTUM WELL INTERFACES

M.H. Meynadier*

Bell Communications Research

Red Bank, NJ 07701-7020, USA

INTRODUCTION

There has been in the recent years an impressive amount of studies of low dimensional semiconductor systems such as quantum wells (QWs), superlattices and modulation-doped heterojunctions. The importance of doping such structures for device applications has very early led to both theoretical [1-4] and experimental [5-7] works on shallow impurities, showing the impurity states in quasi-2 dimensional systems to possess specific properties unencountered in bulk materials. In particular, the binding energies of impurities are found theoretically to vary with well thicknesses and impurity position within the QW.

Experimentally, just as in three dimensional semiconductors, photoluminescence (PL) is a very powerful tool for the study of shallow donor and acceptor state energies and properties. This is even true of single, isolated quantum wells because of their very high radiative efficiency, which results from the combined effects of confinement and carrier capture efficiency. However, at low temperature, while the photoluminescence of undoped bulk III-V compounds is dominated by extrinsic recombination involving shallow residual impurities, the PL emitted from QWs is mainly the result of excitonic recombination [8]. QW emissions at energies a few meVs below the excitonic bandgap (as given from absorption or photoluminescence excitation spectroscopy) have been understood in terms of exciton trapping on interface structural defects [9]. In doped QWs, however, extrinsic emissions can easily be observed. In particular, the impurity position-dependence of the binding energy has been evidenced by PL measurements on selectively ("δ") Be [6] and Si [5] doped QWs, in which the impurities were incorporated in specific regions along the growth axis.

Extrinsic PL from undoped QWs can also be observed, provided that the excitation power density is kept very low -below saturation of the impurity recombination channel, which is usually much lower than in the bulk. At higher pump regimes, these emissions are masked by excitonic line tails. Molecular beam epitaxy (MBE) grown quantum wells, even of good quality, have been often found to show electron-to-acceptor recombination [7]. The acceptor is believed to be carbon atoms localized at the QW interface. It has been proposed that the trapping of this impurity at the GaAlAs/GaAs interface is due to a difference of carbon solubility in the two materials, and is correlated with interface roughness [10].

The first part of this paper will draw a simple theoretical description of hydrogenic impurities in quantum wells, focussing on the fundamental 1s state and its role in the QW optical properties. We will then present experimental results obtained from MBE-grown isolated QWs showing e-A^0 photoluminescence. This emission is due to the presence of residual acceptors; we will not discuss here intentionally doped QWs. The corresponding linewidth is shown to arise from a distribution of acceptors near the first quantum well interface, which is obtained from a lineshape analysis making use of the impurity position-dependent binding energy. This distribution, and a set of data taken on

QWs grown after prelayers of various Aluminum profiles, are consistent with a model of impurity trapping at the GaAlAs interface resulting from a higher Carbon incorporation with decreasing Aluminum concentration.

IMPURITIES IN QUANTUM WELLS: A SIMPLE MODEL

The main characteristic of quasi-2 dimensional systems such as quantum wells is their lack of translational invariance along the growth axis, and the corresponding localization of the electron (and hole) wavefunction. As a direct consequence, the binding energy and wavefunction of the impurity strongly depend on its position within the quantum well, with a maximum binding energy when the attractive center is at the center of the QW, i.e. where the electron or hole presence probability is maximum. Accordingly, one also expects for a given impurity position a strong dependence of the binding energy on the well width L. Between the two extreme situations $L=0$ and $L=\infty$ (corresponding respectively to the bulk binding energies of the barrier and QW materials), the binding energy is increased by the fact that attractive center and attracted carrier are at a smaller average distance than in the bulk crystal. The sketches displayed in Figure 1 illustrate qualitatively the evolution of the impurity wavefunction with decreasing well width (a) and increasing distance to the well center (b). The evolution in (a) is from a bulk-type 1s wavefunction in large wells to a cosine resembling the fundamental electron (or hole) confined QW state at smaller wells. It evidences the fact that for small well width the binding of the particle along the growth axis is provided more by the barriers than by Coulomb interaction. Moving the impurity away from the well center, which is sketched in (b) for a large well, results in a transformation from a quasi-1s to a quasi-$2p_z$ wavefunction at the well interface.

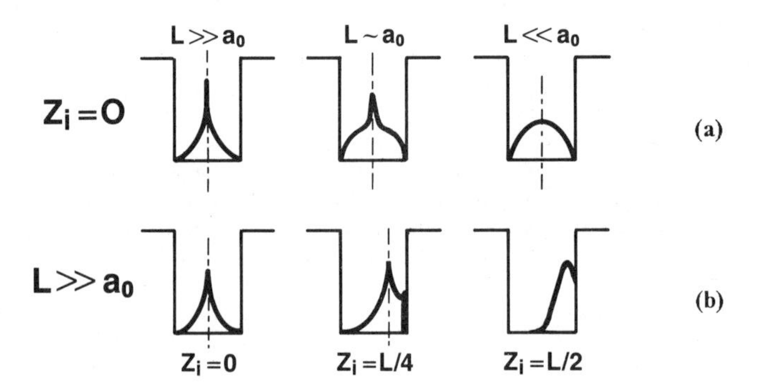

Figure 1. Schematic variation of an impurity wavefunction with (a) quantum well width and (b) position along the growth axis. a^0 is the bulk Bohr radius of the impurity.

An exact theoretical treatment of the coulombic problem in quasi-2D systems is beyond the purpose of this paper. In particular, the complexity of the valence band makes the acceptor problem harder to describe than the donor [4]. The general behavior is however essentially the same, so that we will restrict here to the case of a hydrogenic donor attached to a parabolic conduction band. The discontinuities of the effective mass and dielectric constant at the QW interfaces neglected, the impurity hamiltonian is, in the envelope function approximation [11]:

$$H = p^2/2m^* + V_b Y[z^2-L^2/4] - e^2/\{\varepsilon[\rho^2+(z-z_i)^2]^{1/2}\} \quad (1)$$

where m^* is the effective mass, V_b the barrier potential, Y the step function, ε the dielectric constant, z_i the impurity position along the growth axis and z,ρ cylindrical coordinates. The first two terms of H describe the quantum well in the absence of the impurity and give rise to the usual QW subbands. The impurity coulombic potential couples these subbands together and with the continuum. The problem has no exact solution, but good results can be obtained from a variational method. Several trial wavefunctions have been proposed and give similar results. We have retained here [1]:

$$\Psi(z) = N\,\Psi_1(z)\exp[-(\rho^2+(z-z_i)^2)^{1/2}/\lambda] \quad (2)$$

where λ is the variational parameter, N the normalization constant and $\Psi_1(z)$ the fundamental quantum well wavefunction in the absence of the impurity.

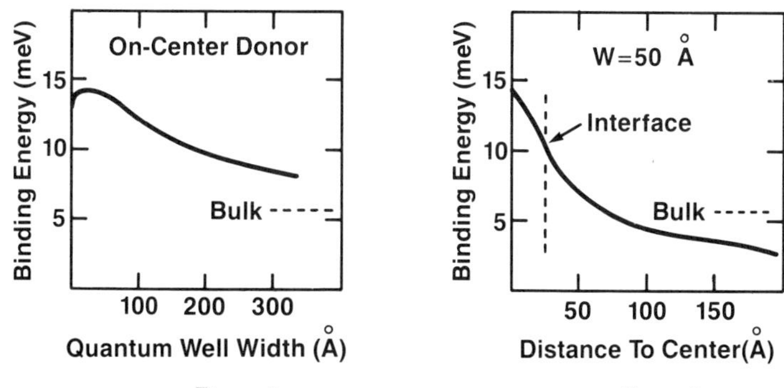

Figure 2 *Figure 3*
Calculated binding energy of a donor in a $GaAs/Ga_{.6}Al_{.4}As$ quantum well. In Figure 2, variation with quantum well width for an on-center donor; in Figure 3, variation with the donor position in a 50A thick QW.

In addition to the bound states attached to the fundamental subband, the impurity also originates short-lived states attached to excited subbands, which are degenerate with the electron continuum. We will not discuss these here, but will concentrate on the fundamental bound state associated with the first electronic subband. For a given quantum well width, the resulting impurity binding energy is a monotonically decreasing function of the distance of the impurity to the well center. Figure 3 illustrates this behavior for a donor in a 50A $GaAs/Ga_{.6}Al_{.4}As$ QW. The on-center donor binding energy is found to increase by a factor of ≈ 2.5 with respect to that of the bulk, which is reached ≈ 75 Å away from the well center. The quantum well width dependence is displayed in Figure 2 for an on-center impurity and the same barrier height. The relevant parameter measuring the amount of 2-dimensionality of the system (and the resulting increase of binding energy) is the dimensionless ratio L/a_0, where a_0 is the bulk Bohr radius of the donor (100Å in GaAs). Significant increases are found for $0.1 < L/a_0 < 1$.

Because these impurity states are of lower energy than the band they are attached to, they are, just as in the bulk, populated in the steady state regime set in a cw PL experiment. Donor-to-hole or electron-to acceptor recombination can thus be observed, with a transition probability given in the electric dipole approximation by the Fermi golden rule. Because of the confinement of both the free carrier and the one bound to the impurity, the optical matrix element is strongly dependent on the impurity position, with a maximum for the on-center impurity. These insights will be discussed below.

In what follows we have applied this model to the case of the acceptor. Using a heavy hole mass of $0.35m_0$ and a dielectric constant of 12.5, we obtain in the bulk limit a binding energy of 27 meV, in good agreement with measured values for Beryllium (26 meV) and Carbon (28 meV) acceptors [12]. We do neglect, however, the complexity of the valence band and in particular the coupling between heavy and light hole states at finite in-plane wavevectors. These should be more important for large well thicknesses and corresponding small valence subband splittings. For the experimental configuration described below, our simple model gives results in good agreement with more detailed treatments [4] and is to our belief adapted to the problems addressed here.

PHOTOLUMINESCENCE LINESHAPE AND ACCEPTOR CONCENTRATION PROFILE

The prediction of strong binding energy variations with impurity location has originated a number of experimental studies on δ-doped and unintentionally doped quantum wells [5-7]. The energy of e-A^0 photoluminescence line peaks, measured or estimated from the E_1H_1 band-to-band edge, was found to be in reasonable agreement with the trends discussed above or, in the case of residual acceptors, gave an estimation of their position along the growth axis. The purpose of our study was to take more fully advantage of this strong dependence, and to extract from the extrinsic photoluminescence lineshape the actual impurity concentration profile along the growth axis of unintentionally doped QWs [13].

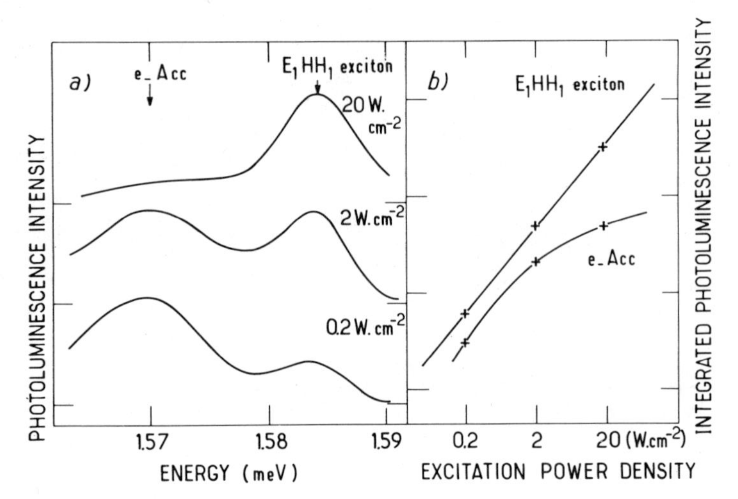

Figure 4. Photoluminescence spectrum of a 50Å thick QWSCH taken at different excitation power densities (left); the integrated photoluminescence intensities of the E_1H_1 exciton and e-A^0 lines are dispayed on the right.

The samples studied are MBE-grown quantum well separate confinement heterostructures (QWSCH) [14]. They consist of a GaAs quantum well embedded in a $Ga_{0.85}Al_{0.15}As$ layer of variable thickness, which is clad between 1000A thick ternary layers of higher Aluminum concentration (≈ 34%). The intermediate, low Aluminum concentration layer thickness is varied from 200 to 3000Å depending on the sample. The samples are unintentionally doped, which corresponds to a p-type background doping in bulk GaAs of ≈ $2x10^{14}$ cm^{-3}. Photoluminescence experiments were performed on these samples at 4K using standard cw excitation and lock-in techniques. Typical spectra are presented in Figure 4 for different excitation power densities. They exhibit two main features, the higher energy one at the same energy as the E_1H_1 exciton as measured by excitation spectroscopy. The energy of the second one (≈ 25 meV below the bandgap) and the saturation of its intensity with increasing pump power, as displayed on the right of Figure 4, strongly suggest that this line involves free electron to acceptor recombination.

The involved average acceptor binding energy is deduced from the energy at the peak of this emission with respect to that of the excitonic line maximum, added to the exciton binding energy. The latter is calculated [15] to be 8.2, 7.8 and 7.18 meV respectively for the 50, 100 and 150Å thick QWs investigated, in fair agreement with experimental determinations of the exciton binding energy for these thicknesses [14]. The resulting binding energies are plotted in Figure 5 together with the results of the hydrogenic model discussed above for an acceptor at the well interface. The on-center acceptor binding energy (dashed line) is more than twice larger. The good agreement obtained indicates that the observed extrinsic PL arises from acceptors mainly distributed at the interface of the quantum well. We will assume in what follows that we are dealing here with the first interface (GaAs on GaAlAs) of the quantum well. Earlier studies on "inverted" and "normal" heterojunctions [16] have shown that this interface is poorer in terms of both purity and crystalline quality. The linewidth (≈ 10 meV FWHM) of the emission observed here cannot be accounted for by quantum well width fluctuations. The latter result in a broadening much smaller for the excitonic line (≈ 3meV), and should besides be lower for the e-A^0 line if it arose from acceptors located exactly at the interface of a well of variable thickness. We have to conclude that the broadening is mainly due to a distribution of acceptors along the growth axis, and we will in what follows estimate the spreading of this distribution on both sides of the interface.

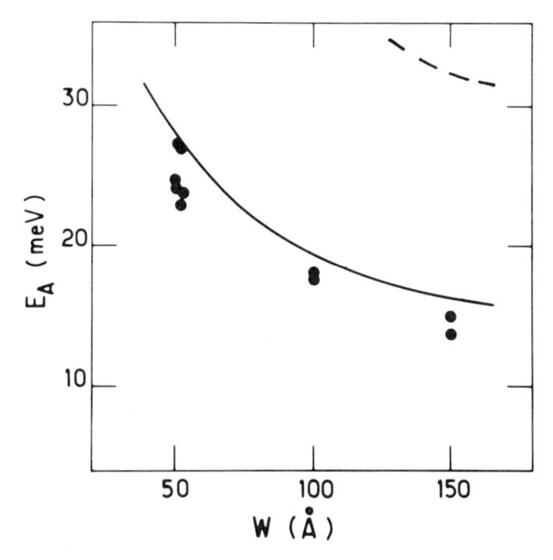

Figure 5. Calculated (solid line) and measured (dots) binding energy of the observed acceptor as a function of QW width. The dashed line corresponds to the on-center acceptor.

Two issues are prominent in determining the e-A^o lineshape: one is the actual acceptor concentration profile along the growth axis, and the corresponding spread in binding energy. The second is the population of holes on these states. Holes are created in the HH_1 subband by absorption; in the steady state regime and for the relatively low pump power densities used, they are expected to thermalize to the lowest valence states, here the closest-to-center acceptor states, before recombination. However, several phenomena do affect substantially this simple Fermi distribution figure, such as saturation effects due to the limited number of acceptor states. Furthermore, the trapping time by the impurity depends on its position with respect to the hole wave function. For impurities far away from the QW, one would expect the hole capture time to be longer than the hole lifetime. As a consequence, recombination of the holes might take place from subband states before thermalization on these distant impurities.

Several experimental features give us insights on the hole distribution in our samples for the excitation regimes investigated. One is that in samples of more mediocre quality, we do observe e-A^o recombination involving acceptors located in the barrier as far as 15Å from the interface. This demonstrates that in the present case, for which the impurities are evidently closer to the interface, the lineshape is not limited by hole capture times. Also, there is no substantial change in the linewidth and lineshape with increasing pump power, which shows that we are in the saturation regime for which all acceptor states involved are filled.

These considerations lead us to assume a constant filling factor of 1 on all acceptor states involved. The electron population is taken to be a Boltzman distribution at the lattice temperature (2K). The lineshape is thus given by the product of the acceptor concentration at a given position multiplied by the radiative recombination probability of the e-A^o transition for this impurity position. The latter is in the envelope function approximation [11] proportional to the square of the overlap integral between electron and hole envelop wavefunctions. Because of the cosine character of the electronic wavefunction, the result is a decreasing function of the impurity distance to the center. The lineshape is:

$$I(h\omega) = \int N_a(z)dz \int 2\pi \, kdk \, \tau^{-1}[\varepsilon_k - \varepsilon_A(z)] * f(\varepsilon_k) \qquad (3)$$

where $N_A(z)$ is the acceptor concentration profile, ε_k and $\varepsilon_A(z)$ the electron and acceptor energies and $f(\varepsilon_k)$ the electron distribution. The transition probability τ^{-1} is given by the Fermi golden rule in the electric dipole approximation.

Numerical simulations of (3) show that the impurity profile is strongly asymmetric with respect to the quantum well interface. It is found to extend about 12 to 30Å into the barrier for about 6 to 8Å in the well, depending on the sample structure. Numerical fits of experimental results are presented in Figure 6b-d for three quantum well thicknesses. The concentration profile functions are taken to be semi-gaussian in the QW and exponential in the barrier, as depicted in Figure 6a. The connection between these two functions, where the concentration is maximum, is an adjustable parameter z_0. Although we do not pretend to reproduce exactly the real profile, the agreement is quite good and indicates the essential correctness of the model. The asymmetry suggests that most of the impurity trapping takes place in the first GaAs layers grown after the interface. This is consistent with models of Carbon accumulation at the GaAlAs interface based upon a Carbon solubility higher in GaAs than in GaAlAs. The existence of a few impurities inside the barrier, as given from our fit, can be the result of out-diffusion. Also, the concentration maximum is found to be 1 to 3Å into the barrier. Either the QW is actually somewhat larger, and we are depicting its real interface, or the interface is graded (or rough) over 1-2 monolayers, with the acceptors trapped in the thicker QW regions. Several approximations in our model restrict however our accuracy to a few Angstroms, so that no conclusions can be drawn.

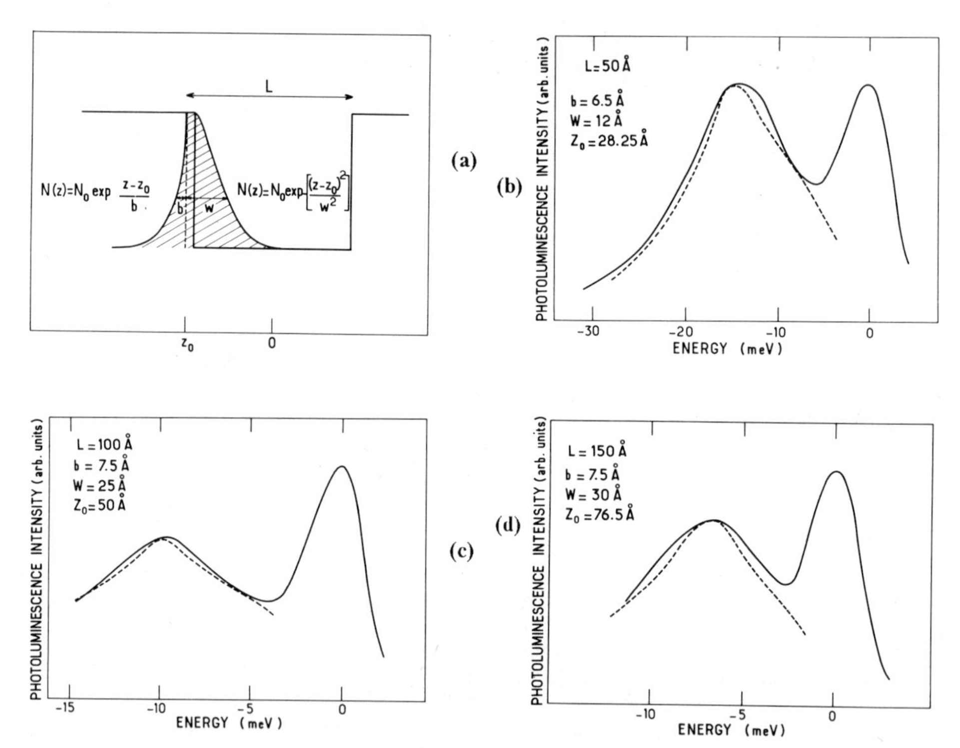

Figures 6a-d. Experimental (solid line) and calculated (dashed line) e-A^0 lineshapes for QWs of thicknesses 50Å (b), 100Å (c) and 150Å (d). The concentration profile functions used for the fit are shown in (a); b and w measure the extension in the barrier and the QW, respectively. z_0 measures the distance between the QW center and the maximum of the impurity distribution.

Since the impurities incorporate at the interface -after "floating" on the GaAlAs surface during growth-, they ought to be reduced for a given quantum well if several interfaces have been grown before. Growth of pre-layers (such as short period superlattices) prior to a structure have been shown to ameliorate device characteristics and PL linewidth [17,18]. Figure 7 shows that this is at least partially the result of a reduction of the acceptor concentration at the interface of interest. The same structure has been grown with 0,1 and 3 pre-QWs of smaller thickness grown 40Å before the well under study. The acceptor luminescence, normalized to the excitonic, is decreased by more than an order of magnitude in the 3 pre-QWs sample. The distribution profiles, not presented here, remain essentially the same.

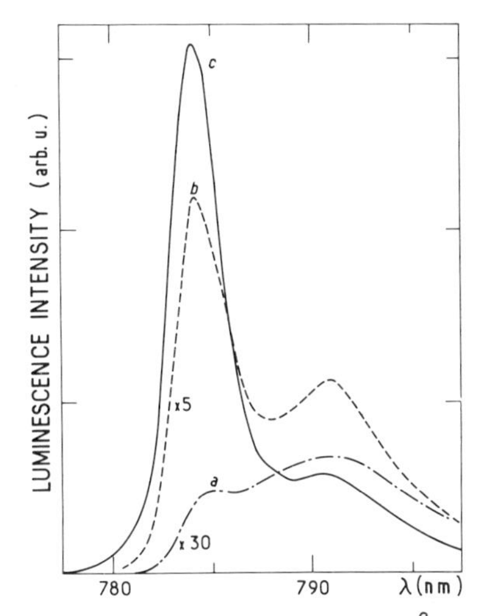

Figure 7. Photoluminescence spectra of a 50Å QWSCH grown with (a) none, (b) one and (c) three prelayers consisting of a 20Å GaAs layer embedded 40Å before the first QWSCH interface.

Finally, one could wonder whether this incorporation at the interface is actually the result of a solubility difference, or for instance of a change in the growth regime. We have studied three samples of identical structure except for the Aluminum profile of thin prelayers grown prior to the quantum well. The stuctures are shown in Figure 8 together with PL spectra of the main QW. These show unambiguously that the impurity trapping efficiency is strongly favored when the pre-QWs have a low Aluminum concentration, in agreement with the solubility model [10]. It is worth noting also that the reduction of the e-A^0 emission goes with a strong increase in the intrinsic, excitonic radiation efficiency. It suggests that Carbon incorporation favors the creation of non-radiative centers. These could be stuctural defects, as it was proposed that Carbon incorporation would inhibit the 2-dimensional growth.

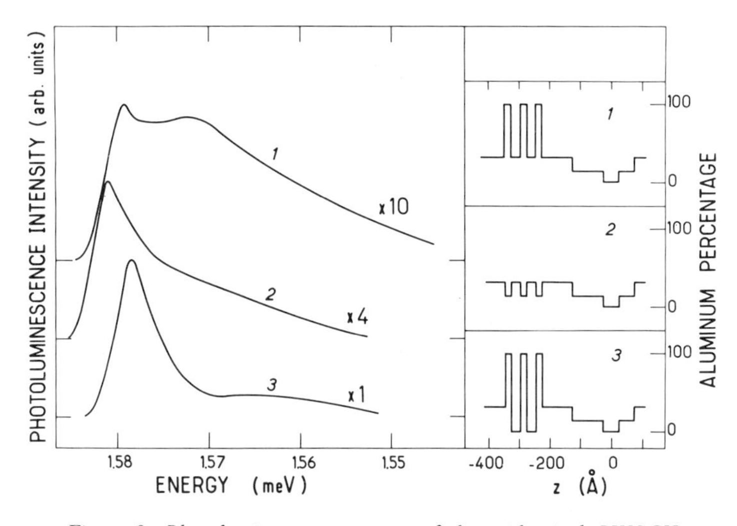

Figure 8. Photoluminescence spectra of three identical QWSCH (50Å QWs) grown after different sequences of prelayers. The Aluminum profile of the prelayers is shown on the right.

CONCLUSION

We have presented a simple treatment of the shallow, hydrogenic impurity problem in quantum wells, accentuating the differences with the bulk situation. In particular, the lack of translational invariance along the growth axis allows one to determine experimentally the location of the impurities participating in radiative recombination. In a more detailed analysis, we have been able to extract from the e-A^O lineshape the concentration profile of acceptors along the growth axis of an unintentionally doped quantum well. The results show that most of the acceptors incorporate in the first few monolayers of GaAs grown after the GaAlAs/GaAs interface, with only a limited tail of the distribution in the barrier. By growing several interfaces before the quantum well investigated, and varying their Aluminum profile, we have proven that the acceptor incorporation is enhanced in low Aluminum concentration materials. This enhancement of the trapping before the quantum well under study goes with a reduction of the number of impurities in the QW. We also show that on-edge acceptor incorporation is linked to the creation of non-radiative centers. Our set of results strongly favors a proposed model for impurity incorporation in heterojunctions, based upon a variation of the impurity solubility with ternary composition.

Aknowledgements I wish to thank my collaborators J.A. Brum, C. Delalande, G. Bastard and M. Voos for their contribution to this work. I am also grateful to F. Alexandre for providing me with good quality samples well adapted to this study. I finally gratefully aknowledge the support of R.E. Nahory during the preparation of this manuscript.

* Work done while at Groupe de Physique des Solides, 24 rue Lhomond, F75231 Paris.

References

[1] G. Bastard, Phys.Rev.B **24** , 4714 (1981).
[2] C. Mailhot, Y.C. Chang and T.C. McGill, Phys.Rev.B **26** , 4449 (1982).
[3] R.L. Greene and K.K. Bajaj, Solid State Commun. **45** , 825 (1983).
[4] W.T. Masselink, Y.C. Chang and H. Morkoc, Phys.Rev.B **28** , 7373 (1983).
[5] B.V. Shanabrook and J.C. Comas, Surface Sci. **142** , 504 (1984).
[6] R.C. Miller, J.Appl.Phys. **56** , 1136 (1984).
[7] R.C. Miller, W.T. Tsang and O. Munteanu, Appl.Phys.Lett. **44** , 374 (1982).
[8] C. Weisbuch, R.C. Miller, R. Dingle, A.C. Gossard and W. Wiegmann, Solid State Commun. **37**, 621 (1981).
[9] G. Bastard, C. Delalande, M.H. Meynadier, P.M. Frijlink and M. Voos, Phys.Rev.B **29** ,7042 (1984).
[10] P.M. Petroff, R.C. Miller, A.C. Gossard and W. Wiegmann, Appl.Phys.Lett. **44** , 217 (1984).
[11] G. Bastard, Phys.Rev.B **24** , 5693 (1981).
[12] Landolt-Bornstein, Vol.17a (springer-Verlag, Berlin Heidelberg New York, 1982).
[13] M.H. Meynadier, J.A. Brum, C. Delalande, M.Voos, F. Alexandre and J.L. Lievin, J. Appl.Phys. **58** , 4307 (1985).
[14] M.H. Meynadier, C. Delalande, G. Bastard, M. Voos, F. Alexandre and J.L. Lievin, Phys.Rev.B **31** , 5539 (1985).
[15] following R.L. Greene and K.K. Bajaj (ref.3) with the Luttinger parameters $\gamma_1 = 7.65$ and $\gamma_2 = 2.41$.
[16] T.J. Drummond, J. Klem, D. Arnold, R. Fischer, R.E. Thorne, W.G. Lyons and H. Morkoc, Appl.Phys.Lett. **42** , 615 (1983).
[17] W.T. Masselink, Y.L. Sun, R. Fischer, T.J. Drummond, Y.C. Chang, M.V. Klein and H. Morkoc, J.Vac.Sci.Technol.B **2** , 117 (1984).
[18] P. Dawson and K.K. Woodbridge, Appl.Phys.Lett. **45** , 435 (1984).

MOLECULAR BEAM EPITAXY OF $Ga_{0.99}Be_{0.01}AS$ FOR VERY HIGH SPEED HETEROJUNCTION BIPOLAR TRANSISTORS

J.L.Liévin*, F.Alexandre, C.Dubon-Chevallier

Centre National d'Etude des Télécommunications
196, rue H.Ravera
92220 Bagneux - France

Abstract

In this paper, we present the molecular beam epitaxy of very high *p*-type doping levels into GaAs using beryllium as a dopant. Very low resistivity, multilayer-compatible material is achieved with doping levels as high as $2 \times 10^{20}cm^{-3}$. We study the beryllium incorporation during growth, and discuss, based on experimental results, how such layers can significantly improve the speed of heterojunction bipolar transistors.

1 Introduction

The advent of molecular beam epitaxy (MBE) makes possible the growth of III-V semiconductors such as GaAs with excellent uniformity and accurate control of the electrically active impurity incorporation. For *p*-type doping, beryllium is now commonly used since it allows an easy active incorporation control over a wide doping range. Heavily doped *p*-type GaAs in particular eases ohmic contact fabrication and is very desirable in device structures such as tunnel diodes or heterojunction bipolar transistors (HBT).

This paper is divided into two distinct parts. First, we present a material study of beryllium incorporation into GaAs where we show how very high doping levels can be achived by varying the conditions of growth. Low resistivity layers are demonstrated despite the fact that holes are the majority carriers. A tentative model for the dopant incorporation is presented to explain the data collected during the course of this study. Secondly, we use the low resistivity *p*-type layers for the fabrication of heterojunction bipolar transistors. Based on the analysis of the device behavior, we try to answer some of the specific questions raised by the incorporation of very high doping levels into multilayer structures. We also demonstrate the application of the heavily doped material to significantly improve the speed of the HBTs.

2 MBE Growth of $Ga_{0.99}Be_{0.01}As$

2.1 Experiments

Beryllium incorporation in GaAs produces a shallow level $\simeq$ 19meV above the valence band. The impurity has a unity sticking coefficient [1]. Its diffusion coefficient D_{Be} during growth is relatively low, although J.N. Miller et al. have reported a concentration dependent behavior involving a substitutional-intersticial mechanism [2]. As a consequence, considerable diffusion tails can be observed for doping levels above $5 \times 10^{19}cm^{-3}$ [3]. Moreover, surface morphology

*Present Address: Laboratoires de Marcoussis, CGE, Route de Nozay, 91460 Marcoussis - France.

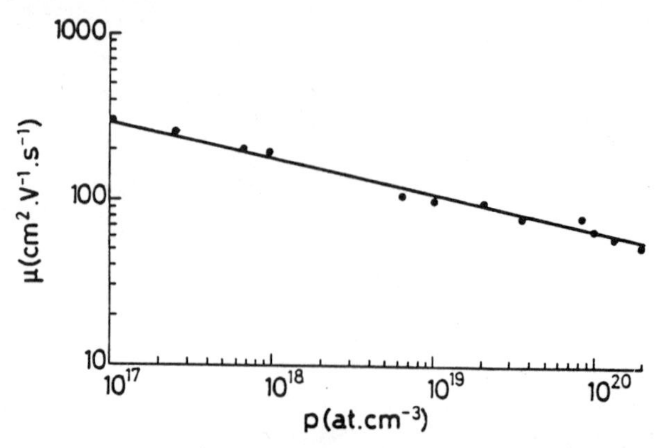

Figure 1. Hall mobility as a function of hole concentration

and luminescence properties were commonly reported to degrade at such high doping levels. We present in this part of this paper information on the influence of MBE growth conditions on Be-doped GaAs in order to control impurity concentrations as high as possible while minimizing undesirable effects.

The epitaxial layers were grown in a Riber 2300 ultra high vacuum system on (001) oriented Cr-doped substrates. Information on surface reconstructions and crystal quality were obtained from reflection high energy electron diffraction (RHEED) experiments. The substrate temperature T_S was monitored with a calibrated infrared optical pyrometer and the As_4/Ga flux ratio was measured with a Bayard-Alpert nude ion gauge. The substrate temperature and the flux ratio, referred as the growth conditions, have been seen by optical microscopy to have an influence on the Be-doped epilayer morphologies. These observations were consistent with a roughening of the surface when increasing the substrate temperature or reducing the flux ratio [4]. Since T_S can be controlled and varied in a more acurate manner than the flux ratio, we have fixed As_4/Ga to 15 approximately (beam equivalent pressure) and we have varied the substrate temperature only. By reducing it to unusually low growth temperatures such as 450°C, we have been able to increase the incorporation of Be into the epilayers up to 2×10^{20}cm^{-3} without any degradation of the layer morphology as seen by optical microscopy or RHEED reconstructions [5].

Secondary ion mass spectrometry (SIMS) analysis carried out on 1.4μm-thick Be-doped GaAs grown on Cr-doped GaAs substrates has allowed us to demonstrate that no dopant diffusion occurs at such high doping levels within the SIMS equipment resolution (better than 300Å). The excellent morphology associated with the weak Be diffusion permits the incorporation of the p^{++}-GaAs epilayers into multilayers structures [5].

Van der Pauw-Hall measurements were performed on different samples taking into account the actual conducting film thickness as known from SIMS profiles. These experiments confirmed an electrically active Be incorporation of 100%. The maximum doping concentrations correspond to a Be/Ga ratio of about 1%, leading to $Ga_{0.99}Be_{0.01}As$ alloy formation. The Hall mobility as a function the hole concentration is shown in Fig. 1. As expected, the mobility decreases as the doping level increases, but the overall resistivity ρ of the layers is minimum for $p = 2 \times 10^{20}$cm^{-3} : $\rho = 5.9 \times 10^{-4}\Omega$.cm. As a comparison, the lowest resistivity reported for n-type GaAs is $\rho = 4.8 \times 10^{-4}\Omega$.cm [6]. An interesting feature related to the high doping levels is that double diffraction X-rays analysis shows a significant impurity-related mismatch $\Delta a/a$ (Fig. 2). A Vegard law expression can be fit to this data yielding $\Delta a/a$ as a function of

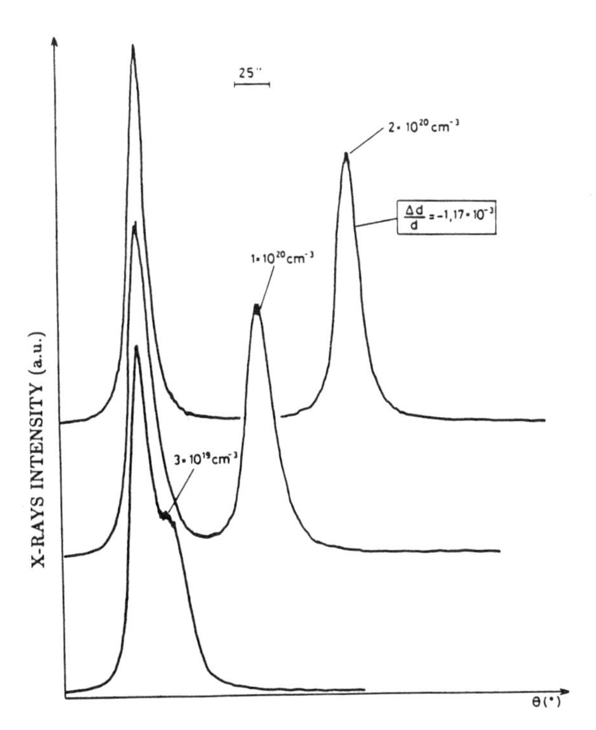

Figure 2. CuK_{α} (400) rocking curves of Be-doped GaAs with high impurity concentrations

the impurity concentration N_B:

$$\frac{\Delta a}{a} = -5.99 \times 10^{-24} N_B \tag{1}$$

A significant mismatch (above 10^{-3}) can therefore result from a dopant incorporation greater than $2 \times 10^{20}\text{cm}^{-3}$.

2.2 Beryllium Memory Effects

Although we have demonstrated the ability of molecular beam epitaxy to grow sharp interface profiles while using very high doping levels, a memory effect due to a slow evacuation process of the impurity out of the growth environment may induce a higher background impurity level. In order to estimate the undesirable effects of a high Be doping level, we have grown a succession of undoped and intentionally doped layers as summarized in Table 1. The carrier concentrations have been measured by Hall experiments. Two conclusions follow from these experiments. First, a memory effect is detected and increases as the intentional doping level in the intermediate layer increases. The background doping level remains however in the 10^{14}cm^{-3} range. Secondly, the memory effect is very limited in time. The 6716-6717 sample sequence shows that one resting day causes the memory effect to disappear. These results therefore confirm that high doping levels using Be can be safely employed during the course of high quality sample fabrication.

2.3 Discussion

Since the substrate temperature appears to play a key role in the incorporation of the Be impurities during the MBE process, we have carried out the following experiment. Keeping

Table 1. Memory effect of Beryllium. The samples not separated by an horizontal line were grown during the same day. Sample 6717 was grown two days after sample 6716.

Sample		$N_B(\text{cm}^{-3})$
6071	undoped GaAs	2×10^{14}
6072	Be-doped GaAs	6×10^{19}
6073	undoped GaAs	4×10^{14}
6716	Be-doped GaAs	2×10^{20}
6717	undoped GaAs	2×10^{14}
6718	Be-doped GaAs	2×10^{20}
6719	undoped GaAs	7×10^{14}

a constant As_4/Ga ratio, we have grown three 100nm-thick Be-doped layers ($N_B = 5 \times 10^{19}$) spaced by 300nm-thick undoped layers. The substrate temperatures during the Be incorporation were 600°C, 500°C and 435°C, respectively. The beryllium SIMS profile shown in Fig. 3 presents two interesting features. First, the temperature-dependent Be-diffusion is very clear. The interface width indicated in the figure is calculated as the thickness between 10% and 90% of the maximum impurity concentration. Since the growth temperature was lowered during the epitaxy, the influence of the growth time after each Be spike is negligible. Secondly, a Be-segregation tail is seen on the SIMS profile. This phenomenon has already been reported by D.L.Miller et al. [7]. These authors suggested the presence of a different incorporation mechanism responsible for the segregation phenomenon since its amplitude is several orders of magnitude lower than the Be plateau. We speculate that the second incorporation mechanism involves the formation of a (Be,As) chemical compound that requires two or more Be atoms. Growth conditions increasing the probability of having two or more Be atoms in the same neighborhood would therefore facilitate the formation of such a compound. This is consistent with our observations since higher substrate temperatures or lower arsenic overpressures are known to enhance the surface mobility of adatoms [8,9,10]. The exponential segregation tail is characterized by a length L_{SEG} and an initial level N_0 that appears to be constant: $N_0 \simeq 10^{17}\text{cm}^{-3}$, and can be interpreted as a solubility limit. The total amount of Be that has segregated, N_{SEG}, is therefore:

$$N_{SEG} = N_0 \cdot L_{SEG} \tag{2}$$

This quantity increases with the substrate temperature and can be expressed as a function of T_S:

$$N_{SEG} = N_0 \cdot L^0_{SEG} \cdot \exp\left\{-\frac{E_a}{kT_S}\right\} \tag{3}$$

The activation energy E_a is then directly related to the impurity mobility at the surface. From our measurements on Fig. 3, we find:

$$E_a = 0.34\text{eV},$$

$$L^0_{SEG} = 10.5\mu\text{m}.$$

In this model, the morphology degradation is associated with a critical concentration of a (Be,As) chemical compound on the growing front. Since a lowering of the growth temperature allows the incorporation of more active beryllium, the incorporation limit is presumably close to $2 \times 10^{20}\text{cm}^{-3}$ because growth temperatures under 450°C have been found to produce poor electrical activation in MBE grown GaAs [11].

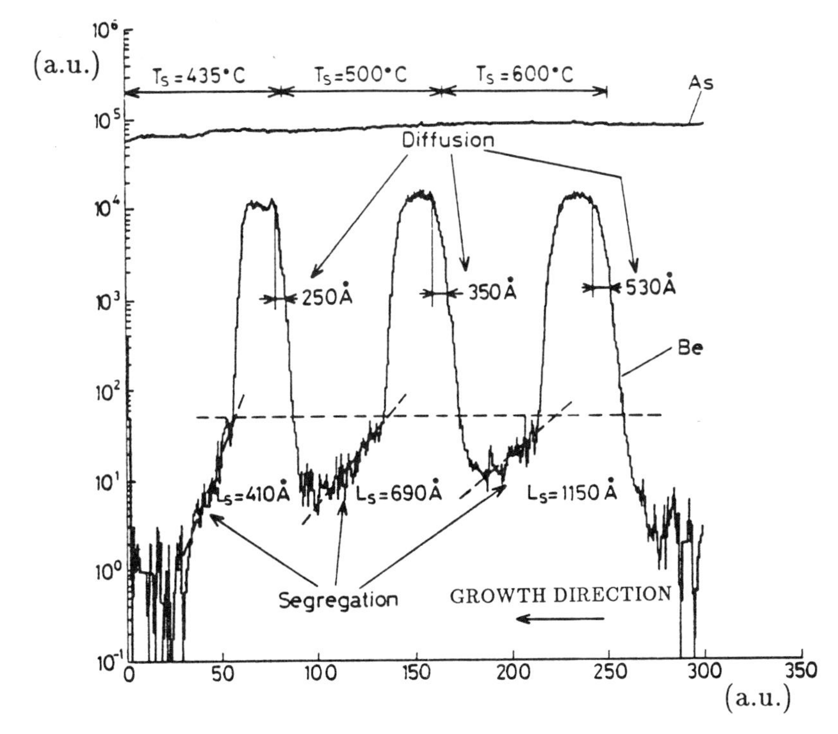

Figure 3. Beryllium and arsenic SIMS profile. The Be doping level is $5 \times 10^{19}\text{cm}^{-3}$.

3 $Ga_{0.99}Be_{0.01}As$ incorporation into HBT structures

3.1 Interest of high p-type doping levels for HBT

Heterojunction bipolar transistors have shown great promise for high-speed microelectronic applications [12]. AlGaAs/GaAs HBTs offer potential advantages over Si bipolars transistors in terms of higher cutoff frequency, lower base resistance, and lower emitter capacitance [13]. The current gain of the device $\beta = I_C/I_B$ can be written as a function of the injection efficiency γ and the base transport factor δ:

$$\beta = \frac{\gamma\delta}{1-\gamma\delta} \tag{4}$$

where the injection efficiency γ, taking the notations of Figure 4, is defined as a function of the hole current density from the base to the emitter, J_t, the electron current density in the base, J_e, and the recombination current density at the heterojunction, J_R:

$$\gamma = \frac{J_e}{J_e + J_t + J_R} \tag{5}$$

The base transport factor δ is defined as:

$$\delta = \frac{J_e - J_{R'}}{J_e} \tag{6}$$

where $J_{R'}$ represents the recombination current density in the base region. δ can be expressed as a function of the electron diffusion length in the base material L_{eB} and the base thickness W_B:

$$\delta = \frac{1}{\cosh(W_B/L_{eB})} \tag{7}$$

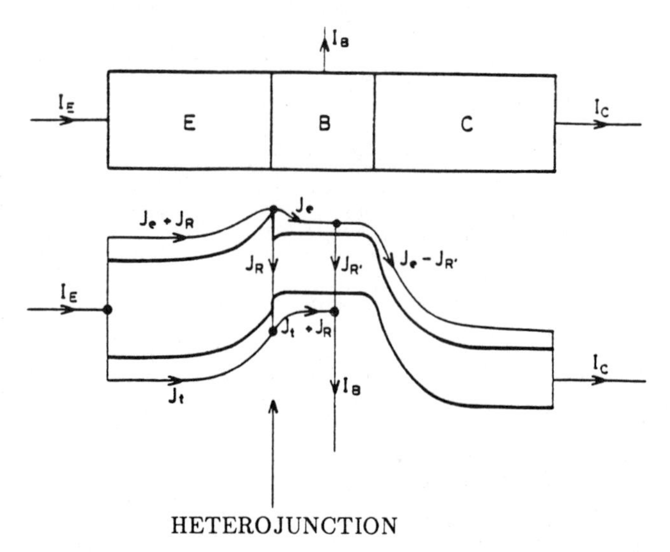

Figure 4. Schematic band diagram of an HBT and current flows involved. The arrows indicate the electron flow direction.

The schematic band diagram of Figure 4 illustrates that, contrary to homojunction silicion bipolar transistors, a forward bias of the emitter-base junction does not result in a significant hole current towards the emitter since the heterojunction creates a potential barrier for the holes greater than the one for the electrons. As a consequence, the injection efficiency γ should be close to unity when the current density through the device is high enough to neglect the recombination currents at the heterointerface. In the approximation where $W_B \ll L_{eB}$, and if the injection efficiency γ is close to unity as should be the case for HBT structures, the current gain of the device is:

$$\beta = 2 \cdot \left(\frac{L_{eB}}{W_B} \right)^2 \tag{8}$$

The overall speed of the device is usually discussed in terms of its maximum oscillation frequency f_{max} [14]:

$$f_{max} = \frac{1}{2} \cdot \left(\frac{1}{2\pi\tau_{EC}} \cdot \frac{1}{2\pi R_B C_{TC}} \right)^{\frac{1}{2}} \tag{9}$$

where τ_{EC} is the total transit time of the electrons from the emitter to the collector, R_B is the base resistance and C_{TC} is the collector capacitance. As can be seen from Equation 9, reducing the base resistance will improve the speed of the device. The base resistance per square can be written as a function of the base material resistivity ρ_B and the base layer thickness W_B as:

$$R_{B\square} = \frac{\rho_B}{W_B} \tag{10}$$

Since we have seen in Section 2.1 that $Ga_{0.99}Be_{0.01}As$ exhibits a very low resistivity, it appears as an attractive alternative to the usual doping levels employed for HBTs. A doping level of 3×10^{19}cm^{-3} gives $R_{B\square} = 600\Omega/\square$ for a typical structure where the $Ga_{0.99}Be_{0.01}As$ gives $R_{B\square}$ of about $150\Omega/\square$ for a base thickness of $500\AA$.

The question that we want to address in the next sections is whether the incorporation of high doping levels alters the ideal HBT behavior as assumed for the derivation of Equation 8.

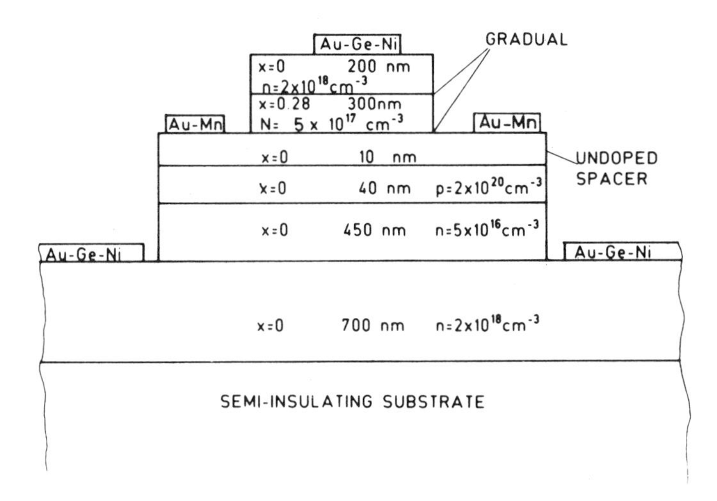

Figure 5. Schematic diagram of the layers comprising the transistor.

3.2 DC device characterization

The detailed dc performance of the $Ga_{0.72}Al_{0.28}As/Ga_{0.99}Be_{0.01}As$ transistor has been reported elsewhere [15]. A typical structure is shown in Figure 5. The role of the undoped spacer layer between the base and the emitter is to partially offset the diffusion of Be into the emitter. The presence of such spacer layers has been shown experimentally to lead to a significant current gain increase [16]. A maximum current gain of about 10 was measured at a base current density of 60kA/cm^2. Assuming an ideal behavior for the HBT as in Equation 8 allows to calculate an apparent diffusion length L_{eB} for the minority carriers in the base:

$$L_{eB} = 0.11\mu\text{m}$$

This method provides an interesting characterization of minority carrier transport into a heavily doped layer. In fact, the HBT base layer has already been used as a means to study perpendicular transport in resonant tunneling structures [17] or to observe the transition from hopping to Bloch conduction regimes in $Ga_{1-x}Al_xAs/GaAs$ superlattices [18]. We give in Figure 6 the L_{eB} values as calculated with the method described above for varying base doping level HBTs fabricated in our laboratory. To have additional and independent information on the minority carrier transport properties, we have performed cathodoluminescence measurements on Be-doped GaAs test samples grown with growth conditions similar to those used in the HBT structures [19]. The results are also plotted in Figure 6. Good agreement is seen between the two methods for doping levels lower than 10^{19}cm^{-3} while a discrepancy is observed for the heavily doped base layers.

Starting from Equation 7, we can calculate the base transport factor δ using $L_{eB}(1 \times 10^{20})$ as given by the cathodoluminescence experiments, $L_{eB} = 0.25\mu\text{m}$. We find: $\delta = 0.97$. It follows from the measured current gain β that the injection efficiency is:

$$\gamma = 0.94$$

and is thus smaller than the ideal unity value. The lattice mismatch $\triangle a/a$ between the emitter and the base layers is not likely to have a significant contribution to the injection efficiency degradation since the base is thin enough to be grown in a pseudomorphic manner [20,21]. The hole to electron current density ratio J_t/J_e on the other hand is found to be negligible in view of the large aluminum concentration used for the emitter. Moreover, a significant bandgap shrinkage ($\triangle E_g = 94\text{meV}$) results from the base material degeneracy [22]. The origin of the injection efficiency degradation is then likely to be related to a beryllium diffusion towards the

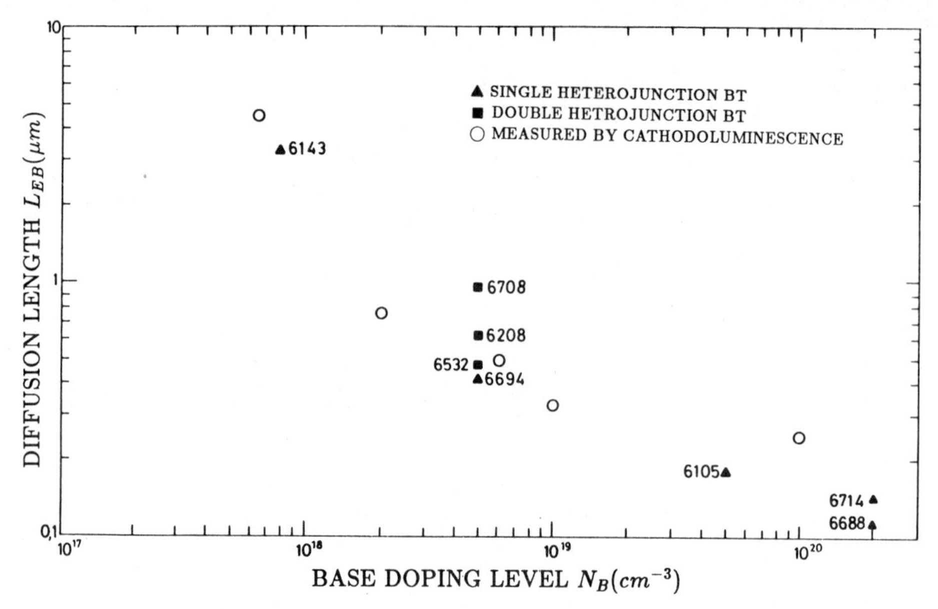

Figure 6. Apparent diffusion length as calculated from current gains of different HBT structures and comparison with cathodoluminescence measurements (○).

emitter, and current gain improvement can be expected from a thicker spacer region between the emitter and the $Ga_{0.99}Be_{0.01}As$ base.

3.3 Towards very high frequencies

From the preceding results, the superiority of $Ga_{0.99}Be_{0.01}As$ as a base material for the HBT as compared to lower doping levels can easily be stated. Assuming an optimized epitaxial structure that ensures an injection efficiency close to unity, one can compare the base resistance $R_{B\Box}$ and therefore the potential speed of the device for different doping levels by using Equation 8 and the electron diffusion length measured by cathodoluminescence L_{eB} (Fig. 7). The current gain was assumed to be equal to 50 to carry out the calculation. It is clear from this calculation that the best doping level-base thickness compromise resides in the heavily doped region.

This prediction has been recently confirmed by the measurement of the highest maximum oscillation frequency values ever reported ($f_{max} \geq 100$GHz) on self-aligned HBT structures using base doping levels above $10^{20}cm^{-3}$ [23].

4 Conclusion

In this paper, we have presented the study of MBE growth conditions in order to incorporate higher active beryllium concentrations into GaAs. As a consequence, doping levels as high as $p = 2 \cdot 10^{20}cm^{-3}$ were obtained, leading to a very low resistivity $Ga_{0.99}Be_{0.01}As$ alloy. In addition, we have discussed an impurity incorporation model consistent with our experimental observations. $Ga_{1-x}Al_xAs/Ga_{0.99}Be_{0.01}As$ heterojunction bipolar transistors were fabricated and their dc behavior was discussed in view of the anomalous dopant quantity present in the base. Based on minority carrier diffusion length measurements, we have shown the interest, as now confirmed by experimental results, of very high doping levels in order to increase the maximum oscillation frequency of the device.

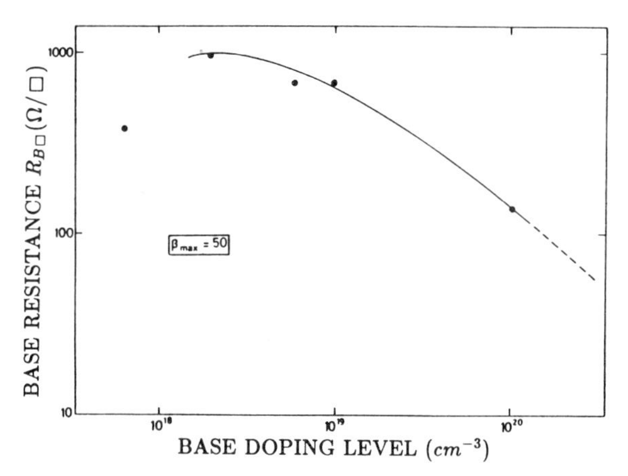

Figure 7. Base resistance per square as a function of doping level, assuming a current gain of 50.

Aknowledgements: The authors gratefully aknowledge Prof. C.G. Fonstad for critical reading of the manuscript, Dr. A. Mircea for encouragements and discussions during the course of this work, P. Daste and S. Godefroy for SIMS analysis, and A.C. Papadopoulo and B. Akamatsu for cathodoluminescence experiments. Part of this work was supported by ESPRIT Project 522.

References

[1] M. Ilegems, J. Appl. Phys. **48**(1977), 1278.

[2] J.N. Miller, D.M. Collins, N.J. Noll, J.S. Kofol, Electronic Materials Conference (1983), unpublished.

[3] P. Enquist, L. Lunardi, G.W. Wicks, L.F. Eastman, Int. Conf. on Molecular Beam Epitaxy (1984).

[4] J.L. Liévin, Thèse de Doctorat, Université de Paris XI (1986), unpublished.

[5] J.L. Liévin, F. Alexandre, Electron. Lett. **21**(1985), 413.

[6] R. Sacks, H. Shen, Appl. Phys. Lett. **47**(1985), 374.

[7] D.L. Miller, P.M. Asbeck, J. Appl. Phys. **57**(1985), 1816.

[8] J.H. Neave, P.J. Dobson, B.A. Joyce, J. Zangh, Appl. Phys. Lett. **47**(1985), 100.

[9] J.M. Van Hove, P.R. Pukite, P.I. Cohen, J. Vac. Sci. Technol. **B3**(1985), 563.

[10] A. Madhukar, S.V. Ghaisas, Appl. Phys. Lett. **47**(1985), 247.

[11] T. Murotani, S. Shimanoe, S. Mitsui, Journ. of Cryst. Growth **54**(1978),302.

[12] For a recent review, see N. Moll, in Proceedings IEEE/Cornell Conf. on High Speed Semicond. Dev. and Circ. (1985), 35.

[13] H. Kroemer, Proc. IEEE **70**(1982), 13.

[14] S.M. Sze, Physics of Semiconductor Devices, 2nd Ed., Wiley-Interscience (1981).

[15] J.L. Liévin, C. Dubon-Chevallier, F. Alexandre, G. Le Roux, J. Dangla, D. Ankri, IEEE Electron. Dev. Lett.**EDL-7**(1986), 129.

[16] R.J. Malik, F. Capasso, R.A. Stall, R.A. Kiehl, R.W. Ryan, R. Wunder, C.G. Bether, Appl. Phys. Lett. **46**(1985),600.

[17] F. Capasso and R.A. Kiehl, J. Appl. Phys.**58**(1985),1366.

[18] J.F. Palmier, C. Minot, J.L. Liévin, F.Alexandre, J.C. Harmand, J. Dangla, C. Dubon-Chevallier, D. Ankri, Appl. Phys. Lett.**49**(1986), 1260.

[19] For a description of the method, see B. Akamatsu, P. Henoc, A.C. Papadopoulo, in Scanning Electron Microscopy IV(1983), 1579.

[20] R. People and J.C. Bean, Appl. Phys. Lett. **47**(1985), 322; **49**(1986), 229.

[21] J.L. Liévin and C.G. Fonstad, accepted for publication in Appl. Phys. Lett., oct.1987.

[22] H.C. Casey, F. Stern, J. Appl. Phys. **47**(1976), 631.

[23] M.C. Chang, P.M. Asbeck, K.C. Wang, G.J. Sullivan, N.H. Sheng, J.A. Higgins, D.L. Miller, IEEE Electron. Dev. Lett. **EDL-8**(1987), 303.

PROGRESS REPORT ON MOLECULAR BEAM EPITAXY OF III-V SEMICONDUCTORS -

FROM FIBONACCI TO MONOLAYER SUPERLATTICES

L. Tapfer, J. Nagle, and K. Ploog

Max-Planck-Institut für Festkörperforschung
Heisenbergstr. 1
7000 Stuttgart-80, FR-Germany

ABSTRACT

We present two examples for the unique capability of molecular beam epitaxy to modify the bulk properties of semiconductors through band-gap (or wavefunction) and structure-factor engineering in artificially layered semiconductors, i.e. GaAs/AlAs short-period and ultrathin-layer superlattices as well as Fibonacci, Thue-Morse and random GaAs/AlAs superlattices. In the first example the concept of artificially layered semiconductor structures is scaled down to its ultimate physical limit normal to the crystal surface. In the second example we consider artificially layered structures that range between periodic and random systems. These quasi-periodic structures are obtained by imposing an incommensurate modulation of composition on the direction of layer sequence.

1. INTRODUCTION

In the past two decades, precisely controlled epitaxial growth techniques have emerged, including molecular beam epitaxy (MBE) and metalorganic chemical vapour deposition (MOCVD), which have provided a variety of new opportunities for the fabrication of artificially layered semiconductor structures / 1, 2 / . In such highly refined quantum well heterostructures (QWHs) and superlattices (SLs)made of III-V, II-VI, or IV-VI compound semiconductors or of group-IV-elements novel physical phenomena due to quantum confinement occur which have opened up new horizons in semiconductor research. The combination of signal transport by photons as well as by electrons in these ultrathin multilayer structures has led to the realization of a variety of new device concepts.

In this review paper we present two distinct examples for the unique capability of MBE to modify the bulk properties of semiconductors through band-gap (or wavefunction) and structure-factor engineering in artificially layered semiconductors; i.e. GaAs/AlAs short-period and ultrathin-layer superlattices as well as Fibonacci, Thue-Morse and random GaAs/AlAs superlattices. In the first example the concept of artificially layered semiconductor structures is scaled down to its ultimate physical limit normal to the crystal surface, as each constituent layer in, e.g., $(GaAs)_1(AlAs)_1$ monolayer superlattices has a spatial extent of less than the lattice constant of the respective bulk materials. In the second example we consider artificially layered GaAs/AlAs structures that range between periodic and random systems. The quasi-periodic structures are obtained by imposing an incommensurate modulation of composition on the direction of layer sequence. In Fibonacci superlattices (FSL), e.g., the incommensurability ratio of the two basic GaAs/AlAs building blocks is given by the golden mean $\tau = (1 + \sqrt{5}) / 2$. The Thue-Morse sequence, on the other hand, is not quasi-periodic but automatic and thus more closely connected to the random systems / 3, 4 / .

2. SAMPLE PREPARATION

The $(GaAs)_m(AlAs)_n$ short-period, ultrathin-layer, Fibonacci- and Thue-Morse superlattices are grown on (001)oriented GaAs substrates in a three-chamber MBE system of the quasi-horizontal evaporation type. The contamination-free substrate surface is prepared by the cleaning method described in detail in Ref. / 5 / . In brief, the etch-polished GaAs wafer is treated with H_2SO_4, and the stable surface oxide for passivation is generated by heating the wafer to 300 °C in air under dust-free conditions for about 3 min. The surface oxide is then thermally desorbed by heating to 600 °C in a flux of arsenic under ultrahigh vacuum (UHV) conditions. A clear (2 x 4) surface reconstruction is observed in the reflection high-energy electron diffraction (RHEED) pattern when this desorption process is finished. On the heat-cleaned substrate first a 0.2 µm thick GaAs buffer layer is grown to provide an atomically flat starting surface. The substrate temperature is measured by an infrared pyrometer and calibrated through the desorption temperature of the passivating surface oxide from the GaAs substrate (taken to be 580 °C). This temperature is reproducible from run to run for the substrate preparation technique described before. The growth temperature is kept below 550 °C to prevent interdiffusion of Ga and Al atoms between adjacent layers. To compensate for the reduced surface mobility of the adsorbed species on the growing surface, we use growth rates as low as 1 monolayer per 5 to 10 s (≙ 0.34 - 0.17 µm/hr) for both GaAs and AlAs. At each interface, the growth is interrupted for 5 to 10 s under constant arsenic flux. For large-period GaAs/AlAs superlattices grown under the same conditions we have shown previously / 6 / that a 10-s growth interruption results in a sharpening of the X-ray diffraction satellite peaks thus indicating a smoothening of the interface. A further increase of the interruption time does not improve the results.

Prior to the growth of the superlattices the growth rates of GaAs and AlAs are accurately determined from the period of the intensity oscillations, monitored by a photodiode, of the specular beam in the RHEED pattern (Fig. 1). This periodic intensity oscillation provides direct evidence that MBE growth occurs predominantly in a two-dimensional (2D) layer-by-layer growth mode / 7, 8 / . The period corresponds exactly to the time required to grow a monolayer of GaAs or AlAs, i.e. a complete layer of Ga or Al plus a complete layer of As. Although the fundamental principles underlying the observed damping of the amplitude of the oscillations (Fig. 1) are not

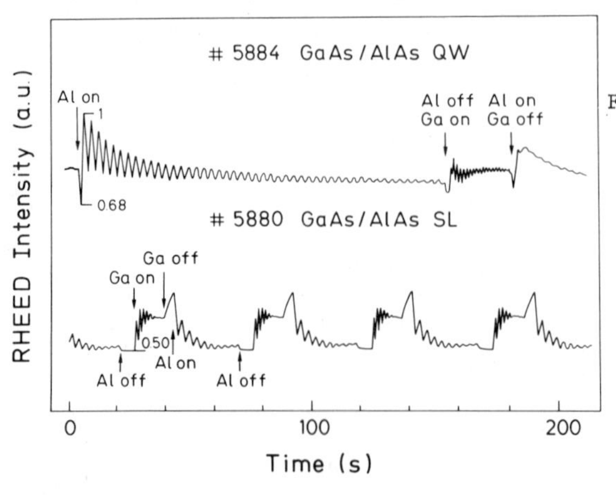

Fig. 1 RHEED intensity oscillations of the specular beam obtained in [100] azimuth on (001) GaAs during growth of GaAs/AlAs heterostructures.

completely understood, the method is now widely used to monitor and to calibrate absolute growth rates in real time with monolayer precision. For this application of the RHEED intensity behaviour the following three assumptions are made:

(i) under specified diffraction conditions (i.e. azimuth, angle of incidence, etc.) the maximum of the specular beam intensity corresponds to the maximum smoothness,

(ii) the maxima in the intensity oscillations correspond to the completion of a monolayer, and

(iii) the damped oscillations indicate that a steady-state distribution of the terrace width has been reached and the growth mode has changed from two-dimensional to step propagation (but not to three-dimensional growth).

The decay of the amplitude of the RHEED intensity oscillations results probably from the loss of coherence between statistically independent areas on the growing surface / 9 / .

The thickness of the GaAs and AlAs layers in the superlattice can thus be controlled by synchronising the actuation of the respective effusion cell shutters with the observed RHEED intensity oscillations. The low growth rate of 1 monolayer per 5 to 10 s. that we employed for the present study, ensures that the crystal growth occurs indeed in a 2D layer-by-layer growth mode. In Fig. 2 we show the RHEED intensity sequence recorded during growth of the m = 1, 2, and 3 ultrathin-layer $(GaAs)_m(AlAs)_m$ superlattices. The intensity oscillations remain constant for the entire growth run of the superlattice indicating the stability of the growth conditions and the excellent periodicity of the layer sequence. The synthesis of each superlattice period requires four steps. At first a monolayer (or bi- or trilayer) of GaAs is deposited. In the second step the crystal growth is stopped for 5 s and the surface is only exposed to the As_4 flux. The third step involves the deposition of a monolayer (or bi- or trilayer) of AlAs. Finally, the fourth step is the same as the second one. The shutters in front of the Ga and Al effusion cells, respectively, are opened at the time when the intensity of the selected feature of the RHEED pattern has reached its maximum value (i.e. maximum reflectivity of a completed monolayer) and closed after one (or two or three) period(s). The second and the fourth step of our growth sequence ensures that each constituent layer of the superlattice is completed before the next one starts. These four steps are repeated several hundred or even thousand times, depending on the required total thickness of the superlattice. The cap layer of the superlattices is always formed by GaAs.

For the growth of the larger-period as well as of the Fibonacci, Thue-Morse, and random superlattices the latter are consisting of sequences of two building blocks A (GaAs) and B (AlAs) of thicknesses of about $d_A = d_B$

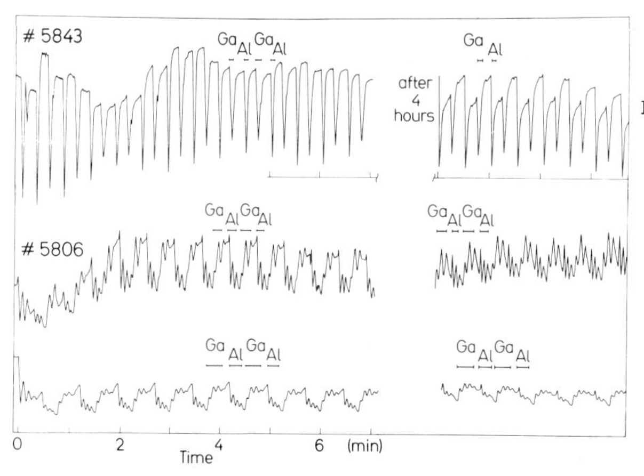

Fig. 2 RHEED intensity oscillations of the specular beam from (001) surface in [100] azimuth during growth of $(GaAs)_m(AlAs)_m$ superlattices with m = 1, 2, 3.

= 2 nm) the growth rate is increased to 1 monolayer per 2 s (≙ 0.56 μm/hr), and the growth interruption at each interface is also increased to 10 s.

3. GaAs/AlAs SHORT-PERIOD AND ULTRA THIN-LAYER SUPERLATTICES

The research activities on GaAs/AlAs short-period superlattices (SPS) were initiated by some detrimental structural, electrical, and optical properties of the ternary alloy $Al_xGa_{1-x}As$. First, the interface roughness for growth of binary GaAs or AlAs on the ternary alloy mentioned in Ref. 8 yields inferior excitonic and transport properties. Second, the electrical properties of n-type $Al_xGa_{1-x}As$ for $0.2 < x < 0.45$ are controlled by a deep donor ("DX center") in addition to the hydrogen-like shallow donor due to the peculiar band structure of the alloy / 10 - 12 / . Third, the X minimum of the conduction band becomes the lowest one when $x > 0.43$ (at 4K), and thus an indirect bandgap of the alloy results / 13 / . The investigations of ultrathin-layer $(GaAs)_m(AlAs)_n$ superlattices (UTLS) with (m, n) from 10 down to 1 were motivated by the possibilities to shift the confined-particle states of the Γ, L and X valleys of the conduction band (and also of Γ of the valence band) to high enough energy to create radiative size-determined "direct-indirect" transitions exceeding in energy the bulk direct-indirect transitions of the ternary alloy $Al_xGa_{1-x}As$ at $x_c \simeq 0.43$ / 14 / . The all-binary GaAs/AlAs short-period and ultrathin-layer superlattices are thus considered as possible substitutes for the random ternary alloy in advanced device structures. The electronic properties of those superlattices, however, which are in the transition between the extremes of quantum well behaviour for period length > 8 nm (i.e. m,n > 15) and of a possible alloy-like behaviour of monolayer superlattices, are not completely understood.

Confinement layers composed of short-period superlattices A_mB_n with $5 \leq (m, n) \leq 10$ play an important role for highly improved optical properties of GaAs, GaSb, and $Ga_{0.47}In_{0.53}As$ quantum wells / 15 - 17 / . The effective barrier height for carrier confinement in the quantum well is adjusted by the appropriate choice of the layer thickness of the lower-gap material in the SPS. A grading of the effective bandgap of the barriers can be achieved by gradually changing the width of the wells (lower-gap material) in the SPS / 18 / . We have studied in detail the process of carrier injection and vertical transport in the SPS towards the quantum well and the process of radiative electron-hole recombination / 19 / . The SPS barriers consist of all-binary GaAs/AlAs for GaAs quantum wells, of all-binary GaSb/AlSb for GaSb quantum wells, and of all-ternary $Ga_{0.47}In_{0.53}As/Al_{0.48}In_{0.52}As$ for $Ga_{0.47}In_{0.53}As$ quantum wells lattice matched to InP. The observed improvement of the optical properties of SPS confined quantum wells is due (i) to a removal of substrate defects by the SPS layer, (ii) to an amelioration of the interface between quantum well and barrier, and (iii) to a modification of the dynamics of photoexcited carriers in the SPS barrier. In particular for GaSb and for $Ga_{0.47}In_{0.53}As$ quantum wells we have provided the first direct evidence for intrinsic exciton recombination by application of SPS barriers / 16, 17 /. Detailed studies of the dynamics of photoexcited carriers sinking into SPS confined enlarged GaAs quantum well have clearly demonstrated an efficient vertical transport of electrons and holes through the thin AlAs barriers of the SPS / 19 - 21 / . In addition, the recombination lifetime of photoexcited carriers in GaAs quantum wells is significantly improved by the application of SPS barriers / 22 / . The improved dynamics of injected carriers and their efficient trapping by the enlarged well make the SPS confined quantum wells very attractive for application in newly designed heterostructure lasers with separate superlattice waveguide and superlattice barriers. Recently, graded-index waveguide separate-confinement heterostructure (GRIN-SCH) laser diodes with a very low threshold current density have been fabricated / 23 / , where the graded-index waveguide was constructed by all-binary GaAs/AlAs SPS with gradually changed GaAs layer thicknesses (from 1.8 to 3.3 nm at eight intervals of six periods with a constant AlAs barrier width of 1.9 nm). When n- or p-doped SPS confinement layers are required, this can be achieved by selevtively doping the GaAs regions of the SPS /24/.

A distinct example for the removal of substrate defects by SPS buffer layers and for the improvement of the interface between quantum well and SPS barrier is given by a modified selectively doped GaAs/$Al_xGa_{1-x}As$ hetrostructure with high-mobility 2D electron gas (2DEG) which we have developed recently / 25 /. A 10-period GaAs/AlAs SPS prevents propagation of dislocations from the substrate so that the thickness of the active GaAs layer containing the 2DEG can be reduced to 50 nm. The SPS confined narrow active GaAs layer is of distinct importance for transistor operation, because the electrons cannot escape too far from the 2D channel during pinch-off. This implies a higher transconductance for the high electron mobility transistor (HEMT). In addition, the growth time of the complete heterostructure is reduced to less than 15 min. An additional 15 min for wafer exchange and heat and cool time makes a total of 30 min through-put time per 2-in high-quality heterostructure wafer grown by MBE. We have further used GaAs/AlAs SPS in one-sided selectively doped $Al_xGa_{1-x}As$/AlAs/GaAs multi quantum well heterostructures which exhibit an enhanced electron mobility of more than 6 x 10^5 cm^2/V_s at 4K with AlAs spacer as narrow as 4.5 nm / 26 / .

The recent progress in the control of interface quality using RHEED intensity oscillation (Fig. 2) and growth interruption has led to the successful growth of ultrathin-layer $(GaAs)_m(AlAs)_m$ superlattices with m = 1, 2, 3, and 4 / 27, 28 / . In Fig. 3 we show schematically the layer sequence of a $(GaAs)_1(AlAs)_1$ monolayer superlattice grown in [001] direction. Growth of these superlattices is achieved by monitoring each deposited GaAs and AlAs monolayer from the RHEED oscillation period, interrupting the group-III-element flux at m = 1, 2, 3, 4 and allowing the RHEED intensity to recover almost to its initial value, and then depositing the next layer. In the ultrathin-layer superlattices the electron states have lost their 2D character and become extended throughout the entire superlattice. The mixing of different k-states due to zone folding effects can no longer be neglected, and the electronic structure can therefore not be described by a combination of the two constituent semiconductors in terms of the effective mass theory. The calculations of the energy band structure have thus to be performed by treating the superlattice crystal as a whole / 29, 30 /.

The well ordered periodic layer-by-layer arrangement of Ga and Al atoms on the appropriate lattice sites in [001] direction manifests itself in the appearance of distinct satellite peaks around the (002) and (004)

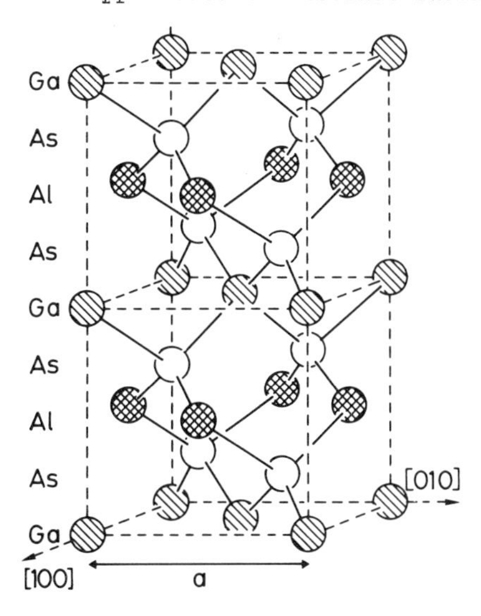

Fig. 3 Schematic arrangement of Al, Ga, and As atoms in a $(GaAs)_1(AlAs)_1$ monolayer superlattice ("ordered $Al_{0.5}Ga_{0.5}As$ alloy")

Fig. 4
X-ray diffraction patterns of $(GaAs)_m$ $(AlAs)_m$ superlattices with m = 1, 2. 3 obtained in the vicinity of the (002) and (004) Bragg reflections using CuKα radiation.

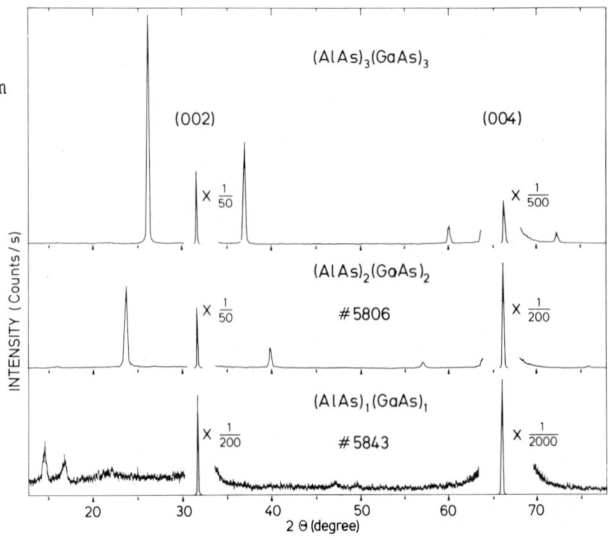

reflections of the X-ray diffraction patterns, shown for the three prototype ultrathin-layer superlattices in Fig. 4. The intensity and sharpness of the satellite peaks confirm the smothness of the heterointerfaces. The nominal and the actual period lengths (i.e. values of m and n) of 10 representative $(GaAs)_m(AlAs)_n$ superlattices are summarized in Table 1. Considering the precision of our X-ray measurements with a double-crystal spectrometer (1% absolute precision for the mean Al composition and less than 1% relative precision for the period), it is important to indicate the measured m and n values with two digits after the decimal point, in particular for the ultrathin-layer superlattices. Also included in Table 1 are the results of low-temperature (4K) photoluminescence, photoluminescence excitation and ellipsometric measurements of the various $(GaAs)_m(AlAs)_n$ superlattices. The position of the intrinsic luminescence peak is deduced from the evolution of the photoluminescence spectrum using the Kr^+ 476-nm line with a power density between 25 mW/cm^2 and 70 W/cm^2. The term "intrinsic" excludes any impurity related and phonon assisted transitions but includes bound excitons. The corresponding gap may thus be underestimated by an

TABLE 1 Period lengths and optical transitons of $(GaAs)_m(AlAs)_n$ superlattices. The E_o gap values were determined by spectroscopic ellipsometry / 31 /

Sample	nominal (m-n)	X-ray m	X-ray n	PL intrinsic peak (eV)	PLE peak or threshold (eV)	E_o gap (eV)
# 5867	100X(7-7)	7.18	7.38	1.831 (X)	1.95	
# 5868	350X(2-2)	2.01	2.06	2.070 (X)	2.19	2.18
# 5869	467X(2-1)	2.06	0.75	1.815 (Γ)	1.83	1.84
# 5870	100X(7-7)	6.99	7.07	1.835 (X)	1.97	1.94
# 5871	350X(3-1)	3.02	0.87	1.776 (Γ)		1.82
# 5872	467X(1-2)	0.74	1.96	2.101 (X)	2.48	2.53
# 5873	350X(1-3)	0.69	3.08	2.127 (X)	2.59	2.60
# 5894	377X(11-6)	10.49	6.29	1.769 (Γ)	1.79	1.77
# 5897	233X(4-2)	3.83	2.02	1.905 (Γ)	1.92	
# 5902	700X(1-1.5)	0.89	1.61	2.086 (X)	2.35	2.27

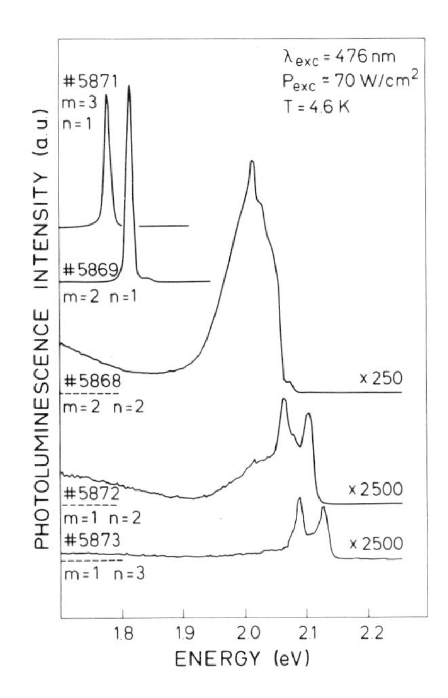

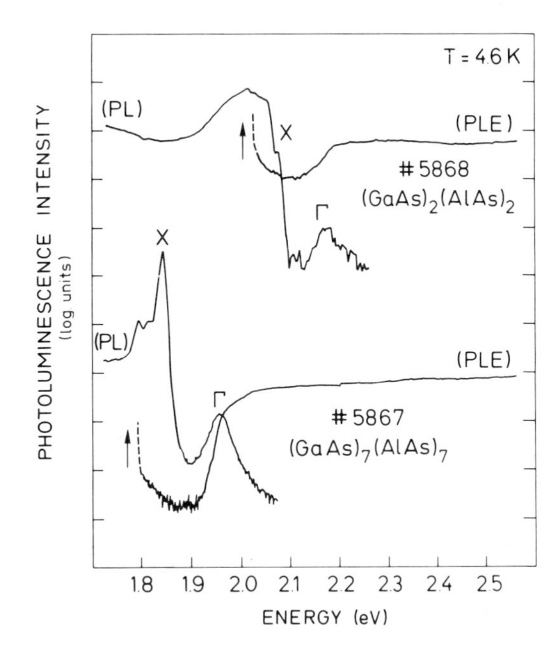

Fig. 5 Photoluminescence spectra of $(GaAs)_m(AlAs)_n$ superlattices having different period lengths and different average Al compositions (i.e. m, n-values)

Fig. 6 Photoluminescence and photoluminescence excitation spectra of two $(GaAs)_m(AlAs)_n$ superlattices having different period lengths but the same average Al composition.

amount equal to the binding energy of the exciton plus an additional binding energy.

The samples labelled Γ in Table 1 exhibit a strong luminescence (see also Fig. 5) and an excitonic peak in the excitation spectra with moderate Stokes shift similar to the behaviour of conventional GaAs quantum wells. The energy levels involved in these optical transitions are obviously Γ-like states. The other samples (labelled X) exhibit a large energy shift between the luminescence and the excitation threshold, as shown in Fig. 6. We attribute the luminescence to X-like states mainly localized in the AlAs layers /32, 33 / and the luminescence excitation threshold to the onset of the Γ-like transition (see Fig. 7 for illustration). This interpretation is confirmed by the results of the E_o gap determination through

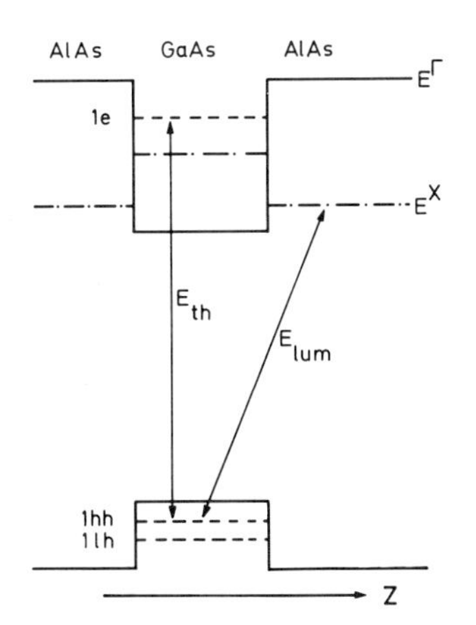

Fig. 7 Schematic real-space energy band diagram of a section of GaAs/AlAs UTLS indicating the transition for luminescence (E_{lum}) and for absorption (E_{th}). The solid lines indicate the conduction and valence band edges of GaAs and AlAs at Γ, the dashed line the bottom of the conduction band at X, and the dash-dotted line the Γ electron and hole subbands in the GaAs quantum well.

ellipsometry. These ellipsometric measurements provide the first gap having a significant absorption, i.e. the Γ-like gap at k = o. The agreement for the two measurements of $E_{g\Gamma}$ given in Table 1 is very good, if we take into account the typical errors of the determination of the luminescence excitation threshold (± 10 meV) and the E_o gap (± 20 meV). The luminescence and the excitation spectra obtained from the (m, n) = 2 and the (m, n) = 7 superlattice and depicted in Fig. 6 clearly demonstrate that in these indirect-gap samples the luminescence associated with the Γ states is directly detectable. The other luminescence features at lower energy which are not labelled in Fig. 6 are probably due to phonon assisted transitions riding over and impurity related background. These features are certainly not caused by slightly different period lengths existing in the samples, as their excitation spectra are similar and indicate the same absorption threshold. The occurrence of GaAs-like and AlAs-like phonon sidebands in the photolumescence spectra is characteristic for short-period $(GaAs)_m(AlAs)_n$ superlattices and extions of the electron wave function beyond the GaAs quantum well. Finally, it is important to note that the $(GaAs)_m(AlAs)_n$ superlattices labelled "indirect (X)" in Table 1 do not necessarily exhibit a weak luminescence.

We next compare our experimental results with calculations using a simple Kronig-Penney model. As a first approximation the energies of the confined Γ and X states are calculated separately / 33 / using the following parameters for GaAs: $E_{g\Gamma}$ = 1519 meV, E_{gX} = 1983 meV, $m_{e\Gamma}$ = 0.0657(m_o), m_{hh} = 0.377, m_{eXl} = 1.3, m_{eXt} = 0.23, and for AlAs: $E_{g\Gamma}$ = 3113 meV, E_{gX} = 2228 meV, $m_{e\Gamma}$ = 0.228, m_{hh} = 0.478, m_{eXl} = 1.1, m_{eXt} = 0.19. The curves shown in Fig. 8 are calculated with a valence-band offset of 558 meV (correspondig to a 65:35 band-gap discontinuity) and they include a simple non-parabolicity correction for the Γ electrons. Inspection of Fig. 8 reveals a surprisingly good agreement between theory and experiment for superlattices with (m + n) ≥ 6 taking into account that the calculation neglects any interaction and mixing of X and Γ states. In addition, as pointed out by Ihm / 34 / , the X_z state of this model (where z represents the direction of layer sequence) is always located at lower energy than the X_{xy} state due to the lower transverse effective mass at the X point. The data of Fig. 9 demonstrate that for the case of ultrathin-layer superlattices, i.e. (m + n) ≤ 6, the alloy approximation is probably the most adequate model which predicts the position and the type of the band gap quite accurately. For the calculated curves of Fig. 9 we use the expression of Casey and Panish / 35 / for the evaluation of the Γ gap and the following expression for the X gap of the ternary alloy: E_{gX}(eV) = 1.986 + 0.113 x_{Al} + 0.129 x_{Al}^2. The Γ and X gaps of the ultrathin-layer $(GaAs)_m(AlAs)_n$ superlattices with (m + n) ≤ 6 are very close to those of the corresponding ternary alloy. In particular, the (m, n) = 2, 3, and 4 superlattices exhibit optical properties which are characteristic for indirect-gap III-V semiconductors. However, other interesting electronic properties of these

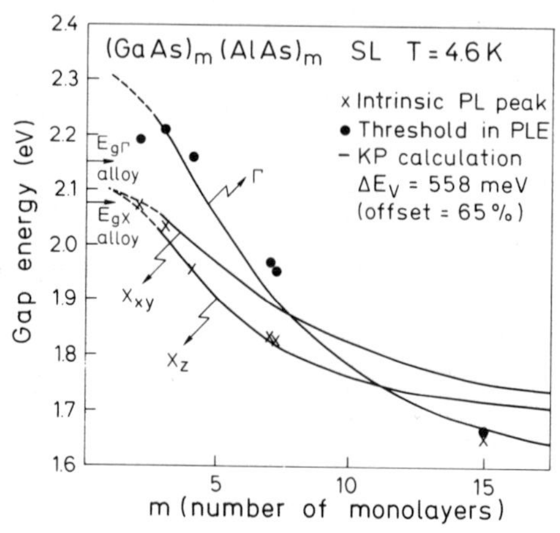

Fig. 8 Comparison of experimental results for $(GaAs)_m(AlAs)_m$ superlattices and calculations based on the Kronig-Penney model (parameters see text). The arrows indicate the position of the band gap of the $Al_{0.5}Ga_{0.5}As$ ternary alloy.

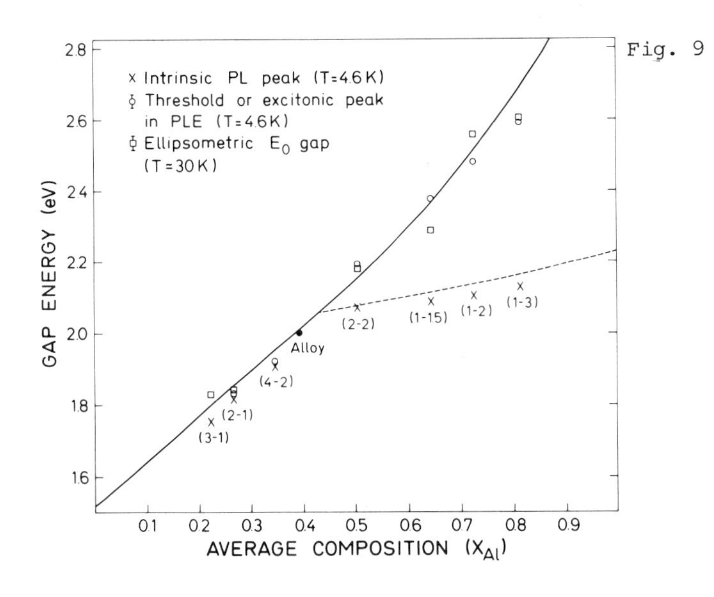

Fig. 9 Comparison between experimentally determined band gaps of several $(GaAs)_m(AlAs)_n$ ultrathin-layer superlattices and the calculated band gaps of the corresponding $Al_xGa_{1-x}As$ alloy. The filled dot indicates the excitonic peak of a ternary alloy determined by excitation spectroscopy.

superlattices, including the suppression of alloy disorder effects / 36 / and the lifting of the valence band degeneracy, do exist.

4. GaAs/AlAs FIBONACCI AND THUE-MORSE SUPERLATTICES

Fibonacci and Thue-Morse superlattices range between periodic and random systems. These one-dimensional (1D) quasi-periodic structures are of great current interest from theoretical and experimental prospects. The two characteristic features of a singular continuous spectrum and critical eigenfunctions should give rise to unusual electrical, optical and magnetic properties / 3, 4, 37 - 40 /. Merlin et al. / 3 / were the first to fabricate quasi-periodic (or incommensurate) semiconductor superlattices by using a Fibonacci sequence to line up the two different materials GaAs and AlAs. In Fig. 10 we show schematically the arrangement of the constituent GaAs and AlAs layers in a Fibonacci sequence and in a Thue-Morse sequence to form the two different quasi-periodic 1D heterostructures which we have studied by means of double-crystal X-ray diffraction / 6, 41 / and Raman scattering / 42 / .

The Fibonacci sequence F_1, F_2, F_3, ... is defined recursively by $F_{n+1} = F_n + F_{n-1}$, where F_n is denoting the n-th Fibonacci number. The arithmetic properties of the Fibonacci sequence are correlated with the golden section number $\tau = (1 + \sqrt{5}) / 2 = 1.618$ .... An algebraic formula useful to determine the type of the n-th layer in a Fibonacci sequence is given by Ω =

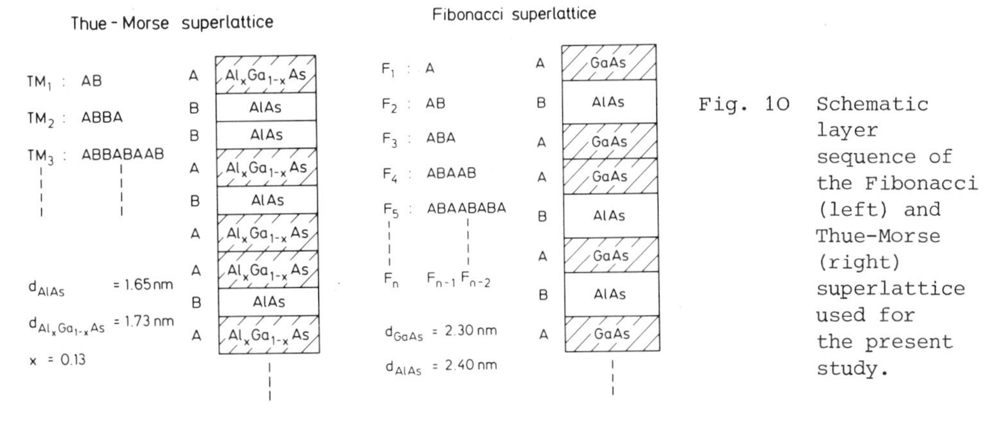

Fig. 10 Schematic layer sequence of the Fibonacci (left) and Thue-Morse (right) superlattice used for the present study.

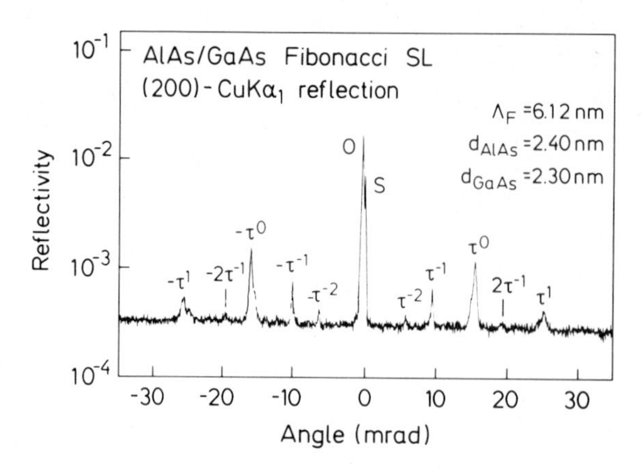

Fig. 11 X-ray diffraction of a GaAs/AlAs Fibonacci superlattice taken around the (200) reflection using $CuK\alpha_1$ radiation.

$INT (n\tau^{-1} + 1) - INT[(n-1)\tau^{-1} + 1]$. If $\Omega = 0$ than layer A (e.g. GaAs), otherwise layer B (e.g. AlAs) is the n-th layer / 41 / . In Fig. 11 we show the measured X-ray diffraction pattern of a GaAs/AlAs Fibonacci superlattice around the quasiforbidden (200) $CuK\alpha_1$ reflection. A dense set of satellite peaks (Cantor set) is a characteristic feature of Fibonacci superlattices. The present Fibonacci sequence is composed of 233 GaAs and 144 AlAs layers. We observe several satellite peaks around the GaAs substrate peak ("S")and the main superlattice peak ("O"). From the angular distance between two different satellite peaks the average Fibonacci superlattice period is determined to be $\Lambda_{SL} = d_{AlAs} + d_{GaAs} \cdot \tau = 6.12$ nm /4/. The Thickness of the constituent layers is found to be $d_{GaAs} = 2.3$ nm and $d_{AlAs} = 2.4$ nm in excellent agreement with the vlaues derived from the growth conditions.

Thue-Morse superlattices can be generated by the substitution rule A → AB and B → BA (see Fig. 10). At present it is still an open question whether the Thue-Morse sequence is a connecting link between the Fibonacci and the periodic superlattices / 43 / or whether the Thue-Morse sequence is not quasi-periodic but automatic and thus more closely related to the random systems / 4, 42 / . The investigation of non-Fibonaccian quasi-

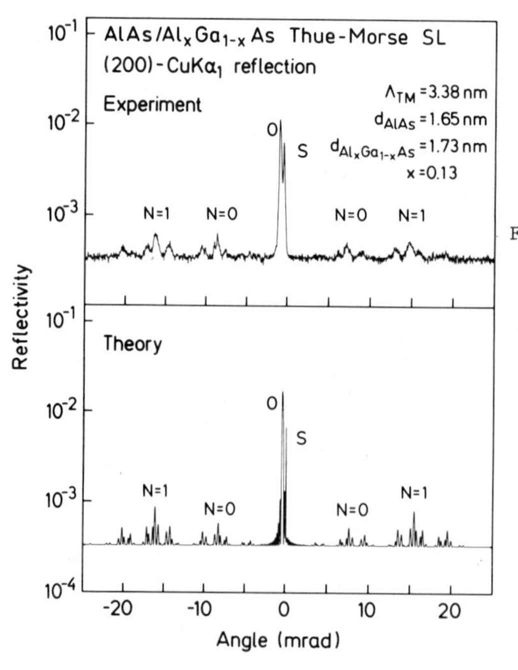

Fig. 12 Measured and calculated X-ray diffraction pattern of a $AlAs/Al_xGa_{1-x}As$ Thue-Morse superlattice around the (200) $CuK\alpha_1$ reflection.

periodic (or automatic) superlattices is thus a promising step towards a unified theory of disordered, quasi-periodic, and periodic systems. In Fig. 12 we show a comparison between the measured and the calculated X-ray diffraction pattern obtained from a $AlAs/Al_xGa_{1-x}As$ Thue-Morse superlattice around the (200) reflection using $CuK\alpha_1$ radiation. The main satellite peaks are labeled by the index N. In the calculated diffraction pattern the main satellite peaks are split into several subpeaks, which are not observed in the experimental curve. This smearing-out of the peaks is probably caused by small thickness fluctuation and/or by an interface roughness of about 1 monolayer. Based on the experimental diffraction data the thickness of the constituent layers of the $GaAs/Al_xGa_{1-x}As$ Thue-Morse superlattice is determined to d_{AlAs} = 1.65 nm and $d_{Al_xGa_{1-x}As}$ = 1.73 nm and the mole fraction in the $Al_xGa_{1-x}As$ layers to $x = 0.13$ in agreement with the data derived from the growth conditions.

In addition, we have investigated nonresonant Raman scattering by acoustic phonons in these Fibonacci and Thue-Morse superlattices / 42 / . The Raman spectra of Fibonacci superlattices exhibit a series of sharp doublet peaks which are centered at frequencies that follow a power-law behaviour. This result implies that the quasi-periodicities originating in the Fibonacci sequence cause zone-folding effects on acoustic phonons. The calculated folded-phonon frequencies using an elastic continuum model are in excellent agreement with the observed frequencies of the doublet peaks. The Raman spectrum obtained from the Thue-Morse superlattice is more complex. A detailed analysis is in progress.

5 CONCLUSION

The unique capability of molecular beam epitaxy to position individual atomic layers one by one allows to produce artificially layered crystals with prescribed or "tailored" properties. We are thus able to fabricate layered materials which test major theories. This is of particular importance for simple prototypes of basic mathematical or physical problems which could never before be tested directly in the laboratory. The whole series of disordered, quasi-periodic, and periodic 1D systems described before may in future form a basic set of structures to study Anderson localization and possible transitions (mobility edges) between localized and extended states.

ACKNOWLEDGEMENT

This work is supported by the Bundesministerium für Forschung und Technologie of the Federal Republic of Germany.

REFERENCES

/ 1 / L.L. Chang and K. Ploog (Eds.): "Molecular Beam Epitaxy and Heterostructures" (Martinus Nijhoff, Dordrecht, 1985) NATO Adv. Sci. Inst. Ser. E 87 (1985)

/ 2 / E.H.C. Parker (Ed.): "The Technology and Physics of Molecular Beam Epitaxy" (Plenum, New York, 1985)

/ 3 / R. Merlin, K. Bajema, R. Clarke, F.Y. Juang, and P.K. Bhattacharya, Phys. Rev. Lett. 55, 1768 (1985)

/ 4 / F. Axel, J.P. Allouche, M. Kleman, M. Mendes-France, and J. Peyriere, J. Physique 47, Colloque C 3, C3-181 (1986)

/ 5 / H. Fronius, A. Fischer, and K. Ploog, J. Cryst. Growth 81, 169 (1987)

/ 6 / L. Tapfer and K. Ploog, Phys. Rev. B 33, 5565 (1986)

/ 7 / T. Sakamoto, H. Funabashi, K. Ohta, T. Nakagawa, N.J. Kawai, T.Kojima, and K. Bando, Superlattices and Microstructures 1, 347 (1985)

/ 8 / B.A. Joyce, P.J. Dobson, J.H. Neave, K. Woodbridge, J. Zhang, P.K. Larsen, and B. Bölger, Surf. Sci. 168, 423 (1986)

/ 9 / F. Briones, D. Golmayo, L. Gonzales, and A. Ruiz, J. Cryst. Growth 81, 19 (1987)

/ 10 / D.V. Lang, R.A. Logan, and M. Jaros, Phys. Rev. B 19, 1015 (1979)

/ 11 / E.F. Schubert and K. Ploog, Phys. Rev. B 30 7021 (1984)

/ 12 / T.N. Morgan, Phys. Rev. B 34, 2664 (1986)

/ 13 / R. Dingle, R.A. Logan, and J.R. Arthur, Inst. Phys. Conf. Ser.33a, 210 (1977)

/ 14 / M.D. Camras, N.Holonyak, K. Hess, J.J. Coleman, R.D. Burnham, and D.R. Scifres, Appl. Phys. Lett. 41, 317 (1982)

/ 15 / K. Fujiwara, H. Oppolzer, and K. Ploog, Inst. Phys. Conf. Ser. 74, 351 (1985)

/ 16 / K. Ploog, Y. Ohmori, H. Okamoto, W. Stolz, and J. Wagner, Appl. Phys. Lett. 47, 384 (1985)

/ 17 / J. Wagner, W. Stolz, J. Knecht, and K. Ploog, Solid State Commun. 57, 781 (1986)

/ 18 / J.J. Coleman, P.D. Dapkus, W.D. Laidig, B.A. Vojak, and N. Holonyak, Appl. Phys. Lett. 38, 63 (1981)

/ 19 / K. Fujiwara, J.L. de Miguel, and K. Ploog, Jpn. J. Appl. Phys. 24, L 405 (1985)

/ 20 / B. Deveaud, A. Chomette, B. Lambert, A. Regreny, R. Romestain, and P. Edel, Solid State Commun. 57, 885 (1986)

/ 21 / A. Nakamura, K. Fujiwara, Y. Tokuda, T. Nakayama, M. Hirai, Phys. Rev. B 34, 9019 (1986)

/ 22 / K. Fujiwara, A. Nakamura, Y. Tokuda, T. Nakayama, and M. Hirai, Appl. Phys. Lett. 49, 1193 (1986)

/ 23 / Y. Tokuda, Y.N. Ohta, K. Fujiwara, and T. Nakayama, J. Appl. Phys. 60, 2729 (1986)

/ 24 / T. Baba, T. Mizutani, and M. Ogawa, Jpn. J. Appl. Phys. 22, L 627 (1983)

/ 25 / K. Ploog and A. Fischer, Appl. Phys. Lett. 48, 1392 (1986)

/ 26 / K. Ploog, H. Fronius, and A. Fischer, Appl. Phys. Lett. 50, 1237 (1987)

/ 27 / M. Nakayama, K. Kubota, H. Kato, S. Chika, and N. Sano, Solid State Commun. 53, 493 (1985)

/ 28 / T. Isu, D.S. Jiang, and K. Ploog, Appl. Phys. A 43, 75 (1987)

/ 29 / T. Nakayama and H. Kamimura, J. Phys. Soc. Jpn. 54, 4726 (1985)

/ 30 / M.A. Gell, D. Ninno, M. Jaros, and D.C. Herbert, Phys. Rev. B 34, 2416 (1986)

/ 31 / J. Nagle, M. Garriga, W. Stolz, T. Isu, and K. Ploog, Proc. 3rd Int. Conf. Modulated Semicond. Structures (MSS-III), Montpellier 1987, J. Physique Colloque, to be published (1988)

/ 32 / E. Finkman, M.D. Sturge, and M.C. Tamargo, Appl. Phys. Lett. 49, 1299 (1986)

/ 33 / G. Danan, B. Etienne, F. Mollot, R. Planel, A.M. Jean-Louis, F. Alexandre, B. Jusserand, G. Le Roux, J.Y. Marzin, H. Savary, and B. Sermage, Phys. Rev. B 35, 6207 (1987)

/ 34 / J. Ihm, Appl. Phys. Lett. 50, 1068 (1987)

/ 35 / H.C. Casey and M.B. Panish, "Heterostructure Lasers", Part A (Academic Press, New York, 1978) pp. 192 - 193

/ 36 / T. Yao, Jpn. J. Appl. Phys. 22, L 680 (1983)

/ 37 / L. Todd, R. Merlin, R. Clarke, K.M. Mohanty, and J.D. Axe, Phys. Rev. Lett. 57, 1157 (1986)

/ 38 / M. Fujita and K. Machida, Solid State Commun. 59, 61 (1986)

/ 39 / S. Das Sarma, A. Kobayashi, and R.E. Prange, Phys. Rev. B 34, 539 (1986)

/ 40 / M. Karkut, J. Triscone, D. Ariosa, and Ø. Fischer, Phys. Rev. B 34, 4390 (1986)

/ 41 / L. Tapfer and Y. Horikoshi, Proc. 14th Int. Symp. GaAs Related Compounds, Crete 1987, Inst. Phys. Conf. Ser., to be published (1988)

/ 42 / R. Merlin, K. Bajema, J. Nagle, and K. Ploog, Proc. 3rd Int. Conf. Modulated Semicond. Structures (MSS-III), Montpellier 1987, J. Physique Colloque, to be published (1988)

/ 43 / R. Riklund, M. Severin, and Y. Liu, Int. J. Mod. Phys. B 1, 121 (1987)

INTERFACE CHARACTERIZATION OF GaInAs-InP SUPERLATTICES GROWN BY LOW PRESSURE METALORGANIC CHEMICAL VAPOR DEPOSITION

Manijeh Razeghi, Philippe Maurel and Franck Omnes
(invited paper)

Laboratoire Matériaux Exploratoires/LCR, THOMSON-LCR
B.P. N° 10, 91401 Orsay Cedex (France)

INTRODUCTION

Very high quality $Ga_{0.47}In_{0.53}As$-InP heterojunctions, quantum wells and superlattices have been grown by low pressure metalorganic chemical vapor deposition (LP-MOCVD). Excitation spectroscopy shows evidence of strong and well resolved exciton peaks in the luminescence and excitation spectra of GaInAs-InP quantum wells. Optical absorption shows room-temperature exciton in GaInAs-InP superlattices (M. Razeghi et al., 1983)[1]. Quantum wells as thin as 8 A with a photoluminescence linewidth of 9 meV have been grown.

Negative differential resistance at room temperature from resonant tunnelling in GaInAs-InP double barrier heterostructures grown by LP-MOCVD has been observed for the first time by M. Razeghi et al., 1986)[2]. The "edge-on" transmission electron microscopy (TEM) characterization on a resonant tunnelling structure of GaInAs-InP of a 28 A thickness grown under the same conditions shows that the thickness of the triple layer is generally regular (Razeghi et al., 1986)[2].

GROWTH CONDITIONS

The gas panel and reaction chamber have already been described in detail in a previous review paper (Razeghi et al., 1985)[3]. Pressure inside the reactor is maintained at 1/10 atm during the growth while the temperature of the susceptor is fixed at 550°C. We have used pure Arsine (AsH_3) and Phosphine (PH_3) as sources of V elements As and P. Triethylgallium (TEG) is used as III element Ga source. Either triethylindium (TEI) or trimethylindium (TMI) have been used as indium sources. The organometallic group III species are contained in stainless steel bubblers, which are held at fixed temperatures of 0°C for TEG and 33°C for TEI and TMI.

Diethylzinc (DEZ) is used as the Zn source for p type doping, and is maintained at a controlled temperature of -12°C.

Hydride H_2S or silane SiH_4 are respectively used to provide S and Si for n type doping in the modulation doped heterostructures.

Pure hydrogen (H_2) and nitrogen (N_2) are respectively used as bubbler gases in the (TMI, TEG) and TEI sources as well as carrier gases in the reaction chamber . The total flow rate is maintained at 6 l/min during the growth. Both gases are mixed in a 50%/50% proportion if TEI is the indium source. With TMI as indium source, only hydrogen is sent into the reactor.

All the samples under study have been grown on semi-insulating (Fe doped) or n ($2.10^{18} cm^{-3}$, Sn doped) InP substrates. They were generally (100) oriented, 2° off towards (110).

GROWTH AND CHARACTERIZATION OF InP

We have recently reported the growth (Razeghi et al., 1987)[4] of the purest InP ever obtained by any technique, using trimethylindium as In source. Growth conditions are summarized in table 1. Photoluminescence measurements were performed at 2K, using a dye laser as source of excitation. The excitation energy was tuned just above the band gap of InP, at about 1430 meV, with a slitwidth of 50 μm. Free exciton line (X) is located at about 1418,8 meV. Various components of the neutral donor bound exciton (D°X) recombination, identified as $(D^{\circ}X)_n$, are observed. These lines are located between 1417,3 meV and 1418 meV, with linewidth less than 0,1 meV (see fig. 1)..

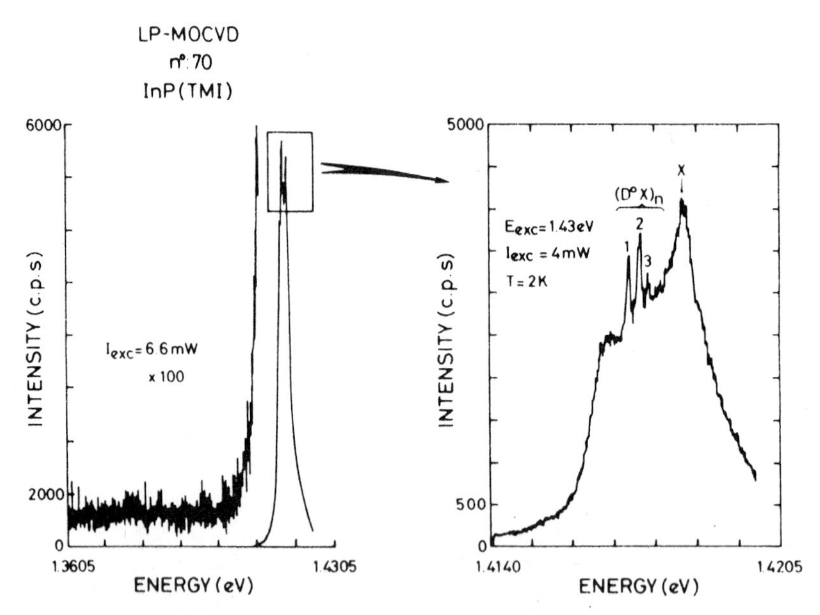

Fig. 1. Photoluminescence spectrum at T = 1.6K of an ultrapure bulk InP grown by LP-MOCVD.

$D^{+}X$ and D°h lines appear at lower energies (1416,6 and 1416,8 meV respectively).

Two peaks related to exciton bound to neutral acceptors (A°X) are generally observed (Dean and Skolnick, 1983)[5] between 1414 and 1414.5 meV. Even by multiplying the signal by a factor of 100 in this region of the spectrum, no trace of such a recombination process is found, thus indicating that the sample is very little compensated. This further demonstrates its high purity, and can explain the exceptionnaly high mobility measured (μ (50K) = 200000 cm $V^{-1}S^{-1}$).

On the other hand, the free exciton peak (X) appear more intense than the donor bound like (D^oX). This is the first time that such a fact is observed in any InP crystal, therefore confirming the very low concentration of donors ($3.10^{13}cm^{-3}$) determined by a polaron profile (figure 2).

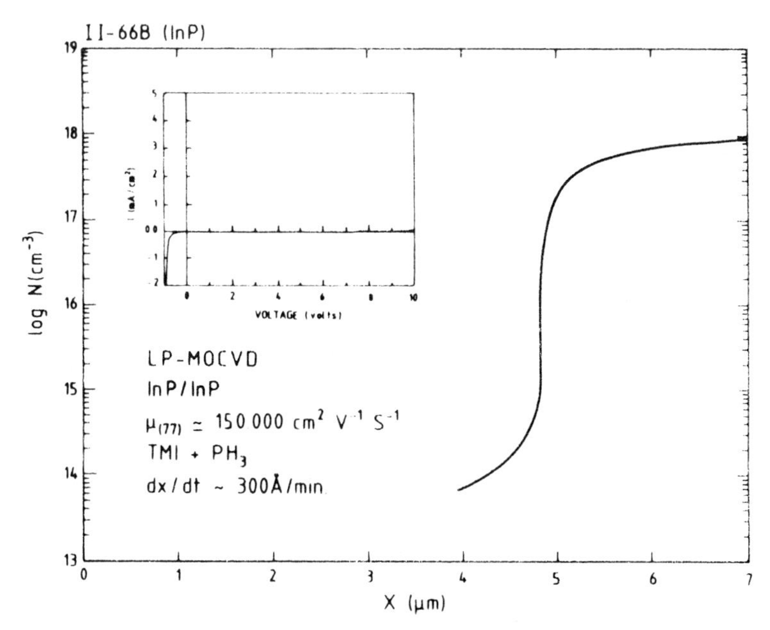

Fig. 2. Polaron profile of an ultrapure bulk InP grown by LP-MOCVD.

To our knowledge, this is the purest InP crystal ever grown, with the highest low temperature mobility reported in the literature.

GROWTH OF $Ga_{0.47}In_{0.53}As$

The growth conditions of $Ga_{0.47}In_{0.53}As$ lattice-matched to InP have already been described in detail (M. Razeghi, 1985)[3]. They are briefly summarized in table 1, using TEI or TMI as Indium source.

The relatively low growth rates of those two compounds makes it possible to obtain very thin layers (quantum wells down to 8 A width) with very sharp interfaces. The growth of the superlattices is controlled by a computer so that the width of the barriers and the wells stays very reproducible.

$Ga_{0.47}In_{0.53}As$, with a bandgap at room temperature of 0.75 eV (λ = 1,66μm) is a very useful material for long wavelength optical communication devices. Moreover, its high mobility and large drift velocity make this material very promising for use in high frequency field effect transistors, or high speed signal processing.

Many high quality $InP/Ga_{0.47}In_{0.53}As$ heterostructures such as modulation doped heterojunctions (MDH), multiquantum wells (MQW) and superlattices (SL) have been grown. We describe now the different characterizations of these structures.

X-Ray diffraction is of much use, particularly in the case of superlattices, to insure the reproducibility of the wells and the barriers. Figure 3 exhibits the X-Ray (400) diffraction pattern of a 10 period $Ga_{0.47}In_{0.53}As/InP$ superlattice, with InP barriers of about L = 250 A and $Ga_{0.47}In_{0.53}As$ wells of L_Z = 200 Å. A synchrotron beam of wavelength λ = 1.2834 A is used as X-Ray source. Satellites corresponding to the artificial crystalline periodicity L_B+L_Z introduced during the growth have been resolved up to n = ± 5. This is rarely observed and underlines the excellent structural quality of the sample under study. The n = 0 satellite, attributed to the (400) Bragg diffraction of the mean parameter of the superlattice appears at the position of the InP substrate peak. This indicates that the lattice parameter of the ternary alloy is well adapted to that of the substrate.

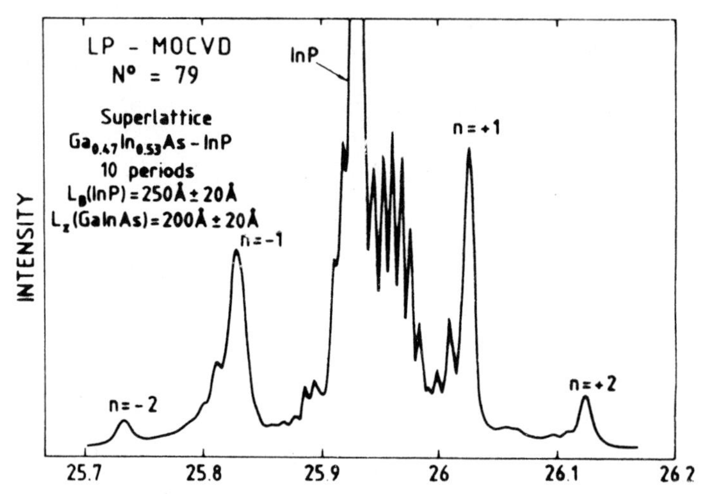

Fig. 3. X-Ray diffraction patern of a 10 periods $Ga_{0.47}In_{0.53}As/InP$ superlattice grown by LP-MOCVD.

From the angle $\Delta\theta$ = 360" between two adjacent satellites, we deduce, using the classical expression[1] : $L_B + L_Z = \lambda / 2\Delta\theta \cos\theta$ the superperiodicity L_B+L_Z = 410 Å ; that corresponds quite well to the expected value from the growth parameters.

The hyperfine structure observed in Fig.3 near the substrate $K\alpha_1$, peak, may be attributed to the interaction between incident and reflected waves through the 5000 A thick epilayer (Pendellosung oscillations) (Bartels and W. Nijman, 1978)[6]. If we use the classical equation : $\delta\theta = \lambda/2L\cos\theta$ with L = 5000 Å, we find $\delta\theta$ = 30" in good agreement with experimental results shown in figure 3. Such behaviour has rarely been observed in III-V superlattices.

SIMS (Secondary Ion Mass Spectroscopy) analyses are performed for quantitative determination of impurities accumulated at the substrate-epilayer interfaces. The analyses are carried out by a modified CAMECA IMS3F. The surfaces of the epitaxial layers are scanned with a focused mass filtered oxygen ion beam ($I_P \simeq 1.5$ Å at 10 keV). The scanned area is 250 μm x 250 μm and the analysed region is 150 μm in diameter. Statistical results of various experiments show that the quantitative results of SIMS are given with an accuracy of ± 20% above a concentration level of 10^{16} at.cm^{-3}. Below this level, results are less accurate, ± 50% at 10^{14}at.cm^{-3}. Talysurf measured depth precision is estimated at ± 10%.

Figure 4 exhibits the SIMS profile of the four majority species (In, Ga, As and P) in a 25 periods $Ga_{0.47}In_{0.53}As/InP$ superlattice with well and barrier thicknesses of 150 A. The indium concentration appears almost constant, because it is the only element present in the barriers as well as in the wells. The signals of the other species are clearly oscillating with the change of chemical composition of the layers, thus evidencing the abruptness of the interfaces and the perfect control of the compositions and of the thicknesses.

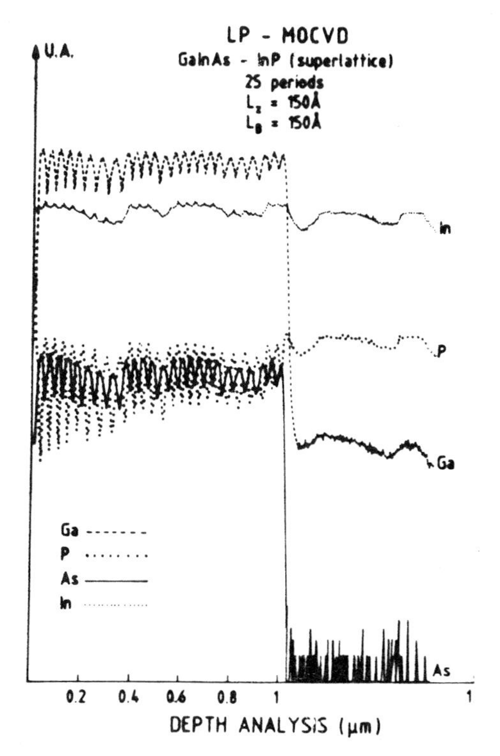

Fig. 4. Ga, In, As and P SIMS profiles of a 25 periods $Ga_{0.47}In_{0.53}As/InP$ L_B = 150 Å, L_Z = 150 Å superlattice grown by LP-MOCVD.

Auger spectra have been performed on chemical bevels. The samples are chemically etched by using a methanol bromine solution (15% Br), in order to obtain a bevel having a mean amplification coefficient of 2100 (measured with a Talysurf). This means that a change of one micron along the surface corresponds to a change of 4.75 A in depth (z direction). By scanning the incident electron beam along the bevel, the phosphorus profile of the previous superlattice has been obtained. It is shown in figure 5. There again the modulation of composition is clearly evidenced, thus confirming the abruptness of the interfaces.

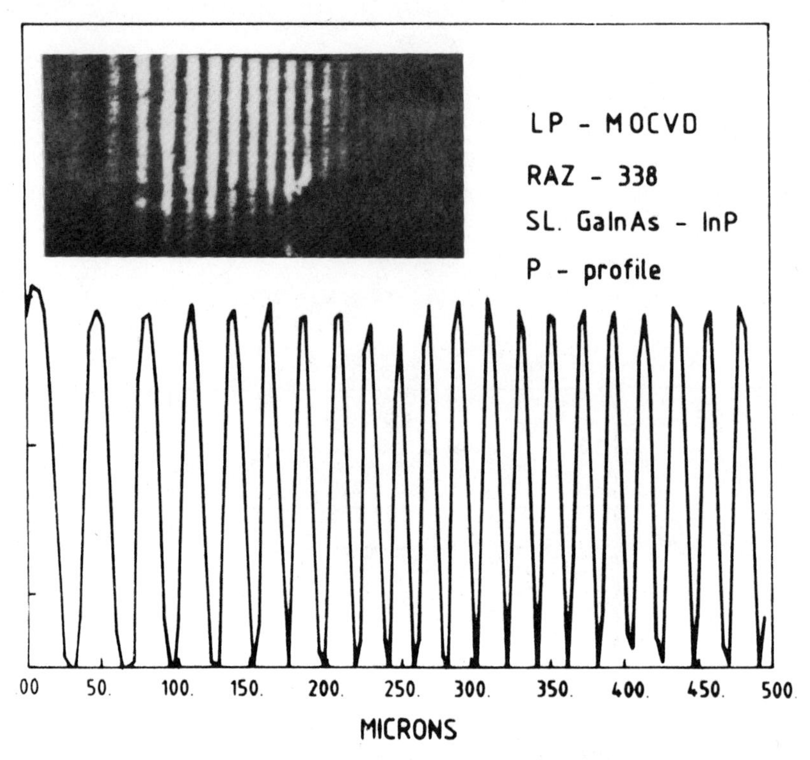

Fig. 5. Phosphorus signal relative to the Auger spectrum of a $Ga_{0.47}In_{0.53}As/InP$ supperlattice grownby LP-MOCVD

The Etch Pit Density (EPD) is also performed on a chemical bevel. The defects are revealed in a H_3PO_4 and HBr solution (Huber et al., 1984)[9]. The etch speed of the acid solution varies with the orientation of the atom planes. Around a dislocation, the atom plane appears slightly disoriented so that it is etched at a higher speed. For a fixed revelation time, the etched thickness is more important around the defects than in the rest of the crystal : etch patterns appear that allow us to determine the density of dislocations.

We show in figure 6 the Etch Pit Density photograph of a $Ga_{0.47}In_{0.53}As/InP$ heterojunction. The epitaxial layers, with a density of dislocations of about $10^4 cm^{-2}$, equal to that of the substrate, show a crystalline quality at least corresponding to that of the InP substrate. No defects are evidenced at the interface.

M. Bulle and T.Viegers(Philips Research Laboratories, Eindhoven) have performed edge-on transmission electron microscopy (TEM) characterization on a 40 period GaInAs-InP superlattice grown by LP-MOCVD. Figure 7-a shows that the thickness of the layers is not regular and there are some dark point defects in the layers. These irregularities are due to the fact that at the time our MOCVD reactor was operated manually, and we had a pyrolysis oven for PH_3. After elimination of PH pyrolysis oven and automatization of the MOCVD reactor (see Razeghi et al., 1986)[13] we improved the thickness homogeneity of the superlattices as shown on figure 7-b. The dark field images at atomic resolution of such a superlattice are shown in figure 7-b. It is clear from this photograph that the interfaces between these layers are perfectly smoth and flat. Looking carefully at the InP-GaInAs interface (fig. 7-c) we can see atomic steps energy 80A, which can be attributed to the 2° misorientation of the InP substrate. The results show that perfect interfaces on an atomic scale can be grown by LP-MOCVD.

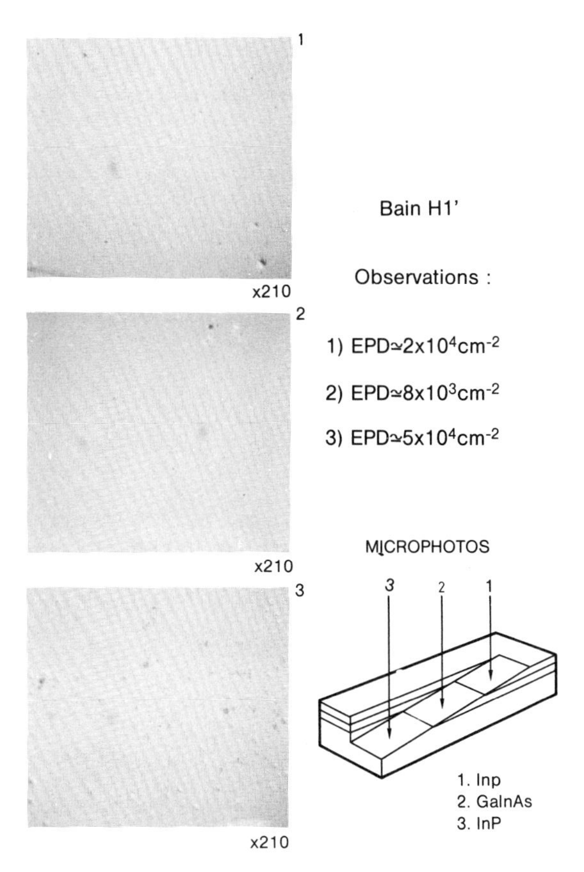

Fig. 6. Etch pitch density photograph of a $Ga_{0.47}In_{0.53}As/InP$ heterojunction grown by LP-MOCVD.

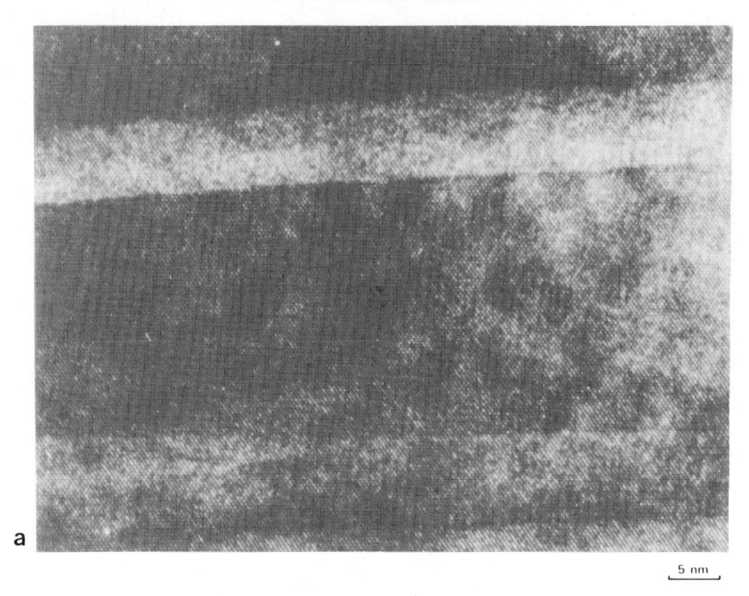

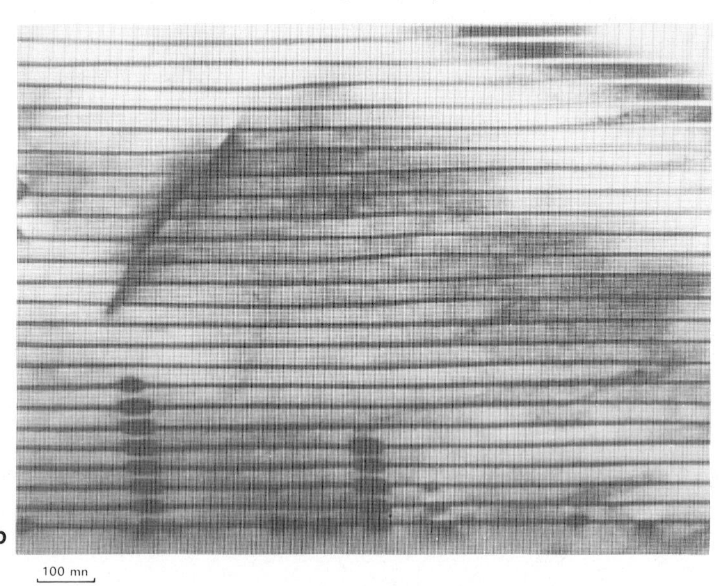

Fig. 7. 7-a : Transmission electron microscope photographe of $Ga_{0.47}In_{0.53}As/InP$ superlattice

7-b : Dark field image at atomic resolution of the same superlattice

LP-MOCVD
n° 640

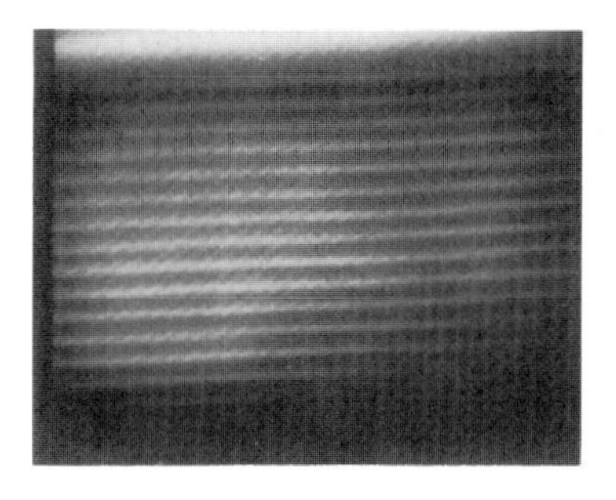

Fig. 7-c : TEM of a $Ga_{0.47}In_{0.53}As/InP$ superlattice grown by automatic MOCVD reactor.

Optical Properties

The quantum size effect appears when the size of the sample is small compared with the intrinsic quantum length scale associated with carriers in the semiconductor. This is given by the de Broglie wavelength $\lambda = h/p$, where h is Planck's constant and p the carrier momentum, typically given by $P^2/2m^* = k_B T$, k_B being the Boltzmann constant.

Figure 8-a shows the optical absorption spectrum of 2 GaInAs-InP superlattices grown by MOCVD. Table 2 indicates the number of periods of alternating layers of InP and GaInAs with thicknesses d_1 and d_2, respectively.

Table 2

sample	d_1(InP)	d_2(GaInAs)	period	$\Delta a/a$
S_1	(200 + 10) Å	(187 + 10) Å	20	2.7×10^{-3}
S_2	(200 + 10) Å	(94 + 5) Å	40	1.3×10^{-3}

The absorption spectra exhibit the step like behavior characteristic of the two-dimensional density of states and excitonic features at the onset of the steps. The magnitude of the absorption coefficient establishes that this system is of the type I, i.e., the layers of the ternary alloy are simultaneously QW for both the conduction and valence-band states (Voisin et al., 1984)[7] These observations show that high quality materials have been achieved.

The structures were evaluated from X-ray double diffraction, using the Cu-$K\alpha_1$ radiation. The standard intensity vs angle plot of (004) diffraction patterns for the two samples reported here (S_1 and S_2) is given in ref. 8 (Razeghi et al., 1985)[8]. A number of high order satellite peaks are resolved, their presence demonstrate the high structural quality of the samples. These patterns show that the structures are actually strained, which indicates a deviation of the alloy composition from the standard composition lattice matched to InP. Taking into account the actual composition and the effect of strains leads to a satisfactory fit of the energy positions of the absorption steps.

Figure 8-b illustrates such a quantum size effect (QSE) for these InP/$Ga_{0.47}In_{0.53}As$ superlattices of 20 and 40 periods. Photoluminescence is performed at 2K, using an He-Ne laser as excitation source. It is analysed in a 60 cm grating spectrometer and detected with a high sensitivity N_2-cooled Ge photodiode.

The QSE can also be evidenced in a multiquantum well structure (MQW) composed of quantum wells of different thicknesses, keeping the barrier width of InP constant. The photoluminescence spectrum of such a sample is shown in figure 9. It is composed of three wells of 400 A, 75 A and 35 A $Ga_{0.47}In_{0.53}As$ separated with 500 A thick InP barriers. The signal corresponding to the 75 A well has been detail enlarged for. It is remarkably thin, with a linewidth of 5 meV. The excitonic nature of the recombination is proved by plotting the photoluminescence excitation spectrum (PLE), using a spectrometer filtered lamp as a tunable source for PLE. It exhibits a series of peaks which can be attributed to the heavy hole to electron (E-HH) and light hole to electron (E-LH) transitions, as shown in figure 9. The indexes of these transitions are associated with the different indexes of the confinement levels of the electron and holes in the quantum well.

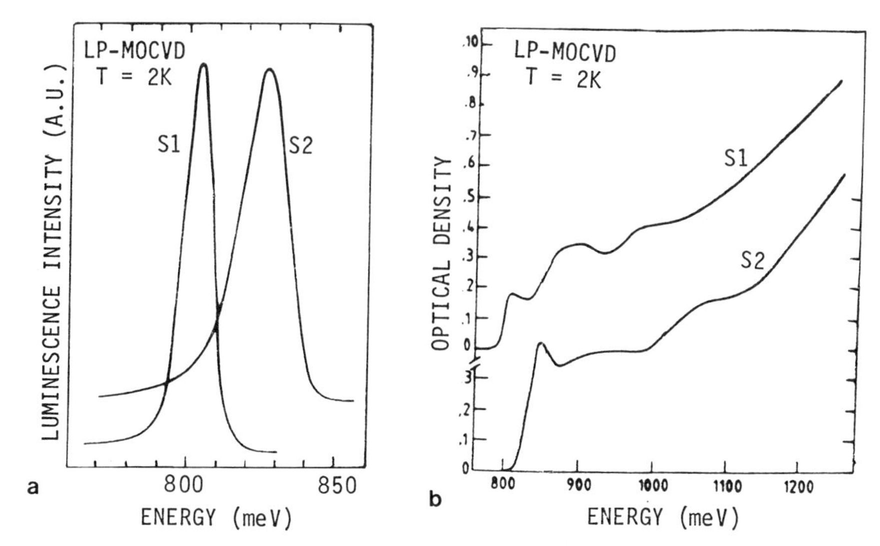

Fig. 8. 8-a : Optical absorption spectrum of two $Ga_{0.47}In_{0.53}As/InP$ superlattices
8-b : T = 2K photoluminescence spectrum the same $Ga_{0.47}In_{0.53}As/InP$ superlattices

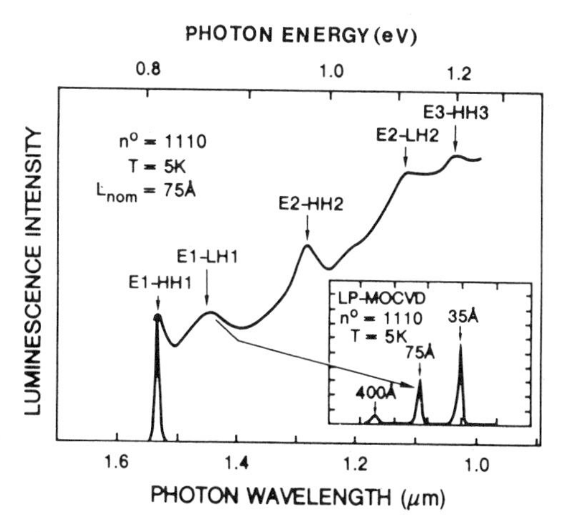

Fig. 9. Photoluminescence and photoluminescence excitation spectra of a $Ga_{0.47}In_{0.53}As/InP$ multiquantum well grown by LP-MOCVD

Figure 10 shows the absorption spectrum at room temperature of a $Ga_{0.47}In_{0.53}As/InP$ superlattice with 40 periods of 100 A thick GaInAs wells and 200 A thick InP barriers together with its T = 5K photoluminescence spectrum. A full width at half maximum of 6 meV is determined. Absorption spectroscopy is performed by polishing the sample on its two faces. It shows that excitonic absorption is present even at room temperature. This result is very important, especially for device applications. LPMOCVD should allow the growth of high quality superlattices for the development of optoelectronic devices for the medium infrared wavelengths such as high speed modulators and optical switches similar to those that have already been demonstrated with GaAs/GaAlAs multiquantum wells and superlattices (Miller and Chemla, 1984)[10].

Tunnelling effect

The dark field image of atomic resolution on a resonant tunnelling structure of GaInAs-InP of a 30A $Ga_{0.47}In_{0.53}As$ layer sandwiched between two InP layers of the same thickness grown under the same conditions, is exhibited in figure 11. It shows that very thin epitaxial layers of GaInAs-InP of accurately controlled thickness and composition can be prepared by LPMOCVD. We therefore observed negative differential resistance at room temperature (figure 12-b). This occurs when the confined level of the well (between the two barriers) is brought below the Fermi level of the injecting contact by applying a potential across the structure. When the potential is increased further the bound level fall below the band edge of the injecting contact and the conductance and the current of the structure are reduced.

This is the first observation of such an effect in the $Ga_{0.47}In_{0.53}As/InP$ system. It can be of great interest for device production such as high frequency oscillators for microwave applications.

Transport properties

We have also pointed out that Ga In As can be very interesting for its transport properties, especially because of its low effective electron mass of about $m^* = 0.047\ m_o$.

Figure 13 shows the Shubnikov-de Haas oscillations at T = 4K of the 10 period superlattice already exhibited in figure 3. They are performed on standard Hall bridge sample with photolithographically defined current and potential contacts, by imposing a constant current (a few μ A) and measuring the potential difference V (see figure 13) while varying the normally applied magnetic field B between 0 and 18 Tesla. Two series of SdH oscillations have been observed, evidency for two occupied electric subbands (Razeghi et al., 1986)[11], the oscillations are periodic in 1/B, which allowed us to deduce their electronic densities, $n\ (1) = 4.4 \times 10^{11} cm^{-2}$ and $n_s(2) = 1.2 \times 10^{11} cm^{-2}$ respectively, following well known equation :

$$n = 2h(e\ \Delta\ (1/B))^{-1}$$

where $\Delta(1/B)$ is the period of the oscillations (Δ (1/B) associated to $n_s(1)$ and $n_s(2)$ are $0.109T^{-1}$ and $0.4T^{-1}$ respectively).

The structure was non-intentionally doped, and the electrons present are supposed to come from the semi-insulating substrate, so that the transport takes place at the first heterointerface between InP and $Ga_{0.47}In_{0.53}As$, as shown in figure 13. A two dimensional electron gas is formed in the quasi triangular quantum well due to the band bending.

The zero field longitudinal resistivity ρ_0 of the sample allows us to deduce the mobility $\mu = 1/ne\rho_0$ of the sample. In that case, we find $\mu = 40000 cm^2 V^{-1} s^{-1}$.

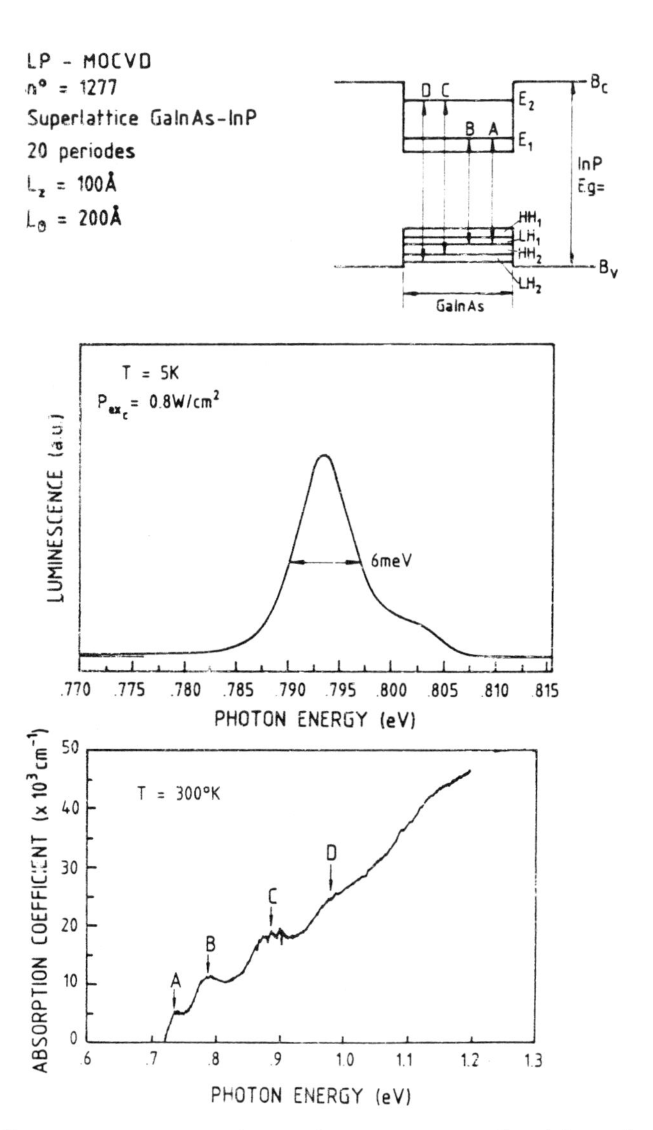

Fig. 10. Room temperature absorption spectrum of a 40 periods GaInAs/InP superlattices grown by LP-MOCVD. Insert T = 5K photoluminescence spectrum of the same superlattice

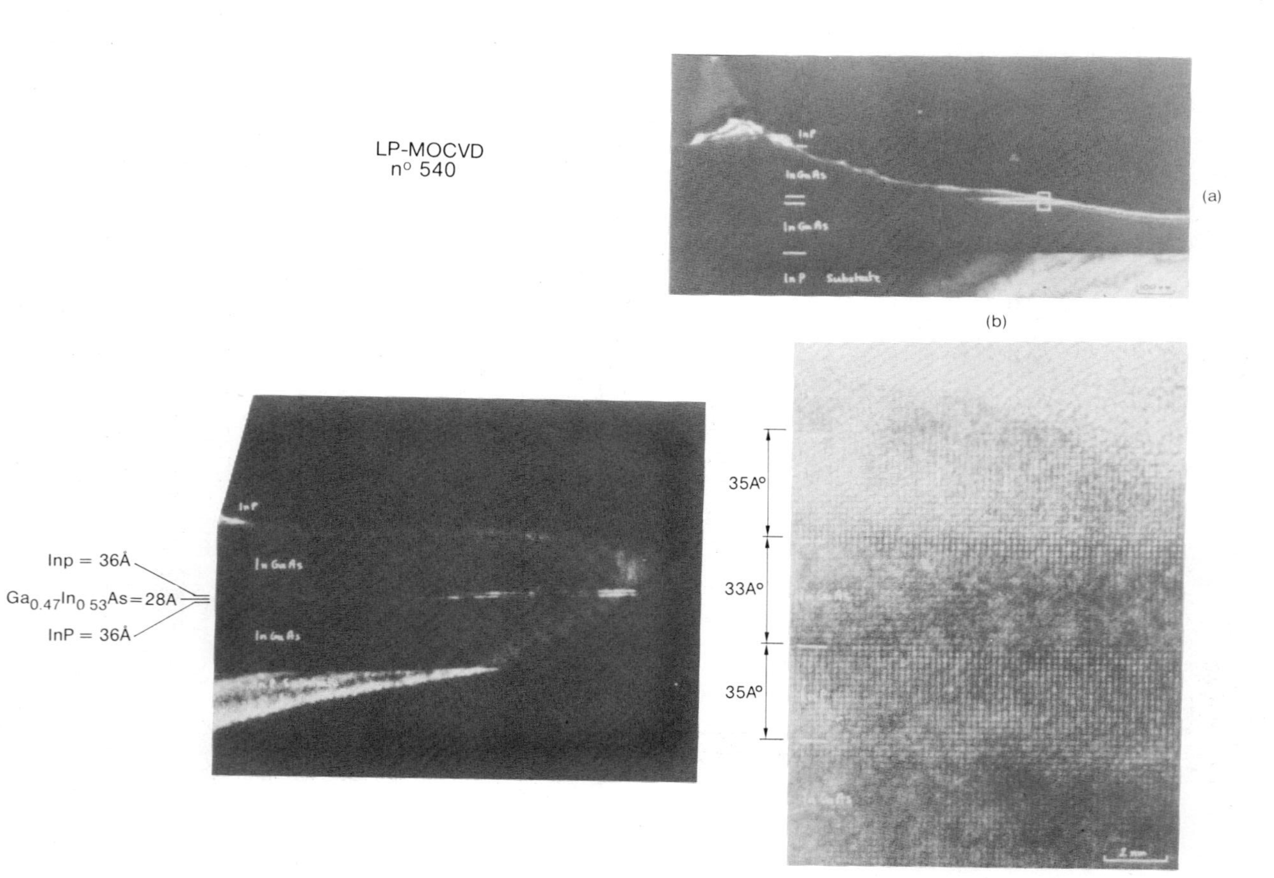

Fig. 11. Edge on transmission electron microscopy (TEM) characterization on a resonant tunneling structure of $Ga_{0.47}In_{0.53}As/InP$ grown by LP-MOCVD

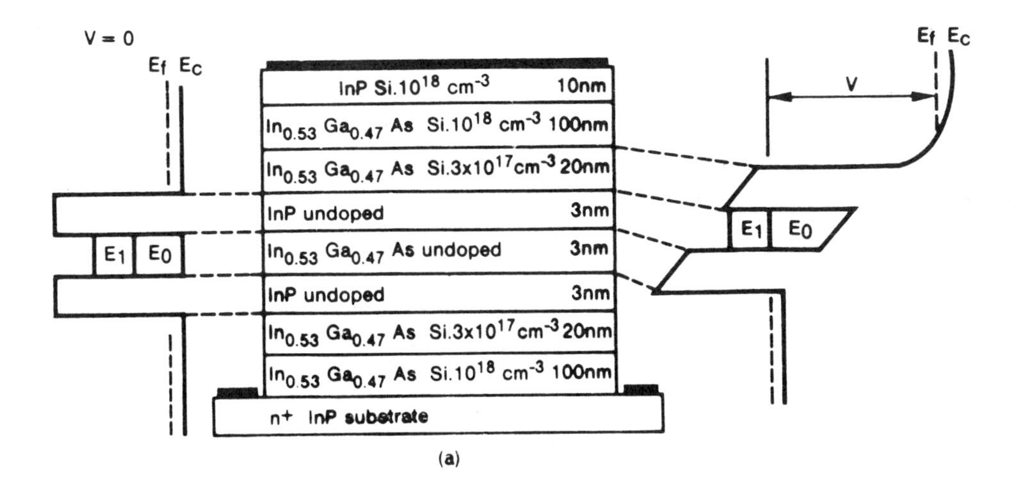

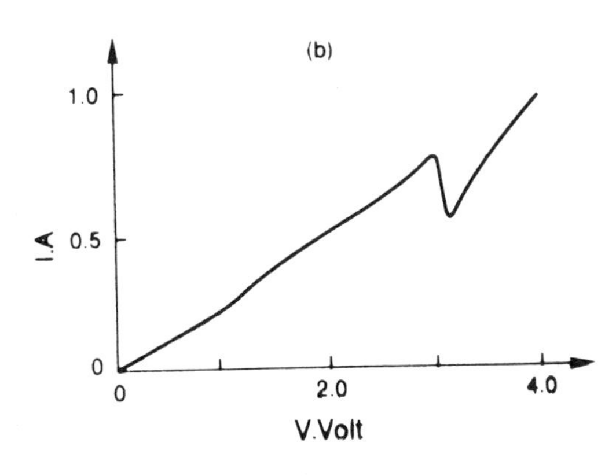

Fig. 12. Negative differential resistance at room temperature of a $Ga_{0.47}In_{0.53}As/InP$ double barrier tunneling structure grown by LP-MOCVD

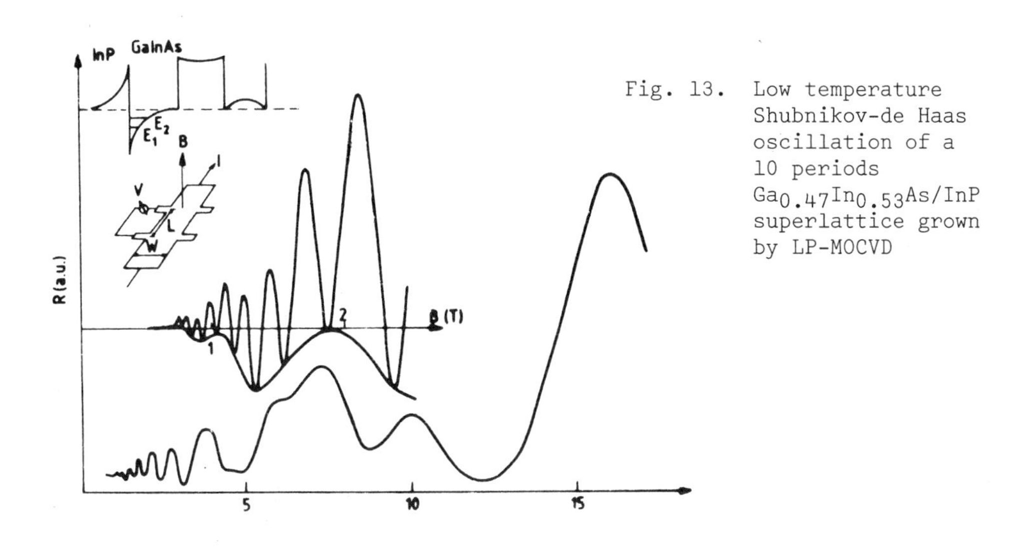

Fig. 13. Low temperature Shubnikov-de Haas oscillation of a 10 periods $Ga_{0.47}In_{0.53}As/InP$ superlattice grown by LP-MOCVD

Much higher mobilities can be obtained in modulation doped single heterojunctions. High gap InP is n doped with silicon or sulphur. Electrons are confined by the band bending in a quasi triangular well at the junction, so that their movements appear two dimensional. They are therefore separated from the donor impurities, and the low temperature mobility can be highly increased.

A T = 4K mobility of about $\mu(4K) = 260.000\ cm^2V^{-1}s^{-1}$ have been obtained. It is the highest value ever reported for this system. It can be well fitted by the alloy scattering due to the random position of the III elements Ga and In (Bastard, 1983)[12].

In figure 14, we exhibit another heterojunction for which we have two series of Shubnikov-de Haas oscillations, associated with two occupied electric subbands. Their respective populations are : $n_1 = 4.3x10^{11}cm^{-2}$ and $n_2 = 1.1x10^{11}cm^{-2}$. A plot of the evaluation of the longitudinal resistivity, with magnetic field applied parallel to the junction (Razeghi et al., 1986)[13] has demonstrated that, in fact, three electric subbands were occupied.

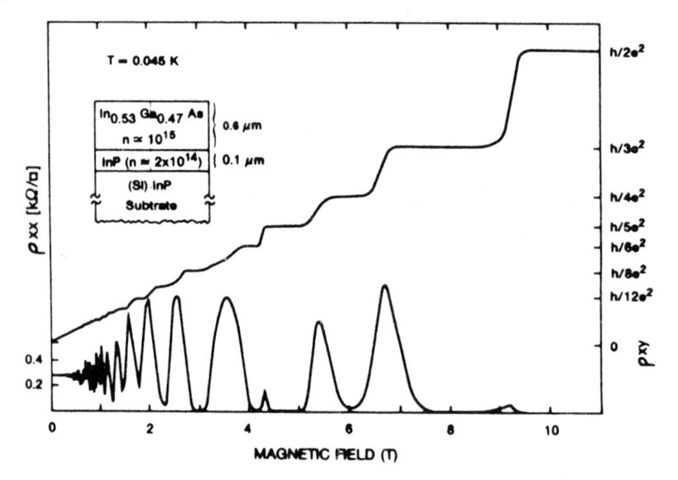

Fig. 14. Quantum Hall effect on a $Ga_{0.47}In_{0.53}As$/InP heterojunction grown by LP-MOCVD with three electric subbands.

We show in figure 14, the quantum Hall effect performed on the same heterojunction at T = 45 mK. Very well resolved Hall plateaus h/ie^2, associated with a quantization of the transverse resistivity ρ_{xy}, can be seen. The relative importance of the odd (i = 2n + 1) plateau, compared to those normally reported in the literature Tsui et al., 1983)[14], is a consequence of the fact that more than one electric subband is occupied (Ando et al., 1982)[15].

Figure 15 shows the cyclotron resonance line of this heterojunction measured at T = 1.6K at an excitation wavelength of $\lambda = 1205\ \mu m$. An effective electronic mass of about $m^* = 0.047\ m_0$ is found, in good agreement with previous work on the subject. Furthermore, quantum oscillations can be seen near the resonance. Theoretically predicted by Ando[15], they are periodic in 1/B. An associated electronic density of $n = 4x10^{11}cm^{-2}$ is found that corresponds well to that of the first electric subband (Maurel et al., 1987)[16]. The quantum oscillations are only very rarely observed on the cyclotron resonance line, and are a further indication of the high quality of the heterojunction under study.

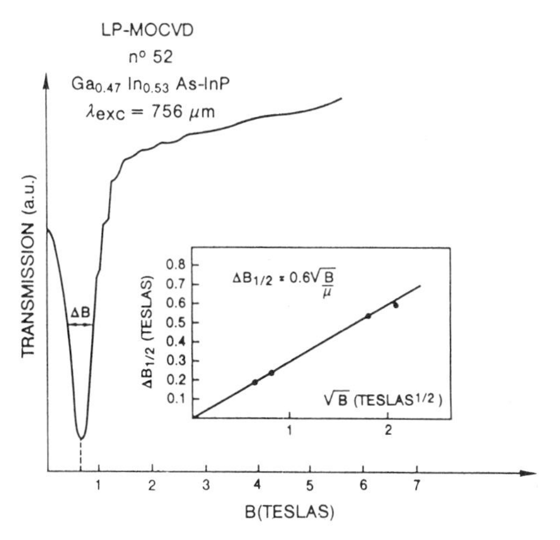

Fig. 15. Cyclotron resonance line at T = 1.6K of a $Ga_{0.47}In_{0.53}As$/InP heterojunction grown by LP-MOCVD with three electric subbands.

Figure 16 shows the quantum Hall effect performed on a two dimensional hole gas obtained at the $Ga_{0.47}In_{0.53}As$/InP interface. The InP was p type doped with Zn at a level of about $N_a - N_d = 10^{17} cm^{-3}$.

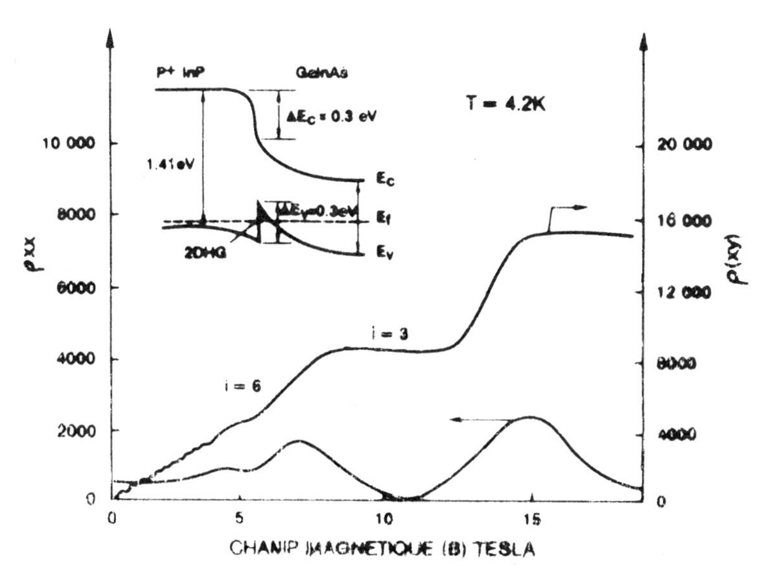

Fig. 16. Quantum Hall effect of a $Ga_{0.47}In_{0.53}As$/InP two dimensional hole gas grown by LP-MOCVD

As for the electrons, band bending occur at the interface, as shown in figure 16. Holes are then confined at the junction plane in a quasi triangular well, so that the system is still two dimensional. That's why quantum Hall effect can be evidenced.

A very well developed $\rho_{xy} = h/3e^2$ add plateau can be seen. It is associated with the lifting of Kramers degeneracy of the heavy hole band, due to the loss of the translation inversion symmetry in the growth direction (Razeghi et al., 1986)[17].

Low field Hall measurements performed on the sample showed a hole mobility $\mu_H(4K) = 10500\ cm^2V^{-1}s^{-1}$, together with a hole density $P_H = 7.10^{11} cm^{-2}$.

To date it is the only observation of Quantum Hall Effect for a two dimensional hole gas in the $Ga_{0.47}In_{0.53}As/InP$ system.

CONCLUSION

We have presented, in this paper, various results in the $Ga_{0.47}In_{0.53}As/InP$ system. High quality quantum wells, superlattices and heterojunctions have been grown by Low Pressure Metalorganic Chemical Vapor Deposition. Results such as excitonic absorption at room temperature, negative differential resistance in a tunneling experiment, two dimensional electron gas mobility as high as μ (4K) = 260.000 $cm^2V^{-1}s^{-1}$, and Quantum Hall Effect for a two dimensional hole gas, are reported. They all necessitate excellent quality of the interfaces, together with a perfect control of thicknesses, doping level or composition of the layers. All of those conditions are now fulfilled by LPMOCVD.

REFERENCES

1. M. Razeghi, J. Nagle, Ph Maurel, F. Omnes, J.P. Pocholle, Appl. Phys. Lett. 49:17 (1986).
2. M. Razeghi, A. Tardella, R.A. Danies, A.P. Long, M.J. Kelly, E. Britton, C. Boothrojd, W.M. Stobbs, Elect. Letters, Vol. 23, n° 3:117 (1987).
3. M. Razeghi, Technology for Chemicals and Materials for Electronic (Howells, London) (1984).
4. M. Razeghi, Ph Maurel, F. Omnes, M. Defour, Submitted to Apllied Physic Letters (1987).
5. P.J. Dean, M.S. Skolnick, J. Appl. Phys., 54:346 (1983).
6. W.J. Bartels, W. Nijman, J. Cryst. Growth, 44:518 (1978).
7. P. Voisin, G. Bastard, M. Voos, Phys. Rev., B29:935 (1984).
8. M. Razeghi, In Feszhörperprobleme, volume XXV, Vieweg, Braunschweig, (1985).
9. A.M. Huber, M. Razeghi and G. Morillot, Inst. Phys. Conf. Ser. n° 74:223 (1984).
10. P.S. Chemla, D.A.B. Millers, P.W. Smith, A.C. Gossard and W. Wiegmann IEEE, J.Q. Electron QE-20: 265 (1984).
11. M. Razeghi, J.P. Duchemin, Semiconductor Science and Technology, (1986).
12. G. Bastard, Appl. Phys. Lett., 43:6 (1983).
13. M. Razeghi, J.P. Duchemin, J.C. Portal, L. Dinowski, G. Remeni, R.J. Nicholas, A. Briggs, Appl. Phys. Lett., 48:712 (1986).

14. D.C. Tsui, H.L. Stormer, A.C. Gossard, Appl. Phys. Lett., 48:1559 (1983).
15. T. Ando, A. Fowler and F. Stern, Rev. Mod. Phys., 54:437 (1982).
16. Ph.Maurel, M. Razeghi, Y. Guldner, J.P. Vicren, To be published in the J. of Semiconductor Science and Technology, (1987).
17. M. Razeghi, Ph. Maurel, A. Tardella, L. Durowski, D. Gauthier, J.C. Portal, J. Appl. Phys., 60:2453 (1986).

STRUCTURAL AND CHEMICAL CHARACTERIZATION OF SEMICONDUCTOR INTERFACES BY HIGH RESOLUTION TRANSMISSION ELECTRON MICROSCOPY

A. Ourmazd

AT&T Bell Laboratories
Holmdel, NJ 07733
U. S. A.

ABSTRACT

The fundamentals of lattice image formation are reviewed to clarify the possibilities and limitations of High Resolution Transmission Electron Microscopy, with special reference to the Si/SiO_2 and semiconductor/semiconductor interfaces.

INTRODUCTION

Modern growth techniques allow the fabrication of multi-layered systems, with electronic properties that can be tuned to the particular application in mind. These systems necessarily contain many interfaces, whose structure can profoundly influence their electronic characteristics. Since the individual layers are sometimes only a few lattice parameters thick, a full characterization of these interfaces requires atomic resolution both spatially and chemically.

High Resolution Transmission Electron Microscopy (HRTEM) can produce compelling images, which often appear to present a direct two-dimensional projection of the atomic structure of the sample. There is little doubt that such images have helped elucidate the structures of a variety of important systems. But it must also be admitted that the temptation towards a simplistic interpretation of lattice images has, on occasion, led to unjustifiable conclusions. It is the purpose of this paper to present a brief description of the fundamentals of lattice image formation so as to promote a critical appreciation of the strengths and weaknesses of HRTEM. At the same time, I will attempt to high-light some of the recent developments, which now allow the simultaneous structural and chemical characterization of semiconductor interfaces on an atomic scale.

FUNDAMENTALS OF LATTICE IMAGE FORMATION

On the simplest possible level, a transmission electron microscope provides a highly magnified image by illuminating a sample with a parallel beam of electrons, which are focussed onto a screen after passing through the sample. The inadequacy of this purely optical picture of the TEM becomes immediately apparent when it is recalled that, although the electron wavelength at the usually employed energies of a few hundred keV approaches 10^{-2} Å, the resolution of the most modern HRTEMs has not yet reached 1 Å. The severe aberrations of electromagnetic lenses which are responsible for this loss of resolution can seriously complicate the way in which a lattice image is related to the sample structure, even when the relevant information lies within the "resolution" of the TEM. In addition, electrons interact with matter more strongly than photons and are, in general, scattered in ways not characteristic of photon/ solid interactions. In order to understand the basic physics of high resolution image formation, two separate processes, electron/sample and electron/lens interactions, must be addressed.

Fig. 1 illustrates the simplest way electron/sample interaction can be viewed. The part of the electron wavefront that passes through a region of high atomic potential experiences a refractive index higher than the part passing through a region of low potential. The sample thus acts as a diffraction grating, and in exact analogy with optics, the emerging electrons give rise to a diffraction pattern. On this picture, an electron undergoes a single scattering event, and a diffracted beam suffers a phase change of $\pi/2$ with respect to the undeviated beam. Under these so-called kinematical conditions, the diffraction pattern is simply the Fourier transform of the sample structure. Fig. 2, however, shows the way in which the phases and amplitudes of the diffracted electron beams change with sample thickness for InP, a typical semiconductor. The "kinematical" phase relationship is rapidly destroyed with increasing thickness and does not hold even for the smallest achievable thicknesses (~ 30 Å). Under these so-called dynamical conditions, the diffraction pattern is no longer related to the sample structure through a simple Fourier transformation. The kinematical phase relation-ship, however, returns with increasing sample thickness and periodically thereafter, thus manifesting "pendellosung" oscillations [1].

The electromagnetic lens brings the diffracted beams to interference on the screen, thus forming a lattice image. A lattice image, therefore, is first and foremost an electron interferogram. In bringing the beams to interference, a perfect lens with no aberrations would simply carry out the inverse Fourier transformation necessary to form an image of the sample projected potential. However, this is in general not the case with an imperfect lens. The performance of such a lens can be characterized by its Contrast Transfer Function (CTF), which describes how, at a given lens defocus, the phase and amplitude of the transmitted information are changed as a function of the spatial frequency of the information. At the so-called optimum or Scherzer defocus, $(1.5\lambda C_s)^{1/2}$, the CTF consists of a passband followed by damped oscillations (Fig. 3). The relative phases of the spatial frequencies lying within the first zero of the CTF are not changed, while frequencies beyond the first zero can undergo relative phase changes as well as amplitude attenuation. Consequently, within the first zero of the CTF (also known as the point-to-point resolution of the microscope) information is transmitted without significant modification, while outside the first zero the lens aberrations can decisively affect the transmitted information.

Due to the dynamical nature of electron/sample interaction, as well as the aberrations of the lens, a lattice image is, in general, not a simple representation of the sample structure. However, if the sample thickness is chosen such that the kinematical phase relationship between the beams holds, and only the information lying within the first zero of the CTF is allowed to contribute to an image formed at optimum defocus, the resulting image is a faithful representation of the sample structure to within the

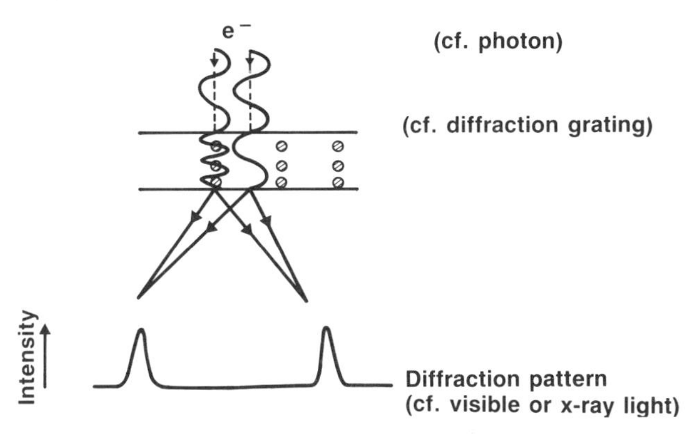

Fig. 1 Simplest representation of electron/sample interaction, giving rise to electron diffraction. In this "kinematic" approximation, the analogy with photons is exact.

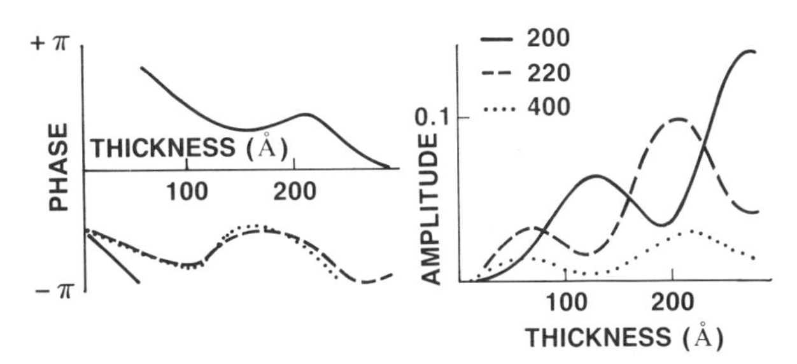

Fig. 2 Variation and phase and amplitudes of beam with thickness for InP. The electron beam is incident along <100>.

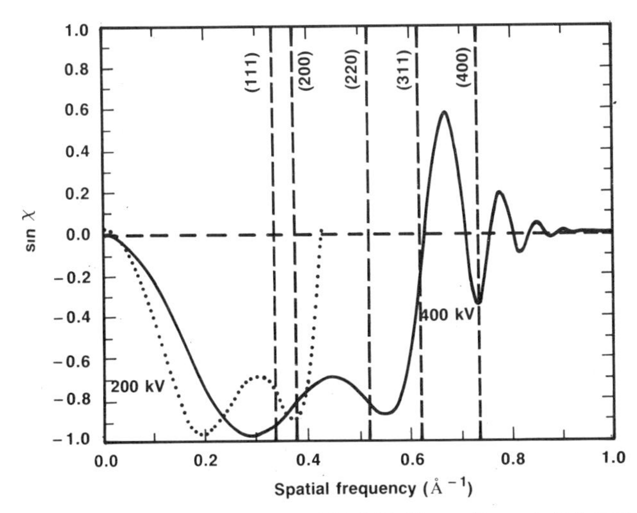

Fig. 3 Contrast Transfer Function at 200 kV (dotted) and 400 kV (solid). The vertical lines represent the various planar spacing of Si.

point-to-point resolution of the microscope. Such an image is known as a *structure* (rather than a lattice) image and, if individual atomic columns are resolved, the term *atomic* structure image is used. The hallmark of a structure image is that atomic columns appear black and the tunnels white on a positive print. Structure images, while being particularly simple to interpret, are sensitive only to the overall geometry of the structure. Images obtained under other conditions, for example at other defocus values or "non-kinematical" thicknesses, can be sensitive to small changes in atom positions, but are not simple representations of the structure. Accurate and reliable structure determination, therefore, requires the fitting of a series of experimental images, where a parameter such as defocus or sample thickness is varied, with a corresponding series of computer simulations. In practice, the structure image is used to deduce the overall structure, which is then refined by fitting the other members of the series.

Although it is in principle possible to deduce information lying well outside the first zero of the CTF, in practice it is difficult to do so reliably. Until the emergence of the latest generation of HRTEMs, only the (111) and (200) planar spacings of semiconductors lay within the first zero (Fig. 3). Consequently, the vast majority of lattice images of semiconductors in the literature are those obtained in the <110> projection, where two sets of (111) and one set of (200) planes can be resolved. Under these conditions, pairs of atom-columns appear as single dark or white blobs and it is not possible even to distinguish between atoms and tunnels without additional information (see Fig. 9a). Fig. 3 compares the CTF of the best 200 kV HRTEM with that of the latest 400 kV instruments, which have recently become available. Most significant is the fact that the (220) reflections of all semiconductors lie comfortably within the first zero of the CTF, and a number of other reflections are within its reasonable proximity. The (220) reflections, which in Si correspond to a spacing of 1.9 Å, are critically important, because their faithful transmission allows the imaging of individual atomic columns in semiconductors. Indeed, this feature has been utilized to obtain the first atomic structure images of Si [2], in the two important orientations <100> and <111>. Additionally, individual atomic columns have been resolved in the <110> orientation, although at defocus values larger than optimum, and lattice images have been obtained in a number of other orientations [2].

These new capabilities have implications beyond the field of HRTEM, because they allow the characterization of semiconductors in ways not previously possible. The ability to resolve individual atomic columns, to do so in more than one orientation, and to obtain lattice or structure images of a given area in different projections makes possible the determination of three-dimensional maps of atom positions around individual features of interest, such as defects and interfaces. Somewhat surprisingly, the aberrations of the lens and dynamical effects impart additional flexibility to HRTEM, allowing the simultaneous structural and chemical characterization of materials. The examples below are intended to illustrate these capabilities.

THE STRUCTURE OF THE Si/SiO_2 INTERFACE

The Si/SiO_2 interface is a corner-stone of semiconductor technology. This, together with its intrinsic scientific interest has stimulated sustained activity for over thirty years. It is thus now generally, but not universally accepted, that the SiO_2 is structurally amorphous and chemically stoichiometric at distances in excess of 10 Å from the interface. No such general consensus exists so far as the interfacial structure itself is concerned. Briefly, the various proposals can be divided into three general categories. a) The $c\text{-}Si \rightarrow a\text{-}SiO_2$ transition is proposed to occur via a stable, bulk phase of $c\text{-}SiO_2$. Due to the structural similarity between Si and cristobalite, this oxide represents the most frequently proposed crystalline interfacial phase. However, it has a lattice parameter 40% larger than Si, and it is difficult to achieve a commensurate Si/cristobalite interface. b) A metastable, sub-stoichiometric oxide is thought to affect the $c\text{-}Si \rightarrow a\text{-}SiO_2$ transition. It is then necessary to postulate a metastable phase of remarkable stability, able to exist at the relatively high temperatures employed in Si oxidation. It is possible that the special conditions present at the interface may stabilize such a phase. c) It has been shown that the $c\text{-}Si \rightarrow a\text{-}SiO_2$ could occur abruptly, with no intervening crystalline or sub-stoichiometric phase at the interface [3].

HRTEM has naturally been employed to investigate the structure of this interface, and in particular, detect the presence of any crystalline oxide [4]. The resultant lattice images show an abrupt transition from c-Si to "amorphous" material. The important question regards the interpretation of such images. An examination of the Si(100) surface shows, that although the surface is structurally four-fold symmetric, the dangling bonds reduce this to two-fold symmetry. An epitaxial oxide would necessarily be tied to the dangling bonds, which rotate through 90° on crossing a Si surface step consisting of an odd number of atomic layers. This implies that, unless an epitaxial oxide were fourfold symmetric, it would be polycrystalline with a grain size determined by the spacing between the steps on the Si surface. Lattice images of the Si/SiO_2 interface are obtained from cross-sectional samples about 100 Å thick. Thus, for an interfacial oxide phase of less than four-fold symmetry, the image would consist of the superposition of the images of a number of oxide grains rotated by 90° with respect to each other due to the presence of Si surface steps. Consequently, unless the grain size is comparable to, or larger than the sample thickness, the detection of a crystalline oxide phase of less than four-fold symmetry is most unlikely. All that can be surmised from the usual lattice images of the Si/SiO_2 interface is that, if a crystalline interfacial oxide is present, it is less than four-fold symmetric.

This discussion also provides guide-lines for the further investigation of this important system. The interfacial structure can only be uniquely determined by HRTEM if the spacing between Si surface steps is much larger than the sample thickness (~100 Å). Modern MBE growth of Si on Si can produce samples of sufficient perfection [5]. Fig. 4 shows lattice images of the $(001)Si/SiO_2$ [110] and $[1\bar{1}0]$ projections for an MBE grown, oxidized Si sample. The presence of an interfacial oxide layer is immediately apparent, while the difference between the two lattice images shows the oxide to be indeed less than four-fold symmetric. Since we have been able to obtain lattice images of this phase in three projections and diffraction patterns in four projections, we have determined a three-dimensional map of atom positions at the Si/SiO_2 interface (Fig. 5) [5]. Modelling of lattice images indicates the interfacial oxide to be tridymite, a stable, bulk phase of SiO_2. This crystalline phase forms a strained, commensurate, epitaxial oxide layer on the Si substrate, and is thus characterized by a critical layer thickness, beyond which strain relaxation must occur. In this way the small thickness of the crystalline oxide (~5 Å) and the production of amorphous SiO_2 can be understood as simple consequences of strain relaxation [5].

SIMULTANEOUS RESOLUTION AND IDENTIFICATION OF INDIVIDUAL ATOM-COLUMNS IN COMPOUND SEMICONDUCTORS

The ability to utilize the (220), and, in other orientations, the (400) reflections, allows

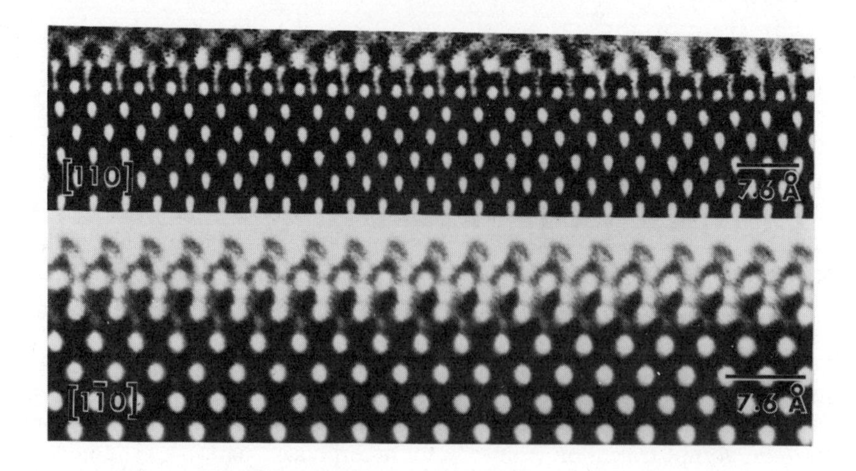

Fig. 4 Lattice images of (100)Si/SiO_2 interface in two orthogonal <110> projections. The Si sample was prepared by MBE growth of a 2000 Å Si layer on a Si substrate. Note the flatness and sharpness of the interface, and the clear presence of an ordered interfacial layer.

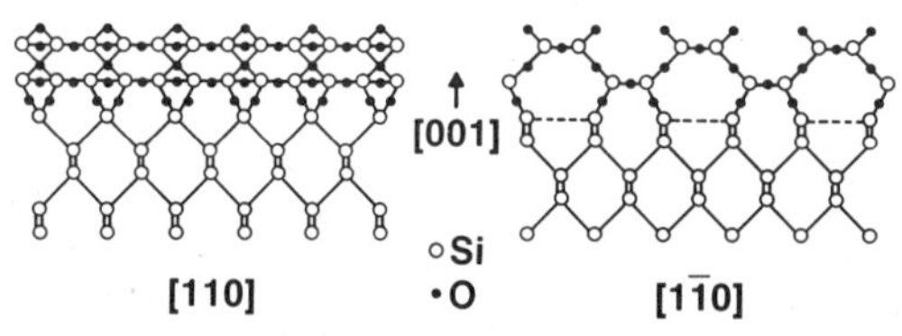

Fig. 5 Schematic representation of the Si/SiO_2 interfacial atomic configuration deduced from lattice imaging. The dashed lines indicate possible dimerisation of unsaturated bonds.

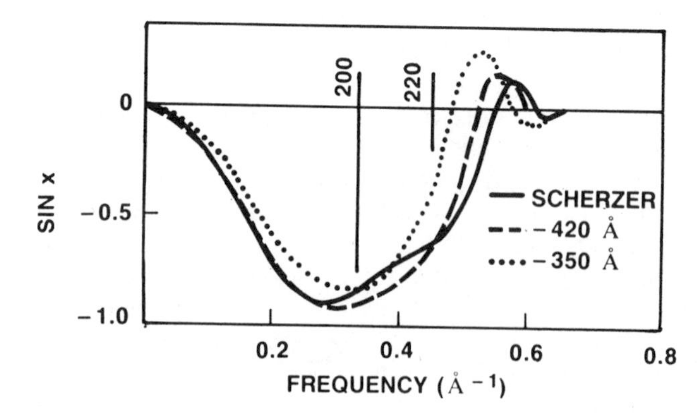

Fig. 6 Contrast Transfer Function for 400 kV HRTEM at different objective lens defoci. The vertical lines indicate the positions of (200) and (220) reflections for InP.

the location of individual atomic columns, and thus a complete structural characterization of elemental semiconductors. In the case of compound semiconductors, however, a full structural characterization requires not only the location of individual atomic columns, but also their identification. In GaAs, for example, it is necessary to determine whether a particular atomic column belongs to the Ga or the As sublattice. In other words, a complete structural characterization of compound semiconductors needs both structural and chemical information.

The (220) reflections, necessary for the resolution of the atomic columns, are relatively insensitive to chemical changes. (200) reflections, on the other hand, are kinematically forbidden in elemental semiconductors, because the contributions from the two fcc sublattices of the diamond lattice cancel out at the (200) position in reciprocal space. These reflections are consequently extremely sensitive to the chemical nature of the occupants of the two sublattices. Thus, in order both to resolve and identify the individual atomic columns, the (220) and (200) reflections must make comparable contributions to the lattice image. In principle, this can be realised in the <100> projection, where four (220) and four (200) beams can be used to form a lattice image, with the advantage that all the contributing beams lie within the first zero of the (CTF). Fig. 2, however, shows the amplitudes and phases of these reflections for InP as a function of sample thickness. In regions where the amplitudes of the (220) and (200) reflections are comparable, their phase relationship is far from kinematical, and a lattice image formed in such a region would not be a simple representation of the sample projected potential. On the other hand, in the kinematical region, the (220) amplitude is dominant and identification of the different chemical species is not possible. This dilemma can be resolved by recognizing that, since the (220) beams lie close to the first zero of the CTF, their contribution to the image can be finely controlled by adjusting the lens defocus or microscope accelerating voltage (Fig. 6). Thus one can choose a sample thickness where the kinematical phase relationship holds, and attenuate the (220) amplitudes to a level comparable with those of the (200) beams. Fig. 7 shows an image of InP obtained by this method, where the lens defocus is used to "tune" the (200) and (220) contributions. Because the defocus deviation from optimum is small, the image is a simple representation of the sample projected potential and the individual atomic columns are directly resolved and identified [6]. As in all structure images, the atomic columns appear black. We have used the same approach directly to resolve and identify the individual atom-columns in GaP and GaAs [6]. This technique can therefore be used for any semiconductor, and, by suitable extension, for other systems, where chemical and structural information is simultaneously required.

CHARACTERIZATION OF INTERFACES IN MULTI-LAYERED SYSTEMS

The *structure* of a perfect semiconductor/ semiconductor interface is well-known; the atoms occupy zinc-blende sites on both sides of the interface. The question of the atomic configuration at such an interface is thus *chemical* and not structural in nature. The complete characterization of the interfacial configuration thus requires the development of a technique capable of yielding chemical and spatial information on an atomic scale. This can again be achieved by utilizing the chemical sensitivity of the (200) reflections. As an example, consider the InP/InGaAs system. Fig. 8 shows the way the amplitudes of the (200) and (220) beams change with thickness for these two materials. Over the thickness range 70 - 150 Å, the (200) beam is dominant in InP, while in InGaAs the (220) beam is stronger. This implies that a lattice image formed in this thickness range should, under suitable defocus conditions, exhibit a strong

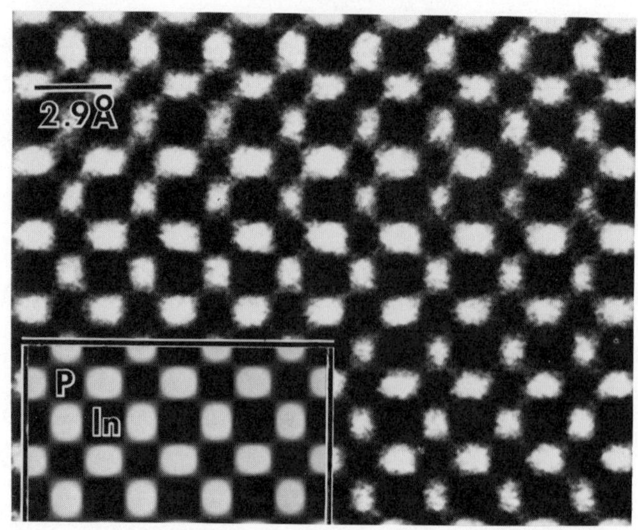

Fig. 7 Atomic structure image of InP in the <100> projection, where the individual atomic columns are directly resolved and identified. The atom-columns appear black, the tunnels white. Bottom inset is the simulated image. Note the heavier In atoms appear larger and darker than the lighter P atoms.

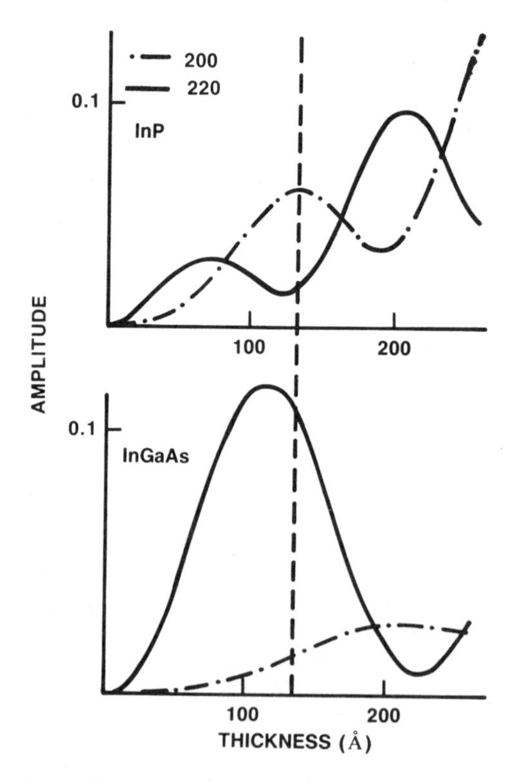

Fig. 8 Variation of amplitudes of (200) and (220) beams with thickness in InP and InGaAs. The vertical lines shows an ideal thickness for maximum change in lattice image periodicity on crossing an interface between InP and InGaAs.

change in periodicity across the interface. Figs. 9a and b represent the usual <110> lattice images of the interface between InP and InGaAs, while the lattice image of Fig. 9c was obtained in the <100> orientation from the same area of the same interface, using the approach just described [7]. The InP gives rise to strong (200) fringes, while the InGaAs is represented by the fine (220) fringes at 45 ° to the interface. The interface, shown by the <110> image to be "abrupt" and "flat", can be seen to be rough on an atomic scale. In addition to the coarser interfacial steps, revealed as strong protrusions of the (220) fringes from InGaAs onto the (200) fringes of InP, a number of weaker (220) "spots" also protrude into the (200) InP fringes. These arise from interfacial steps, whose extent in the interface plane is a small fraction of the sample thickness (~100 Å). It appears, therefore, that this approach can sensitively detect small variations in composition at the interface. This sensitivity is due to the fact that chemical changes in the sample are translated into readily detectable periodicity changes in the lattice image, and, additionally, because "random" effects, such as surface contamination, do not give rise to definite spectral changes.

In Fig. 9c spatial frequency changes of the lattice image reflect compositional changes in the sample. A quantitative analysis of the frequency content of the image is therefore equivalent to a chemical analysis of the sample. This in principle allows the determination of the composition of a given atomic column at the interface. A more detailed discussion of this point can be found in other publications [7].

A complete characterization of the interfacial atomic configuration involves the three-dimensional mapping of atom positions. This requires imaging of the same area in more than one chemically sensitive projection. Since at present chemical sensitivity can only be achieved in <100> projections, and as two different such projections are at right angles to each other, a full three-dimensional characterization of semiconductor/semiconductor interfaces is not yet possible. However, much can be gleaned from a single chemically sensitive image when it is recalled that surface steps run in <110> directions, and that the atoms must occupy zinc-blende lattice sites. These geometrical constraints are sufficient to limit the possible interfacial configurations consistent with experimental data to a reasonably small number. On the basis of such geometric considerations, it is possible to deduce a family interfaces which are consistent with the experimental data, ruling out all other possibilities [7]. Fig. 10a shows one possible three-dimensional interfacial configuration consistent with a two-dimensional experimental profile we obtained. (A detailed and color-coded image of this interface can be found in [7]). As shown in Fig. 10b, a small change in the size of the hole in the arrangement of the islands introduces a significant change in the two-dimensional profile, and in fact destroys the agreement with experiment. It is remarkable that such two similar configurations can be experimentally distinguished.

SUMMARY AND CONCLUSIONS

In this review, I have attempted to emphasize the importance of an elementary appreciation of the fundamentals of lattice image formation, because, under most circumstances, the aberrations of the microscope and the dynamical nature of electron/sample interaction can profoundly affect the recorded information. Only a systematic experimental procedure, augmented by computer simulation, allows the reliable interpretation of lattice images. Nevertheless, the recent past has shown HRTEM to be a powerful tool for the structural analysis of semiconductors. Chemical questions, on the other hand, have not, until recently, begun to be properly addressed.

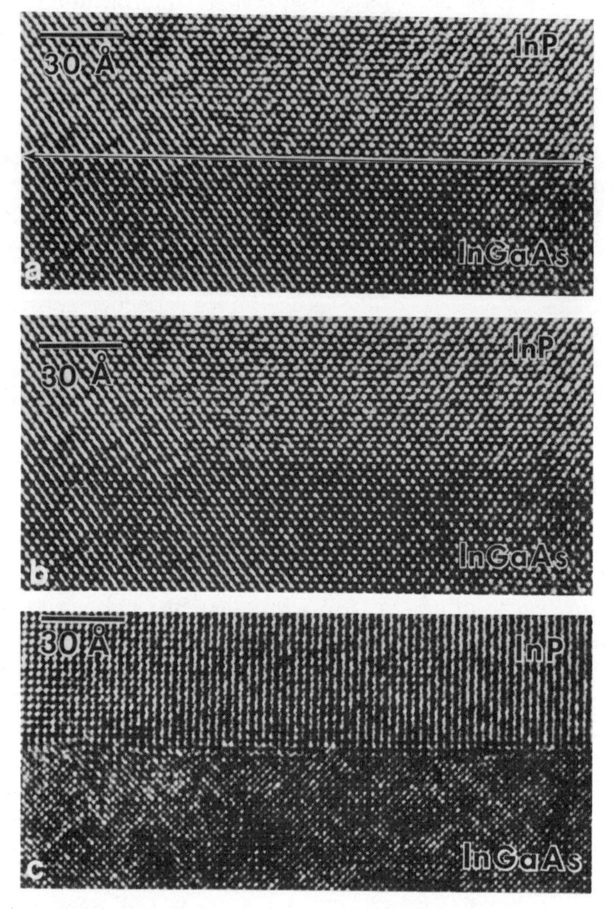

Fig. 9 (a) <110> lattice image of InP/InGaAs interface.

(b) Same image without line drawing attention to the interface, which is now difficult to recognize.

(c) <100> lattice image of the same region of the same interface, obtained as described in the text. Note the protrusion of the finer InGaAs fringer onto InP, as well as the presence of weaker (220) spots in the (200) fringes of InP at the interface. These represent interfacial roughness not revealed by normal lattice imaging.

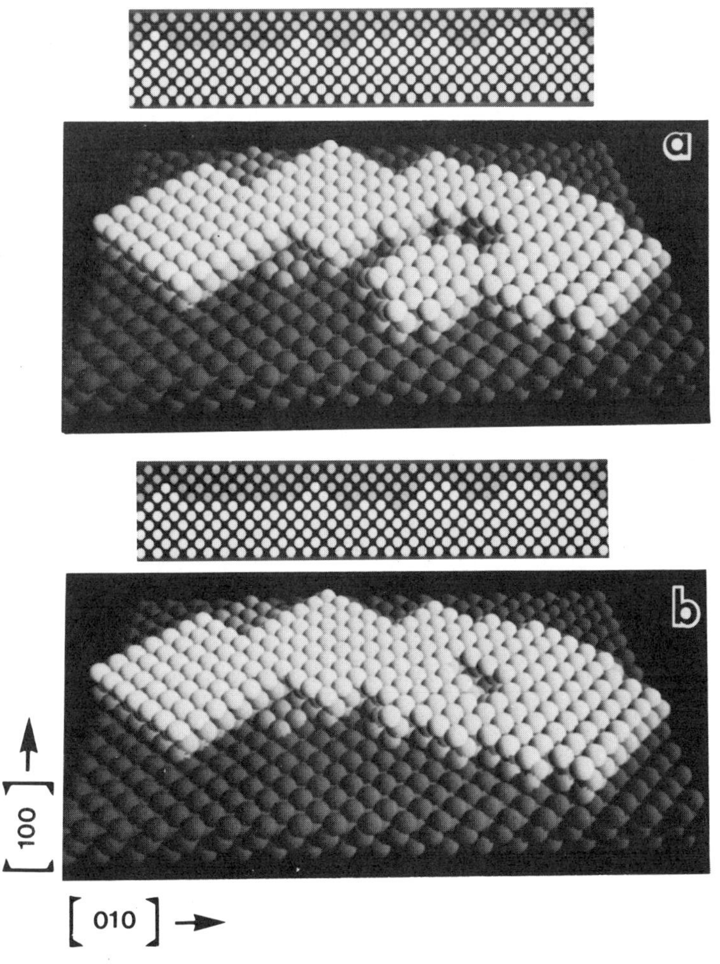

Fig. 10 (a) Possible three-dimensional configuration of the InGaAs surface compatible with the experimental data. The profile of this interface is shown in the top inset.

(b) Three-dimensional configuration produced when a part of the island in (a) is moved by one atomic spacing. Note the change in the profile. This configuration is inconsistent with the experimental data [7].

The improved resolution of the latest generation of electron microscopes now makes possible the structural characterization of semiconductors and their interfaces on a scale not previously possible. The ability to obtain lattice images in several projections by tilting between them, and the possibility to resolve individual atomic columns in three important orientations combine to allow three- dimensional mapping of atom positions around individual features of interest. Intellectually perhaps more appealing, however, is the new capability of simultaneously obtaining structural and chemical information on an atomic scale. Ironically, this is possible because the aberrations of the lens impart to the microscope the character of a band-pass filter, which can be tuned to enhance the chemical sensitivity of lattice imaging.

ACKNOWLEDGEMENTS

This paper contains the results of several collaborative projects, involving a number of expert colleagues. Among them I wish especially to acknowledge the contributions of J. Bevk, L.C. Feldman, J.A. Rentschler, D.W. Taylor and W.T. Tsang.

REFERENCES

1. For a detailed discussion of HRTEM see e.g.: J C H Spence, Experimental High Resolution Transmission Electron Microscopy (Oxford University Press, New York, 1980).
2. A Ourmazd, K Ahlborn, K Ibeh, and T Honda; Appl Phys Lett 47, 685 (1985).
3. S T Pantelides and M Long, in *Physics of* SiO_2 *and its interfaces*, ed. S T Pantelides (Pergamon Press, New York, 1978), p. 339.
4. See, e.g., S M Goodnick, D K Ferry, C W Wilmsen, Z Liliental, D Fathy and O L Krivanek, Phys Rev B **32**, 8171 (1985).
5. A Ourmazd, D W Taylor, J A Rentschler and J Bevk, Phys Rev Lett, **59**, 213 (1987).
6. A Ourmazd, J A Rentschler and D W Taylor, Phys Rev Lett, **57**, 3073 (1986).
7. A Ourmazd, W T Tsang, J A Rentschler and D W Taylor, Appl Phys Lett, **50**, 1417 (1987).

II. DEEP AND SHALLOW IMPURITY STATES

DEEP LEVEL BEHAVIOUR IN SUPERLATTICE

Jacques C. Bourgoin* and **Michel Lannoo****

*Groupe de Physique des Solides de l'École Normale Supérieure+, Centre National de la Recherche Scientifique, Tour 23, 2 place Jussieu, 75251 Paris Cedex 05, France

**Département de Physique des Solides, Institut Supérieur d'Électronique du Nord, 41 boulevard Vauban, 59046 Lille Cedex, France

I - INTRODUCTION

In the old times there were two kinds of physicists working on semiconductors, the ones who were studying the behaviour of electrons in the bands and the other ones who were growing materials and making devices with them. One day, the second ones realized that, if their devices did not work, this was because a kind of physics had been forgotten, the physics of defects, and that it is also important to know the behaviour of electrons when they are localized inside the forbidden gap. It seems that, with the advent of heterostructures and superlattices, the same process starts again: some play with electrons in fancy band structures while others tailor new materials which are supposed to have the virtue of making working devices. However, the question of the influence of defects will arise soon. For instance, those who hope to replace a layer containing defects, such as DX centers in GaAlAs, by a GaAs-GaAlAs superlattice should fear that this superlattice could contain even more defects than a simple GaAlAs layer. There is indeed apparently no transient phenomenum associated with charge trapping on the DX centers in such superlattices, but there are other types of transients (see fig. 1) probably due to a broad continuum of electron states, originating presumably from defects at the interfaces (1). In addition to the point defects present in "bulk" layers grown by the same technique as the heterostructures, such as the EL2 and DX centers in GaAs and GaAlAs respectively, these heterostructures contain distributions of interface states due to the presence of intrinsic defects, as well as impurity contamination (2). Such defect distributions have already been detected at heterojunctions (3,4) as well as in superlattices (1,5).

Defects play an important role because they give rise to localized levels, i.e. located deeply in the forbidden gap. This role is known and qualitatively understood in bulk materials. The aim of this lecture is to extrapolate the knowledge we have on defects in bulk materials to the case of heterostructures and superlattices. We limit ourselves to

----

+Laboratory associated to the Université Paris VII

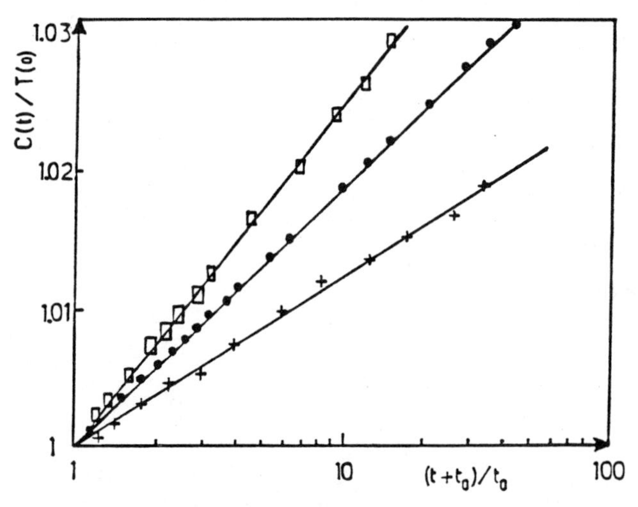

Fig. 1. Variation of the capacitance (normalized to its value at t = 0) of a Schottky barrier made on a 1.7 μ thick 20 - 20 Å, 3 × 10^{16} cm^{-3} Si doped GaAs-$Ga_{0.7}Al_{0.3}As$ superlattice versus time t after a reverse bias (3 V) is applied at different temperatures (50 K (+), 77 K (•) and 100 K (□)).

density of localized states. We first briefly recall the basic characteristics of a defect (a full development can be found in refs. 6 and 7) from which all its properties can be derived, namely its energy level position E_T, the carrier capture cross-sections σ_c and the optical cross-sections σ_0. We then deduce the effects induced on these characteristics by placing the defect in the vicinity of an heterostructure, in a quantum well or in a superlattice. We shall see that, because of the strong localization of its wave function, a deep defect is not affected by the edges of the well in which it is located and that the changes induced in σ_c and σ_0 can be accounted for by the new band structure. As a result, a deep level whose characteristics are known can be used as a probe to characterize an heterostructure or a superlattice.

II - CHARACTERISTICS OF DEEP LEVELS

A defect can bind an electronic charge in its vicinity, introducing a localized level in the forbidden gap. In principle, it is sufficient to know its wave function ψ_T, energy level E_T and total energy to derive its electronic and thermodynamic characteristics which allow to determine all its properties (6,7).

The features common to all defects are first that, because they have a localized state in the gap, they can be present with at least two different charge states (often there are situations with more than one localized level which lead consequently to more than two different charge states). Second, because there is a coupling between the lattice and the defect, which is a function of the defect charge state, its atomic configuration including the distortion and relaxation of the lattice around it, is charge state dependent. A classical illustration of this is provided by the Jahn-Teller effect. This electron-phonon interaction has very important consequences on defect characteristics and behaviour : it can lead to peculiar properties such as bistability (two different confi-

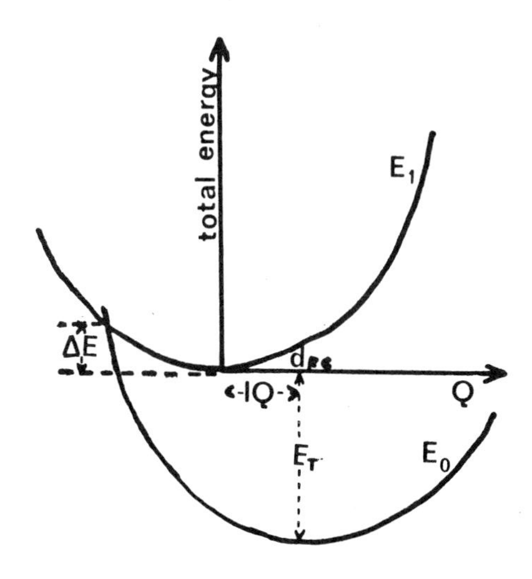

Fig. 2. Configuration coordinate diagram for a neutral (E_0) and ionized (E_1) states.

isolated point defects, i.e. we do not treat the situation of a continuum gurations for the defect in two different charge states), metastability (two different configurations for the same charge state), negative U behaviour (instability of a configuration for certain charge states), charge state dependence of the migration energy, athermal migration induced by carrier capture, etc.

To discuss qualitatively the properties associated with this electron-phonon interaction it is usual to draw what is called a configuration coordinate diagram. This diagram is a representation of the defect energy (i.e. electronic plus vibrational energies) assuming that the defect interacts with only one lattice mode Q. Then, two different charge states E_0 and E_1 of a defect correspond (see fig. 2) to two parabolic branches shifted from one another vertically by the energy E_T representing the ionization of the electron and horizontally by a linear term IQ expressing the electron-phonon interaction. They are :

$$E_1\ (Q) = \frac{1}{2}\ k\ Q^2$$

with the origin of the energy at the bottom of the branch, and

$$E_0\ (Q) = -\ E_T + d_{Fc} + \frac{1}{2}\ k\ Q^2 - IQ$$

where d_{Fc}, the so-called Franck-Condon shift, i.e. the shift of the E_0 branch at Q = 0, is a measure of this electron-phonon interaction.

Such diagram allows to illustrate the mechanism by which free carriers get trapped on the localized level E_0. The classical picture equivalent to the quantum mechanical treatment (valid only for a weak coupling and low temperature, see chap. 6 in ref. 7) of this capture can be viewed as a process in which the free carrier overcome the barrier ΔE (see fig. 2). Then, the probability for this process is given by the Boltzman statistics and the corresponding capture cross-section can be written as :

$$\sigma_c = \sigma_\infty \text{ exo } (- \Delta E/kT)$$

in which ΔE can be expressed in terms of the parameters which describe the branches

$$\Delta E = \frac{1}{2} (k/I^2) (d_{Fc} - E_T)^2$$

This mechanism, which is found to apply for many defects in III-V compounds, is called multiphonon emission.

As to the optical cross-section for a transition between a defect and a band it is given (see chap. 4 in ref. 7) by :

$$\sigma_0 = \frac{\beta}{h\nu} \left[\langle \psi_{bk} \left| - i \hbar \frac{\partial}{\partial z} \right| \psi_T \rangle \right]^2 \rho_b (E_k)$$

where ψ_{bk} is the state of band b for the wave vector k, ψ_T the defect state, ρ_b (E_k) the corresponding density of states and β the oscillator strength. Taking for ψ_T the following simple function :

$$\psi_T = \frac{1}{r} \exp (- \alpha r)$$

allows to compute σ_0 easily because the matrix element of the electronic dipole operator can be taken as a constant or proportional to k. In the first case, one obtains an optical cross-section proportional to :

$$\sigma_0 \propto \frac{\beta}{h\nu} \left(\alpha^2 + k^2\right)^{-2} \sqrt{h\nu - E_T}$$

and the fit of this expression with the experimental curve $\sigma_0(h\ \nu)$ allows to get unambiguously the energy threshold E_T, the Franck-Condon shift, the oscillator strength and the extension in space of the wave function (α^{-1}).

III - DEEP LEVELS IN HETEROSTRUCTURES

In this section we examine the effects that must have on E_T, σ_c and σ_0 the fact a defect is placed in the vicinity of an heterostructure or of a potential well.

III-1 Energy level

Consider a defect at the origin of the coordinate z, situated at a distance z_0 from the potential step of height Δ representing the band off-set of the heterostructure. We want to know the shift ε induced on E_T by the presence of this potential step. For this, we use a treatment in perturbation, i.e. we assume that Δ is small compared to E_T. This can be considered as valid, in case of GaAs for instance, for a level situated in the middle of the forbidden gap if Δ is of the order of few times 100 meV. To first order in the perturbation V(z) (such that V(z) = 0 for $z < z_0$ a,d $V(z) = \Delta$ for $z \geq z_0$) we have :

$$\varepsilon = \langle \psi | V | \psi \rangle / \langle \psi | \psi \rangle$$

In order to get a solution in a simple algebraic form we take for the defect wave function the following function :

$$\psi_T = b \exp(-\alpha r)$$

Because of the geometry around the z axis, the integration of

$$\langle \psi \mid V \mid \psi \rangle = \int \psi V \psi \, dv$$

can be made using cylindrical coordinates :

$$\langle \psi \mid V \mid \psi \rangle = b^2 \Delta \int_0^{2\pi} d\varphi \int_{z_0}^{\infty} dz \int_{|z|}^{\infty} r \; e^{-2\alpha r} dr$$

Such function is integrated by parts. As to the normalization $\langle\psi|\psi\rangle$ its integration is made in a similar fashion, giving as result :

$$\varepsilon = \Delta \exp(-2 \alpha z_0) \left(\frac{1 + \alpha z_0}{2}\right)$$

Fig. 3a which represents the way ε varies with αz_0 shows, as it can be intuitively deduced, that a non negligible value of the shift ε occurs only when the defect is at a distance z_0 of the order of, or smaller than, the extension in space α^{-1} of the wave function; ε/Δ is decreased by a factor 2 for $\alpha z_0 = 0.5$ and is of the order of 10^{-2} for $\alpha z_0 = 2$.

In case the defect is placed in a potential well of height Δ and thickness 2L, the energy shift can be taken as the sum of the shifts induced by the two wells. Thus for a defect situated at a distance z from the middle of the well

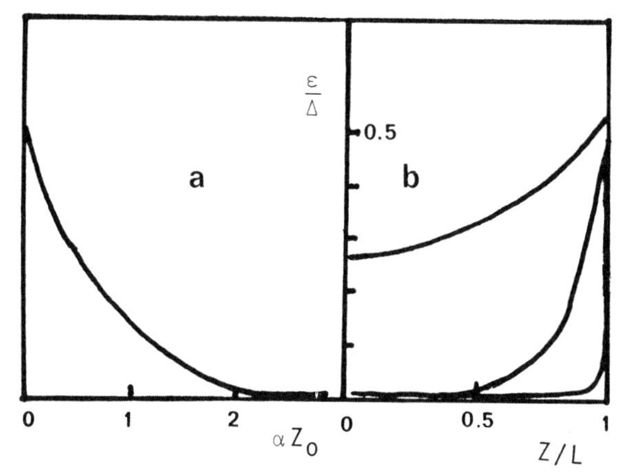

Fig. 3. Variation of the relative energy shift ε/Δ as a function of (a) the distance z_0 between the heterostructure and a defect characterized by an extension in space α^{-1} of its wave function; (b) of the position of the defect in a well of width ZL.

$$\varepsilon = \frac{\Delta}{2}\left\{\exp\left[-2\beta\left(1-\frac{z}{L}\right)\right]\left[1+\beta\left(1-\frac{z}{L}\right)\right] + \exp\left[-2\beta\left(1+\frac{z}{L}\right)\right]\left[1+\beta\left(1+\frac{z}{L}\right)\right]\right\}$$

where : $\beta = \alpha L$. The variation of ε/Δ as a function of the position of the defect in the well is depicted in Fig. 3b for various values of β. It shows once again that a non negligible shift occurs only on the edges of the well when this well is large compared to α^{-1}. It becomes large in the center of the well only when β becomes of the order of 1.

Since the extension in space of a deep level wave function is usually limited to the first neighbors, i.e. to one interatomic distance a, only the defects which are located in the layer adjacent to the heterostructure plane will undergo a shift of their energy level. Is there any experimental evidence for such a shift, limited to the adjacent layer ? In the GaAs-GaAlAs system for instance, the conduction band offset, 270 meV (see next lecture), implies for $\alpha z_o = 1$ a shift ε of the order of 20 meV. Such shift is too small to be detected using a thermal spectroscopy such as Deep Level Transient Spectroscopy (DLTS) (see next lecture) and should be detected by optical means. Because in practice only transition metal impurities give rise to sharp emission lines in GaAs, the existence of such shift has been looked for Mn impurities using luminescence. The Mn associated luminescence in a GaAs well is very similar to the one obtained in bulk material, in particular the width of the peak is the same (8). The only difference consists in a shoulder which appears on the high energy side, at 30 meV above the energy for the maximum of the peak. The ratio of the intensity of this shoulder to the area of the main peak varies linearly with the ratio a/L (9). This demonstrates clearly that this shoulder should be attributed to the luminescence of Mn ions which are located in the two monolayers on each side of the well; the energy shift $\varepsilon \sim 30$ meV is the one expected using the derived expression of ε for a wave function whose extension in space is of the order of a. Thus the binding energy, or energy level, of Mn is constant throughout the well except when this impurity is located at one monolayer from the interface.

The invariance of a defect level position as compared to its original band structure (i.e. in the bulk material) provides a very powerful way to characterize a quantum well or a superlattice, because it is possible to measure its energetic position in the new band structure (10). This is easy to do in doped superlattices using DLTS (11), as it will be developed in the next lecture. The ionization energy of a defect situated at E_{T1} below the band E_{c1} in the bulk material which composes the well of the superlattice becomes (see fig. 4) $E_{S1} = E_{c1} - E_{T1} + \delta$, where δ is the energy difference between the bottom of the superlattice miniband and E_{c1}. For a defect having a level E_{T2} in the material which composes the barrier (of height Δ) the ionization energy $E_{c2} - E_{T2}$ becomes $E_{S2} = E_{c2} - E_{T2} - \Delta + \delta$. Thus, the determination of the energy level positions in both bulk materials 1 and 2 and in the superlattice provides a direct measure of the miniband position δ and of the band offset Δ. Such technique requires of course the knowledge of the layer in which a given defect is present; it can be obtained when one does not consider a single level but a series of levels, because their energy differences are known in the bulk material. This method is illustrated in the next lecture.

III-2 Carrier capture cross-section

We have briefly recalled in section II that one consequence of the electron-phonon interaction, particularly large in compounds semiconductors, is the crossing of the two branches E_0 and E_1 (see Fig. 2) of the configuration coordinate diagram corresponding to the total energies of the defect in its ground and ionized states. This crossing implies an electron recombination through a multiphonon emission process characterized by a capture cross-section σ_c thermally activated with the energy ΔE. Once placed in a superlattice the same defect will have to emit the electron into the new band , the superlattice miniband, shifted by an amount $K = \delta$ or $\delta - \Delta$ depending if the defect is in a well or in a barrier and the energy ΔE becomes

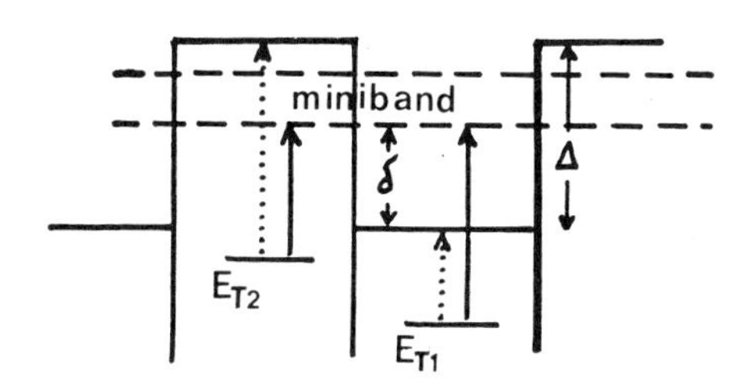

Fig. 4. Representation of the variation of the change of emission rate when the defect E_T is placed in the well (1) or in the barrier (2) of a superlattice.

$$\Delta E_S = \frac{1}{2} (k/I^2)(E_T - d_{Fc} + K)^2$$

Because the defect is strongly localized, on the atomic as well as on the electronic scales, the constants k and I, like the electronic characteristics d_{Fc} and E_T, remain in the superlattice the same as in the bulk materials. Consequently, the energetic barrier which characterizes the evolution of σ_c versus temperature decreases or increases depending whether the defect is placed in a barrier or in a well and σ_c varies accordingly. The changes of σ_c at a given temperature can be rather large since δ or $\Delta - \delta$ can be of the same order of magnitude or even larger than $E_T - d_{Fc}$.

Thus, the value of the capture cross-section of a carrier on a deep defect in a superlattice can be predicted immediately from its value in the bulk material when the band offset and the band structure of this superlattice are known. As illustrated in the next lecture, this prediction seems to be verified at least for some of the cases it has been possible to study up to now.

III-3 Optical cross-section

We have seen in section II that an optical cross-section σ_o is a function of three parameters : the extension in space of the defect wave function α^{-1}, the energy E_T and the Franck-Condon shift d_{FC}. Thus, to first order, the only change which occurs when the defect is introduced in a superlattice is the shift of the threshold energy E_T by the quantity K defined in section III-2. Consequently, once again, when the electronic structure of the superlattice is known, the value that σ_o takes in the superlattice can be predicted.

It must be possible to measure this new cross-section in the same way as in a bulk material by the so-called optical DLTS technique (see chap. 3 in ref. 7) and its variation with the photon energy should provide the density of states in the miniband. However, the attempts which we have made to measure the optical cross-sections of deep levels introduced by electron irradiation in GaAs-GaAlAs superlattices, which have been well characterized in the corresponding bulk materials and in the superlattice themselves by DLTS (11), have failed up up to now. Presumably, the capacitance transients which are observed after optical excitation were dominated by the filling or emptying of interface states.

REFERENCES

1. A. Mauger, S.L. Feng, and J.C. Bourgoin, *Appl. Phys. Lett.* 51:27 (1987).
2. T. Achtnich, G. Burri, M.A. Py, and M. Uegems, *Appl. Phys. Lett.* 50:1730 (1987).
3. D.V. Lang, and R.A. Logan, *Appl. Phys. Lett.* 31:683 (1977).
4. S.R. McAgee, D.V. Lang, and T. Tsang, *Appl. Phys. Lett.* 40:520 (1982).
5. A. Mauger, F. Sillion, J.C. Bourgoin, B. Deveaud, A. Regreny and D. Stiévenard, *in*: "Defects in Semiconductors, H.J. von Bardeleben, ed., Trans Tech Publ., Switzerland (1986) Materials Science, Vol. 10-12:199.
6. M. Lannoo and J.C. Bourgoin, "Point Defects in Semiconductors, I - Theoretical Aspects", Springer, Berlin (1981).
7. J.C. Bourgoin and M. Lannoo, "Point Defects in Semiconductors, II - Experimental Aspects", Springer, Berlin (1983).
8. B. Plot, B. Deveaud, B. Lambert, A. Chomette, and A. Regreny, *J. Phys. C: Solid State Phys.* 19:4279 (1986).
9. B. Deveaud, B. Lambert, B. Plot, A. Chomette, A. Regreny, J.C. Bourgoin, and D. Stiévenard, *J. Appl. Phys.* to be published (1987).
10. J.C. Bourgoin, A. Mauger, D. Stiévenard, B. Deveaud, and A. Regreny, *Solid State Comm.* 62:757 (1987).
11. D. Stiévenard, D. Vuillaume, J.C. Bourgoin, B. Deveaud, and A. Regreny, *Europhysics Letters* 2:331 (1986).

ROLE OF THE Si DONORS IN QUANTUM AND ULTRAQUANTUM TRANSPORT PHENOMENA IN GaAs-GaAlAs HETEROJUNCTIONS

André Raymond*

Service National des Champs Intenses
Centre National de la Recherche Scientifique
166 X, 38042 Grenoble Cédex, France

ABSTRACT

By experiments under hydrostatic pressure (up to 13 Kbar) in high magnetic fields (0-20T) we investigate the role of Si donors of GaAlAs in the quantum and ultraquantum transport regimes of quasi-2D electrons gas of GaAs-GaAlAs heterojunctions with a spacer. In the ultraquantum regime we analyse the metal-nonmetal transition and magnetic freeze-out due to bound donor states as a function of the spacer thickness and of the 2D electron density N_s. In the quantum regime (Quantum Hall Effect) we show the role of these Si donors in the localization phenomenon and study it when the spacer width and N_s vary.
In both investigated regimes we point out that the screening of the Coulomb interaction between 2D electrons and donors is essential to account for the experiments.

I - INTRODUCTION

The impurities in 2D structures have lately become the subject of numerous studies because of their importance in the physical properties of these structures.
The present work concerns the role of Si donors in the Si doped layer of GaAlAs, of modulation-doped GaAs-GaAlAs heterojunctions(GaAs-H) grown by MBE and MOCVD techniques.
The GaAs-H with a spacer have been investigated by means of transport measurements using external magnetic field and hydrostatic pressure P as

* Permanent address : GES associé au CNRS UA 357, USTL, Place E. Bataillon, 34060 - MONTPELLIER-Cédex, FRANCE

variable parameters. The hydrostatic pressure has been used to reduce [1] the 2D electron density N_s in order to obtain, on the same sample, quantum and ultraquantum regimes at available magnetic fields. In the investigated samples the thickness of the spacer d varies between 60 Å and 250 Å. When the thickness of the spacer layer is not too large, $d < 200$ Å, we observed in the ultraquantum regime [2,3] bound magneto-donor states in the two dimensional electron gas (2DEG) of the GaAs-H. These bound magneto-donors states are due to the Coulomb interaction between Si donors in the Si doped layer and the 2D electrons. In the ultra quantum limit, (UQL), the inversion electrons occupy only the lowest Landau level and the magnetic freeze-out of these 2D electrons into the magneto-donors states is observed. In the quantum limit (Q-L) the Quantum Hall Effect (QHE) can be observed. A theoretical [4] and experimental analysis shows that, for GaAs-H with not too wide a spacer layer ($d < 150$ Å), the width of the plateaus increases strongly compared to the theoretical model, when, due to an increase of pressure, N_{so} decreases.

Such a phenomenon which is very small in samples with a larger spacer thickness ($d = 250$ Å) can be interpreted in terms of localization on magneto-donors, whose effect depends on the screening by 2D free electrons, of the Coulomb interaction between the 2 DEG and Si ions.

A theoretical study [5] of magneto-donors states shows that the inclusion of screening of this Coulomb interaction is also essential for the description of the observed binding energies of magneto-donors.

II - EFFECT OF HYDROSTATIC PRESSURE ON Si-DOPED GaAs-GaAlAs HETEROJUNCTIONS

The effect of pressure on bulk III-V compounds is now rather well known. Although the relative variation on lattice constant is rather small, the effect on the electronic properties can be important. Pressure induces a modification of the band structure and in the case of $Ga_{1-x}Al_xAs$ this modification is similar to that of alloying GaAs with Al.

For the samples we investigated the Al content x is close to 0.3 (see table 1). It is well known that for x larger than 0.2 the Si donor states have a deep character [6-8]. Because of their deep character these donor states are strongly pressure dependent (see Fig. 1) : when the hydrostic pressure is increased, this deep-lying level, located in the energy gap of GaAlAs, shifts rapidly downward with respect to the conduction band minimum. As a consequence, donor deionization takes place in GaAlAs, the energy diagram of the GaAs-H varies (as shown in Fig. 2) and N_s decreases. For temperatures smaller than 120K, this deep-lying level is characterized by a metastable occupation [9]. As a consequence, the 2DEG has a metastable character [10], and because of lattice relaxation effects, N_s can be slight modified for a given pressure, depending on the sample cooling speed.

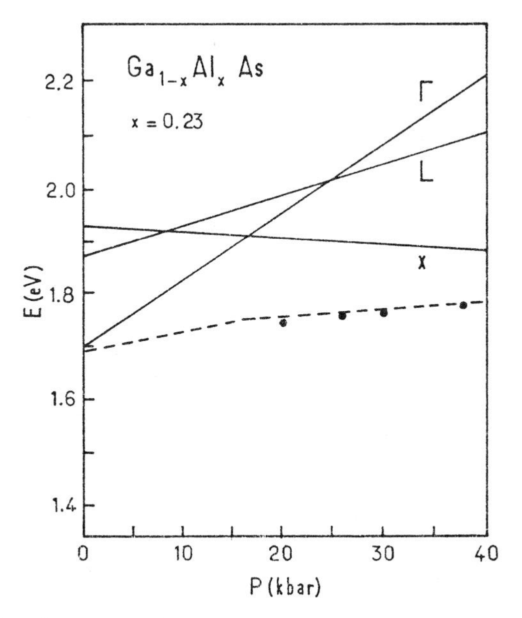

Fig. 1. Pressure dependence of the donor deep level and of the band structure of $Ga_{.77}Al_{.23}$ (Ref. 7).

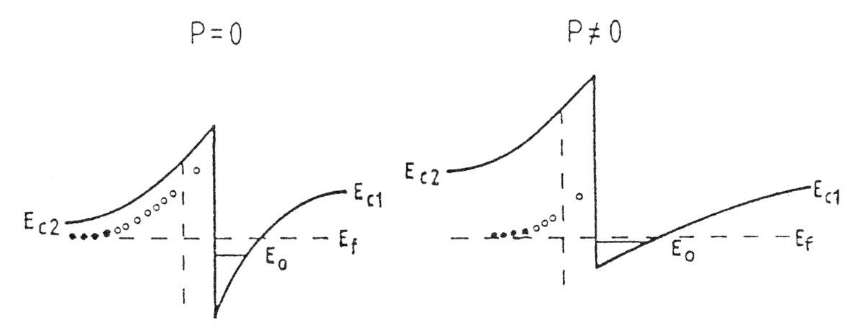

Fig. 2. Schematic variation of the energy diagram of GaAs-H under hydrostatic pressure.

At a sufficiently high pressure one can reduce N_s to values lower than $5x10^{10}$ cm^{-2}, even for highly doped samples.
In Figure 3, we have reported the pressure dependence of N_s on sample 1 at 4.2K ; this linear decrease of N_s with increasing P has been observed at all temperatures for all the investigated samples and has been quantitatively explained[1] assuming a pressure coefficient dE_d/dp = 11 meV/Kbar and E_{do} = 60 meV for the Si donor level in $Ga_{.7} Al_{.3} As$.

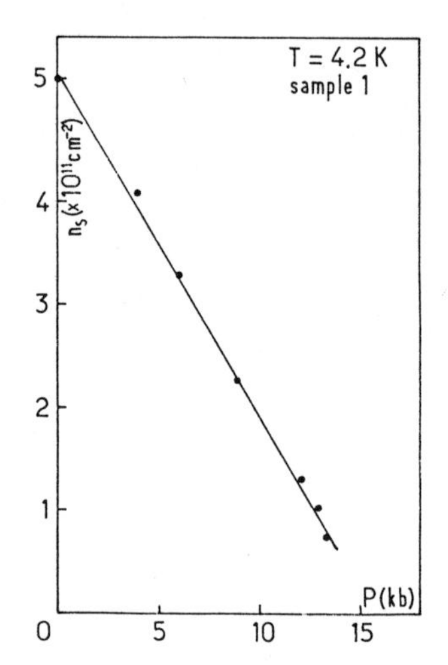

Fig. 3. Typical experimental pressure dependence of the surface electron density N_s (sample 1).

III - STUDY OF MAGNETO-DONORS IN ULTRAQUANTUM LIMIT

III-1- Experimental procedure and experimental results

We have presented in table 1, the samples characteristics at 4.2K (with and without pressure). The values of N_s and of the mobility μ are the Hall values, measured at low magnetic field (B = 0.5 T).

For this analysis of localization the basic problem is the measurement of the surface electron density N_s. The Hall coefficient R_H and the transverse magnetoresistance were measured for two current and magnetic field directions in the temperature range 1.3 - 4.2K. In order to avoid impact ionization effects and to stay within the ohmic regime, we kept the electric field low enough during the experiments.

TABLE 1

Sample characteristics, critical magnetic fields and surface electron densities for metal-nonmetal transition in GaAs-$Ga_{1-x}Al_x$ heterostructures at different hydrostatic pressures. The last column gives the Mott-like criterion for the metal-nonmetal transition.

Sample	x	d (Å)	P=0 T=4.2K $N_{so}(10^{10})$ cm^{-2}	$\mu(10^4)$ cm^2/Vs	P ≠ 0 T = 4.2K p Kbar	$N_{so}(10^{10})$ cm^{-2}	$\mu(10^4)$ cm^2Vs	B_c (T)	$N_{sc}(10^{10})$ cm^{-2}	$\sqrt{N_{sc}}\, L_c$
1	0.3	60	52	7.5	13.3	8.5	0.95	6	6	0.26
2	0.3	90				6.5				
3	0.25	150	24	12.9	8.8	6.5	3	4.8	6.4	0.3
					8.8	5.7	2.36	4.2	5.6	0.29
					8.8	7.8	1.55	3.5	6	0.34
					8.8	5.1	1.84	3.3	4.5	0.3
4	0.27	250	20.8	5.2	5.9	12.5	2.0	10	11	0.27
5	0.3	250	37	41.9	13	6.5	6.82	8	6.5	0.23

If we assume that localized electrons do not participate in the Hall voltage, σ_{xy} being the conductivity tensor components and ρ_{xy} the Hall resistivity, in the relaxation time (τ) approximation we can write :

$$\sigma_{xy} = - \frac{n_s e^2 \omega_c}{m^*} < \frac{\tau^2}{1 + \omega_c^2 \tau^2} > \qquad (1)$$

In high magnetic field limit i.e $\omega_c \tau >> 1$ (ω_c cyclotron frequency) one has

$$\sigma_{xy} \simeq - n_s e/B$$

On the other hand :

$$\sigma_{xy} = - \frac{\rho_{xy}}{\rho_{xx}^2 + \rho_{xy}^2} = - \frac{R_H B}{\rho_{xx}^2 + R_H^2 B^2} \qquad (2)$$

which leads to $N_s \simeq \dfrac{B \rho_{xy}}{e (\rho_{xx}^2 + \rho_{xy}^2)} = \dfrac{R_H B^2}{e (\rho_{xx}^2 + R_H^2 B^2)}$ (3)

where ρ_{xx} is the magneto transverse resistivity.

Thus, in high magnetic field conditions, N_s must be determined by using equation (3). In the magnetic freeze-out regime this expression is never equivalent to the simple expression $n_s = B/\rho_{xy}e$ because in this regime the condition $\rho_{xx}^2 << \rho_{xy}^2$ is never satisfied.
Figure 4 represents the temperature and magnetic field dependence of N_s for sample 1 under a pressure of 13.3 Kbar. It is clearly seen that above a critical field B_c, N_s is thermally activated (magnetic freeze-out). This allows us to determine the critical value of the surface electron density N_{sc} corresponding to the transition between metallic and nonmetallic types of conduction. Both quantities B_c and N_{sc} are reported in table 1. The surface electron density being thermally activated can be written $N_s = N_o \exp(-E_a/kT)$, where E_a is the activation energy. Figure 4 shows a linear dependence of ln (N_s) versus 1/T. Thus, N_o is temperature independent and E_a can be determined from the slopes of the curves.

The experimental values of the binding energy E_a for four different intentionally doped GaAs-H with different values of the spacer thickness d are shown in Fig. 5 : E_a increases with magnetic field and is lower when d is larger. For example, for samples 1 and 3, the pressure has been chosen such as to obtain approximately the same critical density $N_{s.c}$ and practically the same critical magnetic field B_c at the Metal-Nonmetal

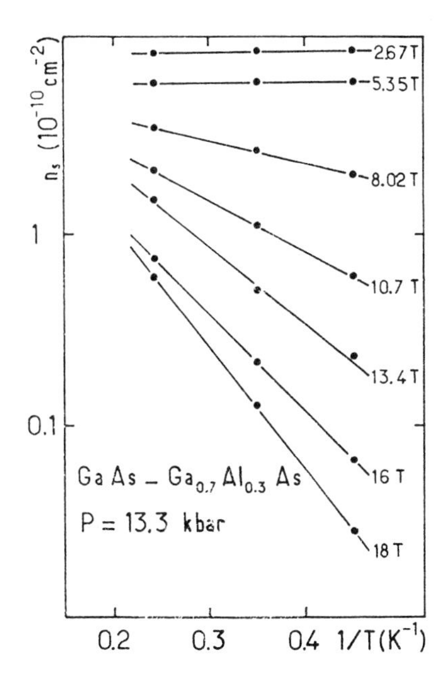

Fig. 4. Temperature dependence of N_s for different magnetic fields. Sample 1.

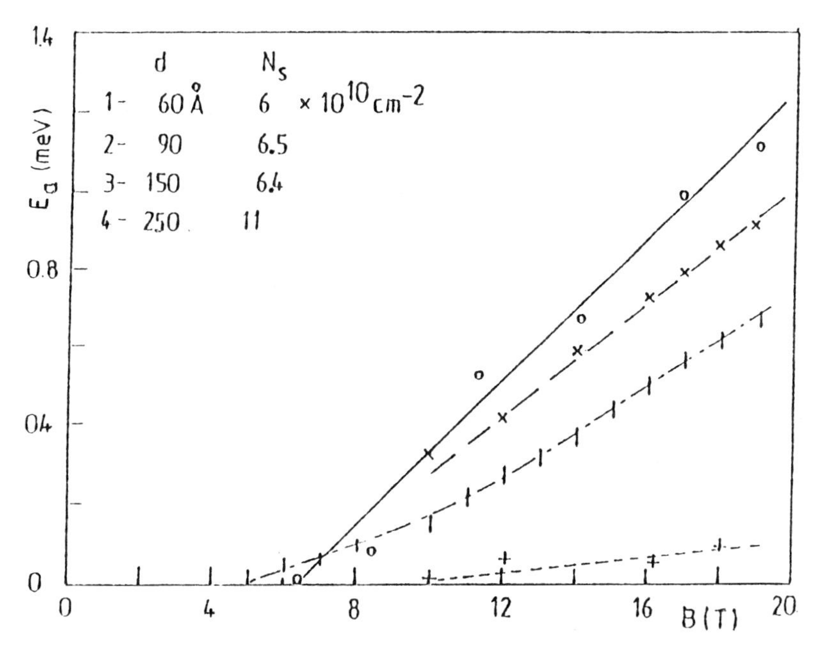

Fig. 5. Experimental magnetic field dependence of E_a for four GaAs-H with a spacer. N_s values at the MNMT are indicated. The lines are drawn to guide the eye.

Transition (MNMT). In these conditions, the overlap of the donor wave functions and the screening of the donor potentials by surface electrons are similar at the transition. Nevertheless Figure 5, shows that the magnetic field dependences of the activation energies are distinctly different for different samples : the activation energy decreases with increasing spacer thickness. This result suggest that the observed localization effect should not be ascribed to the Wigner condensation of a dilute 2D gas, which, in the case of similar electron densities, would lead to the same activation energy value, but indicate that we deal with states related to Si donors on the other side of the spacer.

In Figure 6, we have reported the experimental activation energies in sample 3, measured at the same pressure, for which the zero field electron density N_{so} has been varied by changing the sample cooling speed. In this case one deals with the same spacer thickness and Figure 6 shows that the binding energy E_a increases with decreasing N_{so}, or some other quantity related to it.

III - 2 - Model of magneto-donors in this quasi 2D system

Robert et al [2] first gave a semiclassical model which qualitatively explained the observed phenomena. The basic idea is that a transverse magnetic field shrinks the donor orbit in the plane parallel to the interface so that the electron is on an average closer to the donor ion than without a magnetic field. As a result, the Coulomb binding energy is enhanced by the presence of a magnetic field. This results in a magnetic freeze-out which we do observe experimentally. A characteristic of this quasi 2D situation is related to the interface :

First it is clear that in high field conditions the binding energy of such a magneto-donor should be much smaller than the one in bulk GaAs, since in the latter case the electron is on an average closer to the donor atom. This is in fact what we observe.

Second, the interface also prevents, in the high field regime, the electron from coming closer to the donor as B increases (which is the case in the bulk cf. Ref 11). As a consequence the binding energy should reach a saturation at high fields, its value being determined by the spacer thickness.

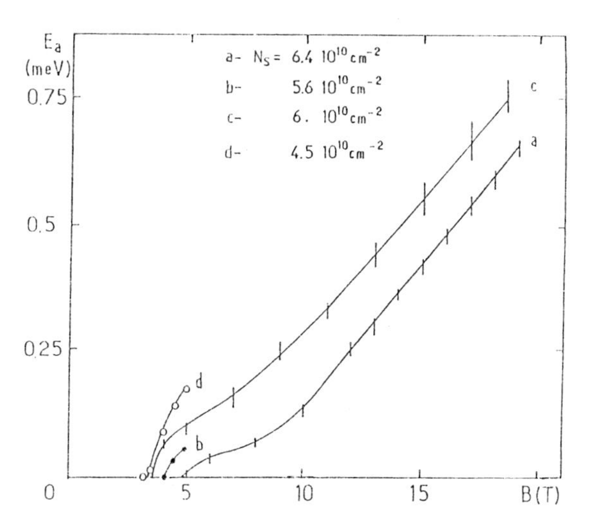

Fig. 6. Experimental activation energies for sample 3 (P = 8.8 Kbar) versus B. Different E_a (B) dependences correspond to various N_s values at the MNMT by different speeds of cooling of the sample. The lines are drawn to guide the eye.

A metal-nonmetal Mott type transition is usually associated with an overlap of the impurity wave functions. As shown by Yafet et al[11] at high magnetic fields, the 3D electron bound to a donor moves on the orbit, whose transverse radius $a_\perp$ is nearly equal to the magnetic length $a_\perp \simeq L = (\hbar/eB)^{1/2}$. Robert et al [2] assumed that, in their semiclassical model, the surface electron moves on the orbit with the same radius. The average distance between surface electrons at the transition being $n_{sc}^{-1/2}$, they proposed that the overlap condition for the Mott transition should be approximately given by $2 L_c \simeq n_{sc}^{-1/2}$, i.e, $n_{sc}^{1/2} L_c \simeq 0.5$. As it appears in table 1, in fact, the product $n_{sc}^{1/2} (\hbar/eB_c)^{1/2}$ is close to that value.

Recently Zawadzki et al[5] used a variational procedure to describe theoreticaly such magneto-donor states. The characteristic parameter for the problem is $\gamma = \hbar\omega_c/2R_y^*$, where R_y^* is the effective Rydberg. For the donor ground state they took the trial function, $F(\rho, z) = \psi(\rho) f(z)$, in which the transverse motion (parallele to the interface) is described by a product of atomic type and magnetic type two parameter function,
$\psi(\rho) = A \exp(-\alpha\rho - \beta\rho^2)$, while the longitudinal motion is described by $f(z) = C z \exp(-b_o z/2)$. They have assumed that the envelope $f(z)$, is the same for the free and bound 2D electron [12] (unchanged value of b_o). They calculated the influence of the donor potential on the motion in the z direction by using first order perturbation theory, and they took for b_o the Stern-Howard expression $b_o = [48 \pi m^* e^2 (N_{de} + 11/32 N_s)/\chi_o \hbar^2]^{1/2}$, where N_{de} is the density of depleted changes. The Price procedure[13] has been used to describe the static screening of the Coulomb potential, (neglecting the effect of the magnetic field on the screening).
The binding energy obtained is, in units of R_y^* : $E_b = \gamma - T(\alpha, \beta) - U(\alpha, \beta)$ where T and U are trial averages of kinetic and potential energies respectively. For the numerical calculations the following parameters have been used : $m^* = 0.07\ m_o$, $\chi_o = 12.56$, $R_y^* = 5.8$ meV.
In Fig. 7, we have reported calculated binding energies of magneto donors for GaAs-H with a spacer of 150 Å, $N_{de} = 6\times10^{10} cm^{-2}$, T = 4.2K and different values of N_s. It appears clearly that at lower N_s, the theoretical binding energy depends very strongly on surface density. This shows that, in the theoretical procedure, the inclusion of screening is essential to reach an even rough agreement with the experimental values (see sample 3 in Fig. 5).
Figure 8 shows the calculated binding energies of magneto-donors for four values of d and N_s corresponding to the experimentally investigated samples (see table 1). A comparison between these values and the data of Fig. 5 shows that, at high magnetic field i.e away from the metal-nonmetal transition, the theory describes quite well the experimental values.

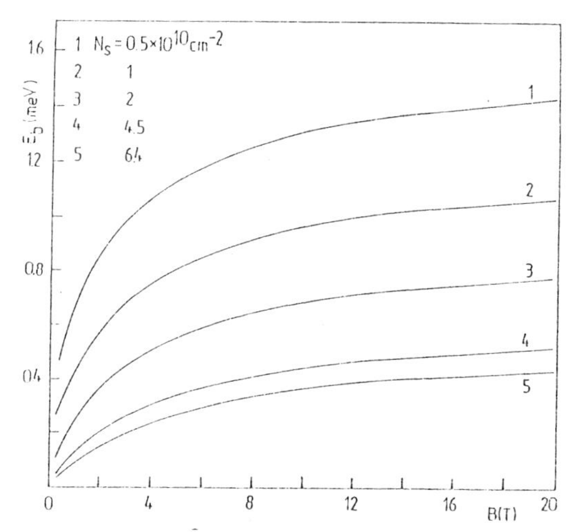

Fig. 7. Theoretical binding energies of magneto-donors in a GaAs-H with d = 150 Å and N_{de} = $6x10^{10}$ cm^{-2} versus B calculated for different surface densities N_s including screening.

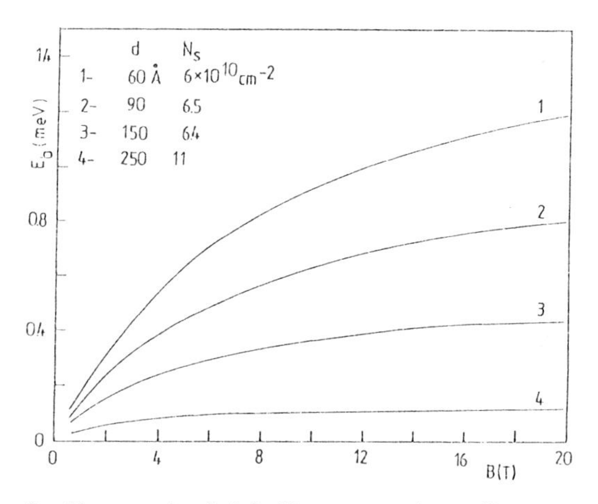

Fig. 8. Theoretical binding energies of magneto-donors in GaAs-H versus B, calculated for d and N_s indicated in Fig. 5 and N_{de} = $6x10^{10}$ cm^{-2}.

Though there exists an uncertainty concerning the real positions of Si donors reponsible for the electron binding and the validity of the one-donor-one electron picture even at higher fields, the agreement should be considered as surprisingly good.

So, both the semiclassical model and this variational procedure show the coherence of the experimental analysis which demonstrates the existence of bound magneto-donor states in quasi 2DEG of GaAs-H. These bound magneto-donor states are due to the Coulomb interaction between Si ions in the Si doped GaAlAs layer and the quasi 2DEG.

IV - STUDY OF THE QUANTUM HALL EFFECT : ROLE OF THE Si DONORS IN QUANTUM LIMIT

IV - A model of the Quantum Hall Effect

The analysis we made of the experimental results in the Ultra Quantum Limit is based on the determination of N_s by using equation (3).

In Fig. 9, we have reported the experimental magnetic field dependence of N_s, determined by using equation (3), on sample 3 for two different temperatures, between 0 and 15 T. The Metal-Nonmetal transition (MNMT) appears clearly at $B_c = 4.5$ T ; between 2.5 T and 4.5T we observe the Ultra Quantum Limit and below 2.5 T the Quantum Limit. We have previously seen that, as long as we believe in this determination of N_s, we can satisfyingly account for the magnetic and temperature dependences of N_s beyond the MNMT in terms of a localization on magneto-donors. If we determine N_s by the same method for $B < B_c$ we clearly obtain oscillations of the 2D electron density, with a period exactly equal to the one of the Shubnikov-de-Haas effect or of the Q.H.E. (Filling factor values $\upsilon = 2$ and $\upsilon = 1$ are reported in Fig. 9). This result shows, if we believe in it, that in the QHE regime the density of mobile electrons (those which participate in the Hall effect) varies when B varies. In that case the oscillations reach 10%.

In normal QHE i.e., for example, for lower values of the pressure far from the MNMT, ρ_{xx} is very much smaller than ρ_{xy} ($\rho_{xx}^2 << \rho_{xy}^2$). In this case equation (3) gives $\rho_{xy} \simeq B/N_s e$. We, then, can think that oscillations of N_s induce variations of ρ_{xy} which can give rise to plateaus.

Raymond and Karrai [4] have preposed a model based on this idea. First Baraff and Tsui [14] have shown that a tunneling of electrons between the ionized donors of the doped layer and the inversion layer could induce the formation of plateaus in the Hall resistivity ρ_{xy}. Raymond and Karrai (R-K) proposed a model which takes into account both the transfer of electron and the role of localized states. Schematicaly they assumed that two kinds of QHE can be observed depending only on the experimental temperature (T) : one at higher temperature involving a transfer and the other one at lower temperature involving localization. Depending on the quality of the sample and on the temperature these two processes may coexist.

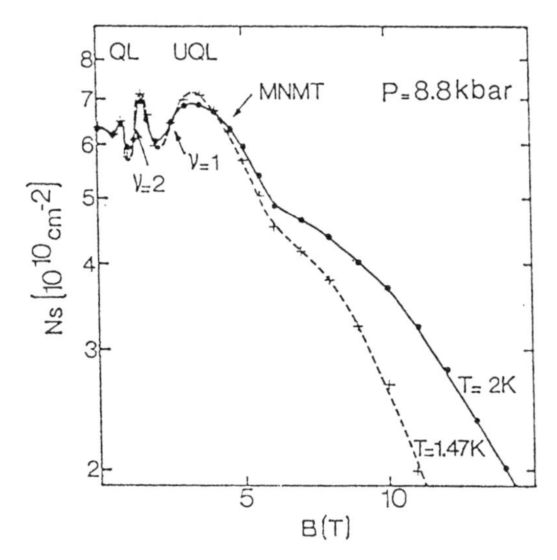

Fig.9. Experimental B dependence of N_s at 1.47 K and 2K for sample 3. N_s is determined from measurements of ρ_{xx} and ρ_{xy} : $N_s = (\rho_{xy} B)/e(\rho_{xx}^2+\rho_{xy}^2)$. The full and dashed lines are a guide to the eye.

Now let's summarize the (R-K) model. We first assume that the density of charges N_s involved in ρ_{xy} is the density of delocalized electrons and that a transfer of electrons is possible between the 2DEG and a reservoir which supplies electrons to this 2D system. If the magnetic field dependence of the density of states (DOS) of the 2DEG is different from the one of the reservoir, then, at the thermodynamical equilibrium, the density N_s of the mobile 2D electrons must be magnetic field dependent. As the magnetic field can modify the charge density, the electrostatic equilibrium of the heterojunction is then B dependent. Two cases are considered :

1) The reservoir is not in the Quantum Well (QW) : the electrons may transfer, for example, by tunneling on ionized donors or on residual impurities or defects in the spacer layer. In this case the equilibrium of the heterojunction must be recalculated.

2) The reservoir is in the QW but the corresponding states are uncorrelated with the 2DEG : the electrons do not leave the QW. They can be localized for example on interface states or other states very close to the 2DEG. In this case, in first approximation, the accumulated charge in the well does not change and the equilibrium does not change either.

The magnetic field dependence of N_s and ρ_{xy} have been calculated for GaAs-H. In the R-K model it is assumed that the Fermi energy E_F, is a constant in the whole structure and does not depend on B since in the undepleted regions of GaAs and of GaAlAs, E_F is in first approximation magnetic field independent (non isolated system). We must specify that E_F-E_o can vary because the lowest electrical subband energy E_o can vary.

As we previously explained, in case 1, when N_s varies the mean electric field F varies and consequently E_o varies too, when, in case 2, the variation of the density N_s of mobile electrons does not change the density of charges in the QW and consequently does not change F and E_o. The calculations have been performed in the Triangular Well Approximation (TWA) assuming than E_o is the only populated subband and considering non interacting electrons in a spherical parabolic energy band. A density of states having the shape of a sum of Gaussian peaks have been used and in the calculations, Raymond and Karrai made the hypothesis that all the 2DEG states are non localized states. This means that in case 2 all the localized states involved in the model are additional states.

The density of mobile electron N_s is then given by :

$$N_s = \sum_s \sum_{n=0}^{\infty} \frac{eB}{h} \left(\frac{2}{\pi}\right)^{1/2} \frac{1}{\Gamma} \int_{-\infty}^{+\infty} \frac{\exp\left(-2\left[\frac{(E-E_o)-(n+1/2+\theta\, s/2)\,\hbar\,\omega_c}{\Gamma}\right]^2\right)}{1+\exp\left[(E-E_F)/kT\right]}\, dE \quad (4)$$

where $\Gamma = \frac{1}{\sqrt{2}\,\sqrt{\mathrm{Ln}2}} \frac{\hbar}{\tau} \simeq \frac{1}{\sqrt{2}} \frac{1}{\sqrt{\mathrm{Ln}2}} \frac{\hbar\, e}{m^*\,\mu}$ is the broadening parameter and μ the zero field mobility, $s = \pm 1$ and $\theta = \frac{g^* m^*}{2\, m_o}$.

In equation (4) N_s is a function of B and in case 1 $(E-E_o)$ depends also on B. The value of the g factor g^* has been calculated taking into account the spin splitting enhancement by the exchange interaction of electrons as developped by Ando and Uemura [15].

In Fig. 10(a) we have reported the results of the calculations using the R-K model for sample 5 at 1.4K. ρ_{xy1} represent the Hall resistivity calculated in case 1 (external reservoir), and ρ_{xy2} in case 2 (internal reservoir). We have used the following parameters $m^*/m_o = 0.07$, $\chi_o = 12.56$, $N_A - N_D = 2x10^{14}$ cm^{-3}(excess donor density of the non doped GaAs layer), and the band bending potential $\phi_d = 1.425$ eV.
In Fig. 10(b) the experimental curves of ρ_{xx} and ρ_{xy} for sample 5 are reported at zero pressure. We can notice for ρ_{xy1} the good agreement obtained without fitting parameter for the width of the plateaus particularly out of the spin splitting regime. A comparative analysis of Fig. 10(a) and 10(b) clearly shows the competition between IQHE and FQHE for the filling factor $\upsilon = 2$ (dashed line). The theoretical oscillatory behaviour of g* is reported in the inset of Fig.10(a).
For all the samples investigated, at zero pressure with mobility varying between $4x10^4$ cm^2/Vs and 10^6 cm^2/Vs, a good agreement is observed between the experimental curves of ρ_{xy} and the theoretical ones in case 1 (ρ_{xy1} for T larger than approximately 1K. Figure 11 shows the effect of temperature between 1.4K and 40K on the theoretical Hall resistivity ρ_{xy1} for sample 5. In this case too, a very good agreement is observed between theoretical and experimental results. Raymond and Karrai deduced that, above such a temperature, the QHE is a consequence of a transfer of electrons due to the statistics, between the 2DEG and for example ionized donors, outside the QW (case 1). This agree well with the recent study of Nizhankovski et al[16].
When the temperature is lowered to values very much smaller than 1K, the widths of the plateaus increase and the model of Raymond and Karrai indicates that the situation is now well described by case 2. They conclude that for very low temperature the QHE is certainly a consequence of the transfer of electrons from mobile states in the broadened Landau levels, into additional localized states. As an example, in Fig. 12, we have reported theoretical (ρ_{xy2} dashed line) and experimental curves of ρ_{xy} given in the litterature[17] at 8 mK for a low mobility sample.
As for sample 5 in Fig. 10 we observe an excellent agreement without fitting parameter between theoretical and experimental results outside the spin splitting regime and a qualitative good agreement inside it.
One argument for the R-K model is that, in the samples we investigated, which do not exhibit clear FQHE, out of the spin splitting regime, in the higher temperature range corresponding to case 1 (transfer with an external reservoir) the width of the plateaus Δ seems to increase with the mobility for similar values of N_s.

We think that, depending on the temperature and the quality of the sample (degree of disorder) both processes may coexist.

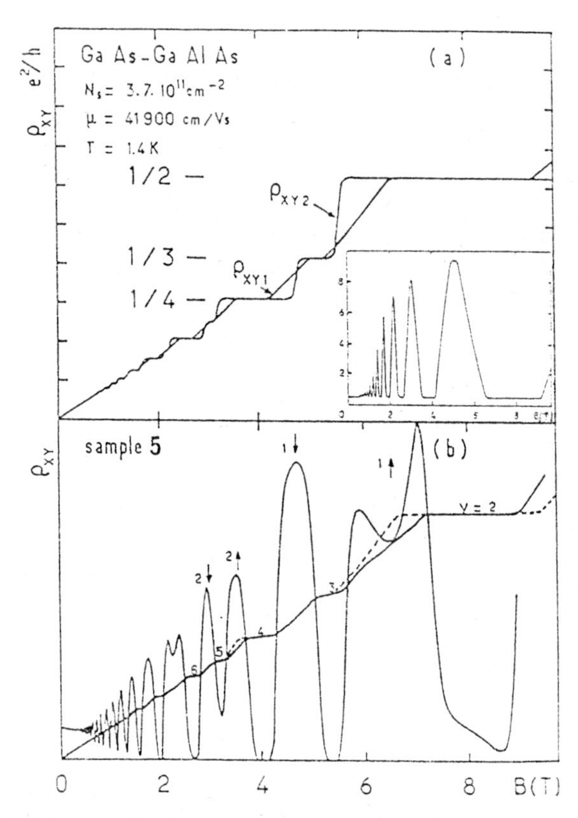

Fig.10. a) Theoretical curves of ρ_{xy} : ρ_{xy1} corresponds to the case of an external reservoir (higher T). ρ_{xy2} corresponds to the localization of non mobile electrons in the well at lower T (internal reservoir). The inset represents the theoretical B dependence of the g factor for sample 5.

b) Experimental curves for ρ_{xx} and ρ_{xy}. Dashed lines correspond to the theoretical curve ρ_{xy1}.

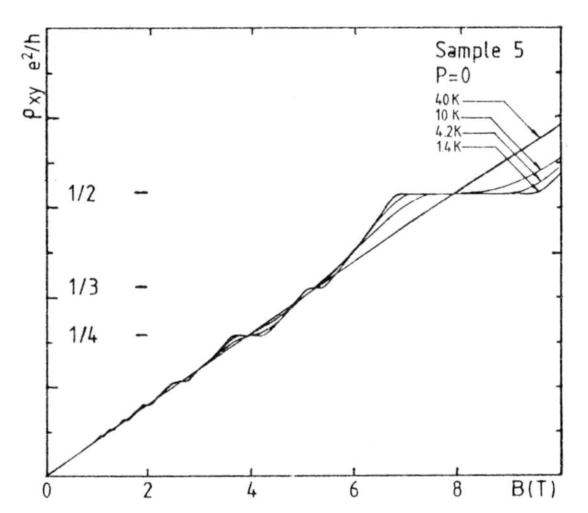

Fig. 11. Theoretical curves of ρ_{xy} (ρ_{xy1}) at four different temperatures for sample 5 at P = 0.

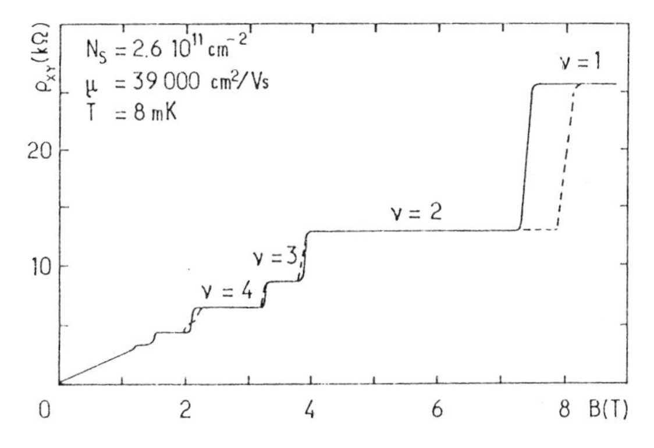

Fig. 12. ρ_{xy} as a function of B for very low T : full line corresponds to the experimental results[17] the dashed line corresponds to the theoretical results in the case of localization (ρ_{xy2}).

Before analysing the role of Si donors of the Si doped GaAlAs layer on the QHE, we would like to discuss briefly the choice we made of the broadening parameter Γ. We assumed that Γ is given by the zero field mobility and is magnetic field independent. This certainly corresponds to the lowest value of Γ. It is well known that the screening by free electrons depends on B because it depends on the relative position of the Fermi energy E_F compared to the Landau levels : when E_F coincides with a Landau level, the screening is maximum and Γ is minimum (abrupt transition between two plateaus at very low T), on the contrary when E_F is between two Landau levels the screening is minimum and Γ maximum. In our calculations we kept Γ constant and equal to its minimum value. This directly increases the calculated plateau width Δ.

We thinks that this is a reason which leads, in high field regime, to theoretical plateaus always larger than the experimental ones.

IV - 2 - Role of the Si donors in the Quantum Hall Effect for GaAs-GaAlAs heterojunctions

We have analysed in the temperature range 1.3-4.2K under hydrostatic pressure i.e for different values of N_s, in the QHE regime, several GaAs-H samples having a spacer layer different from sample to sample. The main result is that for low values of N_s, which means for higher pressure values, the width Δ of plateaus increases, as compared to the R-K model, when the thickness of the spacer layer is small enough to allow the existence of magneto-donor states.

As we said previously in this temperature range for zero pressure or for low pressure values, the R-K model accounts well for the plateau width and shape of ρ_{xy} by the process 1 for all the investigated samples (ρ_{xy1} : external reservoir). As an example we have reported in Fig. 13 the experimental (a) and theoretical (b) QHE for sample 5 at 1.3K under different pressures. When the pressure increases from zero to 11.3 Kbar N_s and μ decreases respectively from 3.7 10^{11} cm^{-2} and 4.19 x 10^5 cm^2/Vs to 1.4 x 10^{11} cm^{-2} and 1.17 10^5 cm^2/Vs. For this sample with a spacer thickness d of 250 Å, in spite of the spin splitting consequences, the agreement between theoretical and experimental results is rather good. Nevertheless we observe on experimental curves, a small increase of Δ for $\upsilon = 2$ when N_s decreases, as compared to the theoretical ones. This effect can be related to a decrease of the spin splitting enhancement, or more probably to the presence of some magneto-donor states which would be less and less screened by free electron when N_s decreases.

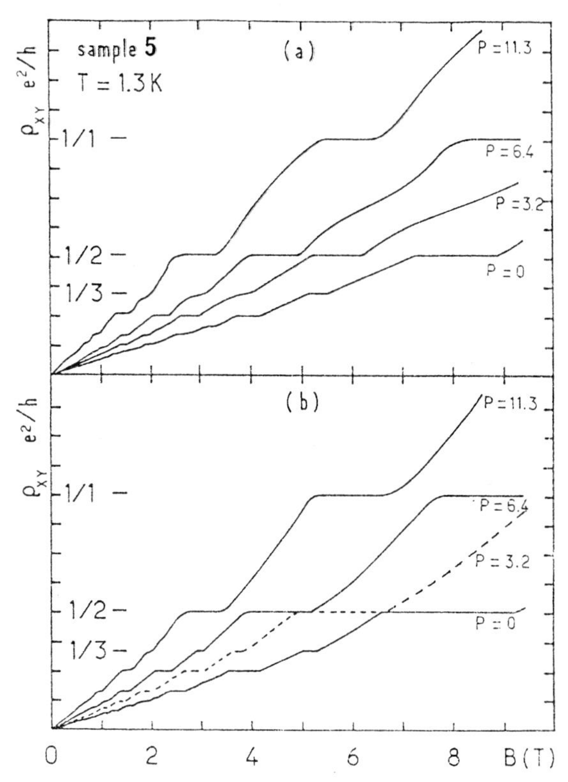

Fig. 13. Experimental (a) and theoratical (b) QHE for sample 5 at 1.3K under different values of the pressure P given in Kbar.

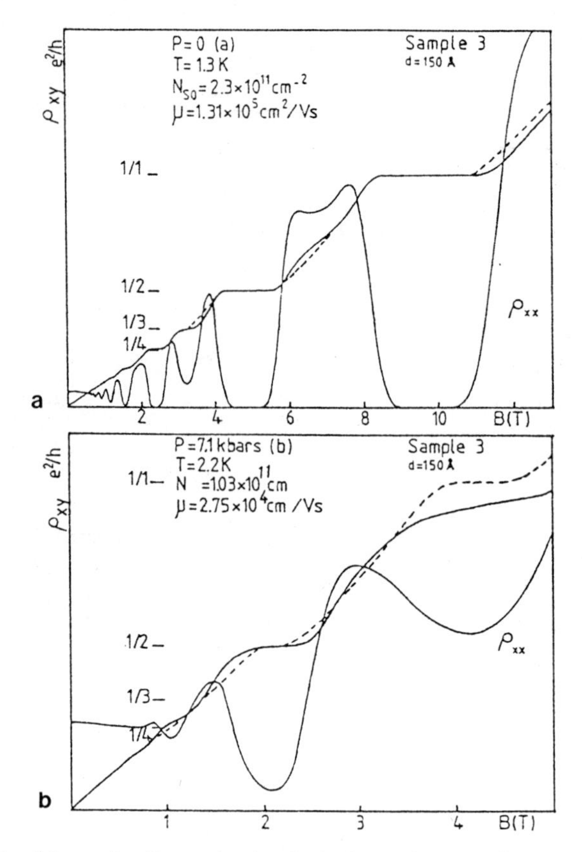

Fig.14. a) Experimental B dependence of ρ_{xx} and experimental and theoretical (dashed) ρ_{xy} curves for sample 3 at zero pressure.

b) Experimental B dependence of ρ_{xx} and experimental and theoretical (dashed) ρ_{xy} curves for sample 3 under a pressure of 7.1. Kbar.

As we explained this effect is much more obvious for GaAs-H with narrow spacer layer, as in sample 1,2 or 3, which clearly present the existence of magneto-donor states.

In Fig. 14 we have reported experimental and theoretical (dashed line) QHE results for sample 3 (d = 150 Å) for two pressure values : (a) P = 0, N_{so} = 2.3 x 10^{11} cm^{-2}, μ = 1.31 x 10^5 cm^2/Vs and (b) P = 7.1 Kb, N_{so} = 1.03 x 10^{11} cm^{-2}, μ = 2.75 x 10^4 cm^2/Vs. The increase of Δ for υ = 2 when N_s decreases from 2.3 x 10^{11} to 1.03 x 10^{11} cm^{-2} is now very important. We can also notice that, for this kind of samples with evidence of magneto-donor states, the quantization of ρ_{xy} in h/i.e^2 disappears for i = 1. This shows that the magneto-donors states which give rise to magnetic freeze-out in the ultra quantum limit can play an important role in the quantum limit (QHE) when the Coulomb interaction, from which they originate is not screened by free electrons. Nevertheless we think the nature of this localization effect is different from the one which is effective in very low temperature in QHE.

Acknowledgements

It is a pleasure to thank Professors J.L. Robert, W. Zawadzki as well as Doctors, C. Bousquet, L. Konczewicz, M. Kubisa, E. Litwin-Staszewska, R. Piotrzkowski, who participate to the study of magneto-donors and Doctor K. Karrai for his collaboration in the QHE analysis. We are also grateful to Doctors J.P. Andre, P.M. Frijlink (LEP), F. Alexandre and J.M. Masson (CNET) for providing the samples.

References

1. J.M. Mercy, C. Bousquet, J.L. Robert, A. Raymond, G. Gregoris, J. Beerens, J.C. Portal, P.M. Frijlink, P. Delescluse, J. Chevrier and N.T. Linh, Surf. Sciences, 142, 298, (1984).
2. J.L. Robert, A. Raymond, L. Konczewicz, C. Bousquet, W. Zawadzki, F. Alexandre, I.M. Masson, J.P. Andre and P.M. Frijlink, Phys. Rev. B.33, 5935, (1986).
3. A. Raymond, J.L. Robert, C. Bousquet, E. Litwin-Staszewska, R. Piotrozkowski, M. Kubisa, W. Zawadzki and J.P. Andre. Proc. 18th Int. Conf. Phys. Semicon. Stockholm 1986, ed. O. Engström, World Scientific, 1261, (1987).
4. A. Raymond and K. Karraï MSS III Montpellier 1987, to be published in Journal de Physique.
5. W. Zawadzki, M. Kubisa, A. Raymond, J.L. Robert and J.P. Andre, to be published in Phys. Rev.B.36, Nov. (1987).
6. N. Chand, T. Henderson, J. Klem, W.T. Masselink, R. Fischer, Y.C. Chang and H. Morkoc, Phys. Rev. B.30, 4481, (1984).
7. A.K. Saxena, J. Phys.C. Solid State Phys. 13, 4323, (1980).
8. H.J. Lee, L.Y. Juravel and J.C. Woolley, Phys. Rev. B.21, 659, (1980).
9. R.J. Nelson, Appl. Phys. Lett. 31, 351, (1977).
10. R. Piotrzkowski, J.L. Robert, E. Litwin-Staszewska and J.P. Andre, to be published in Phys. Rev.
11. Y. Yafet, R.W. Keyes and E.N. Adams, J. Phys. Chem. Solids. 1, 137, (1956).

12. J.A. Brum, G. Bastard and L. Guillemot, Phys. Rev. B30, 905, (1984).
13. P.J. Price, J. Vac. Sci. Technol. 19, 599, (1981).
14. G.A. Baraff and D.C. Tsui, Phys. Rev. B.24, 2274, (1981).
15. T. Ando and Y. Uemura, J. Phys. Soc. Jpn 37, 1044, (1974).
16. V.I. Nizhankovskii, V.G. Mokerov, B.K. Medvedev and Yu, V. Shalden, Sov. Phys. JETP, 63, 776, (1986).
17. K.V. Klitzing and G. Ebert, Physica 117B and 118B, 682, Ed. M. Averous, North-Holland Publishing Company, (1983).

DEFECTS CHARACTERIZATION IN GaAs-GaAlAs SUPERLATTICES

Dominique Vuillaume and Didier Stiévenard

Laboratoire de Physique des Solides,UA 253, CNRS
Institut Supérieur d'Electronique du Nord
41, Boulevard Vauban, 59046 Lille Cedex, France

INTRODUCTION

GaAs-GaAlAs superlattices(SL) have been recently developed , stimulated by their interest in modern electronics. In most SL the GaAlAs layer is large (more than 40 Å) so that the tunneling through the barrier is weak and the SL behaves as an array of quantum wells (QW). However, progress in molecular beam epitaxy (MBE) has now made possible the formation of ultrathin GaAlAs layer where the abrupt interface is of the order of an atomic monolayer[1,2]. In such a case, the conduction electrons tunnel through the GaAlAs barrier and propagate along the direction perpendicular to the layers, in a superlattice conduction miniband common to both barrier and well layers. This behavior of the electronic transport has been recently observed[3], and electrical characterization techniques such as capacitance-voltage (C-V), capacitance transient (C-t) and Deep Level Transient Spectroscopy (DLTS) have been successfully applied to SL[4,5,6,7], in order to probe the materials.

We have used these techniques on periodic 20-20 Å and 40-40 Å GaAs-$Ga_{0.7}Al_{0.3}As$ n-type Si doped ($3x10^{16}cm^{-3}$), 1.7 μm thick, structures in order to detect localized levels in the layers. The localized states associated with deep levels are not affected by the layered structure as soon as the thickness of the layer exceeds the spatial extension of the wave function of the localized states i.e. few angströms (J.C. Bourgoin and M. Lannoo[8]). We have introduced the defects by electron irradiation at 1 MeV whose characteristics and introduction rate are well-known in n-type GaAs [9] but less extensively studied in the GaAlAs [10,11]. When the energy level and capture cross-section[12] of the defects are known in the SL, the comparison with the set of values in both bulk materials therefore provide a direct information on the SL band structure. Notice that a similar idea was already suggested[13], while attempted independantly[14] at the same time, for the measurement of the band edge discontinuity using as bulk reference levels the ones associated with transition metal impurities in the two compounds forming the heterojunction. However, we have pointed out that this approch is weakly reliable in it simple form[4,7]. Two important parameters are known to govern the electronics properties of the SL, (i) the band offset Δ between the conduction bands of both materials of the layer and (ii) the energy location δ of the first conduction miniband of the SL above the conduction band of the GaAs. The values of these parameters are still subject to debate, mainly because they are not directly measured but rather deduced from various analysis of experimental data, which involve approximations and the use of theory. We propose a different method using DLTS measurements. DLTS does not involve any such approximation because it is based on the observation of the carrier emission from two known series of localized states in both types of layers composing the SL[4].

Futhermore, we have studied the native bulk and interface defects in the SL[5,6]. In the temperature range 50-100 K, a logarithmic time decay of the capacitance transient (C-t) is observed. We attribute this time decay to the existence of a continuum of deep levels in the fundamental gap of the superlattice, related to electron traps located at the GaAs-GaAlAs interfaces[5,6]. In the temperature range 300-400 K two localized level traps are detected at 0.06 eV and 1 eV below the first conduction miniband in the SL[5]. These traps are associated with defects in the GaAlAs usually detected in bulk MBE grown $Ga_{0.7}Al_{0.3}As$.

The ouline of this paper is organized as follow. The first section gives a brief summary of the capacitance-voltage behavior of the SL obtained during the course of this work and the second section point out the characterization of the irradiation-induced defects in SL using DLTS. The third section gives the results of the capture cross-section measurements of these defects. Then, in section 4, we show how the SL band structure paramaters (Δ,δ) are obtained using these irradiation-induced defects as a probe. Section 5 is a discussion of these results. Finally, the characterization of the bulk and interface natives defects performed on virgin SL is given in section 6.

CAPACITANCE-VOLTAGE BEHAVIOR OF THE SUPERLATICES

In 1980, H. Kroemer and coworkers have shown that it is possible to determine the band offset at an heterojunction by C-V profiling of the free carriers density[15]. In this case the C-V curve shows a discontinuity which corresponds to the accumulation of free carriers in the GaAs near the interface. It is also the case when the C-V measurements are performed on a multi-quantum wells (MQW). However, this is no longer valid in a SL, with ultrathin GaAlAs layer (typically smaller than 40 Å). In such a case the conduction electrons are delocalized in the growth direction[3],and thus they propagate in a SL conduction miniband common to both type of layers,because the GaAlAs barrier is almost transparent by tunneling process.

We illustrate the C-V behavior of a SL based on GaAs-GaAlAs(30% Al) periodic structure. Both layers, 20 Å thick, are Si doped at about $3x10^{16}$ cm^{-3} .The structure grown by MBE on a n^+-GaAs substrate, is 1.7 μm thick. After a deposition of an Al gate to form the Schottky contact on the top of the SL, the C-V measurements are performed at 1 Mhz by a PAR 410 capacitance meter. The results are digitally acquired with a resolution of 0.25 pF, between 4 K and room temperature, when a reverse bias applied from 0.5 to -5 V on the gate respectively to the substrate. Typical C-V curve is shown in Fig. 1. The behavior is similar to that of the bulk material[7], in contrast with the oscillatory behavior attributed to the free carrier nonuniformity in the thick layer periodic structures such as MQW. The SL exhibits a linear behavior of the C^{-2}-V curve, as expected for a uniformly doped material, even at low temperature (4 K) where the Debye length is still smaller than the width of the layer (Fig. 1). No deviation of the linearity has been detected at the scale of the voltage variation (10 mV) which corresponds to a variation of the space charge region equal to the width of a single layer. Between 100K and 300K the free carrier concentration n(w) at the edge of the depletion zone w is found to be constant $n(w) \sim 3.2 \times 10^{16}$ cm^{-3} ,comparable to the value adjusted during the

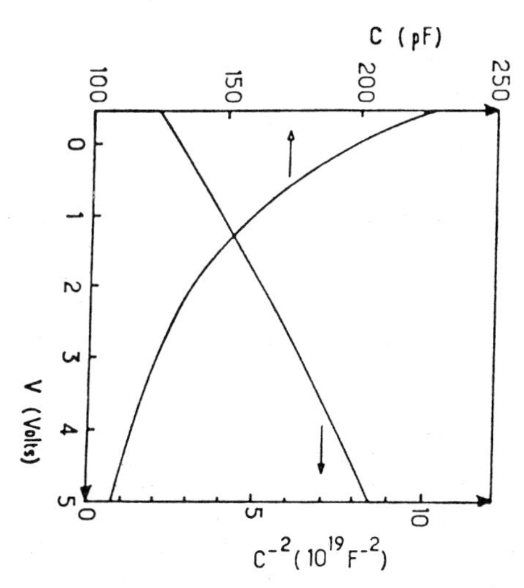

Fig. 1. Typical capacitance-voltage and C^{-2}-V curves recording for a 20-20 Å SL at 4 K. Positive voltages refer to reverse bias (from Ref. 7).

growth process. Only at temperature below 20 K a fraction of the free carriers freezes out and the concentration decreases to $n(w) \sim 2.7 \times 10^{16}$ cm^{-3} at 4 K. Such weak freezing out is probably due to the fact that the doping concentration is near the critical value for the metal-insulator transition in GaAs. Also the frequency used to performe the C-V analysis may induced an error, because the emission and capture rates from the donor impurity level are fast compared to the period at which the measurements are performed, and the electrons cannot be observed localized on donor sites (this has been verified in bulk GaAs). However, preliminary measurements on the 40-40 Å SL, where the donor level is expected to be deeper, show that the freezing is enhanced in this case[16].

The C-V characteristics show a hysteresis with the bias sweep. This indicates a non-equilibrium situation of the free carrier i.e. the presence of deep traps. Consequently, caution has been taken during the C-V measurements to ensure that these traps are always in the same (filled) state. The reported data[4,5,6,7] have been obtained with bias sweep from 0.5 V to -5 V and the voltage was systematically reset to 0.5 V before any change of temperature. The full characterization of these native traps is given elsewhere[5,6], and will be rewieved in section 6 of the present paper.

IRRADIATION-INDUCED TRAPS IN THE SUPERLATTICE

The important role of defects in the behavior of electonic devices is obvious. To obtain a good knowledge of native or related-processes defects which are generally complex defects, the way is to start from an accurate study of primary defects such as those induced by electrons irradiation. This is what has been done in bulk GaAs for example. Moreover, as we shall see, an orignal way to study the SL band structure and to measure the band offset is to use a set of deep levels as a local probe[4]. We shall see how such levels can be detected by DLTS in a SL. For this, we briefly recall how the DLTS works to determine the three main parameters of a deep level[8]: its thermal ionization energy E^t , or energy level; its capture cross-section σ; and the activation energy E^a of the capture cross-section when capture arises via a multiphonons process (MPE)[17].

DLTS is performed on semiconductor devices when a voltage modulable space charge region exists. Usually, Scottky and p-n junction diodes are used to study the bulk traps, and metal-insulator(or oxide)-semiconductor(MIS or MOS) capacitances are used to study the energetical distribution of insulator-semiconductor interface states. Recently, semiconductor heterojunctions were also used for interface studies. In crystalline semiconductor, point defects introduce discrete energy levels in the band gap. The charge state of a defect can be changed by capture and emission of free carriers. The deep level characterization with a transient capacitance technique arises from the observation of the emission of a carrier from a defect level to a band. The measurement technique consists of monitoring the variation of the space charge region induced by a sudden application of an external voltage perturbation. In DLTS measurements, the levels are successively filled and emptied by applying a cycle of external biases. In the first part of the cycle, the diode is biased near the forward regime. The potential drop in the semiconductor is reduced so that the defect level is below the Fermi level and filled by electrons. The filling pulse is applied during a duration t_p. The device is then reverse biased, and the defects situated in the space charge region have now their levels above the Fermi level, and the system relaxes towards equilibrium condition by emission of the trapped carriers. This charge state variation of the defects leads to a corresponding variation of the semiconductor capacitance, or to a transient current associated to the detrapping of the carriers.

Let us considere a single level at an energy E^t below the conduction band. The transient capacitance due to the emptying of this level is approximatively given, in a simple approach, by[18]

$$\Delta C(t,T) = - C \frac{N_T}{2N_D} \exp(- e_n\ t) \qquad (1)$$

where C is the steady-state capacitance, N_T the defects concentration, N_D the n-type doping concentration and e_n is the emission rate of the level which is given by

$$e_n = \sigma_n \, v_{th} \, N_C \, \exp\left(-\frac{E^t}{kT}\right) \tag{2}$$

N_C is the density of states in the conduction band, v_{th} the thermal velocity of the electrons and σ_n the capture cross section of the defect. This formula is the key point of all the transient electrical measurements. The essence of the DLTS technique, as introduced by Lang[19] is the use of a time constant filtering of the transient called "emission rate window" during a temperature scan. This concept is schematically despicted in Fig. 2. The capacitance transients are observed by a fast-response capacitance meter. Eqs. (1) and (2) show the transient behavior. A variation from a slow transient at low temperature (e_n is small) to a fast transient response at high temperature where e_n is great occurs. The repetitive transients are fed into a double box-car analyser[20] with gates set at t_1 and t_2 (Fig. 2). The DLTS signal is simply the difference between the values of the transient capacitances $\Delta C(t_1) - \Delta C(t_2)$ and it goes through a maximum at a temperature for which the emission rate of the trap equals the "emission rate window" fixed by the filtering analysis. In this case, the emission rate window e_{n0} is determined by the two gating times[19]

$$e_{n0} = (t_2 - t_1)^{-1} \, \mathrm{Ln}\left(\frac{t_2}{t_1}\right) \tag{3}$$

Thus, recording the DLTS signal over a temperature range yields a DLTS spectrum. Relative variations of the capacitance $\Delta C/C$ as small as 10^{-5} can be detected. For doping concentration of 10^{15} cm^{-3} for instance, a defect concentration of 1 defect per 10^{12} atoms of the semiconductor may be detected. The fundamental reason of this high sensitivity is the overall small number of charges which are at stake in the semiconductor junction.

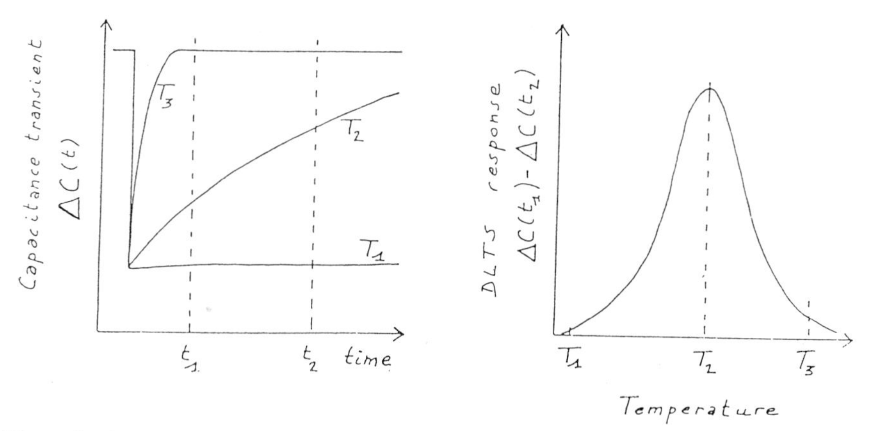

Fig. 2. Transient capacitance behavior under temperature scann and implementation of an emission rate window filtering.

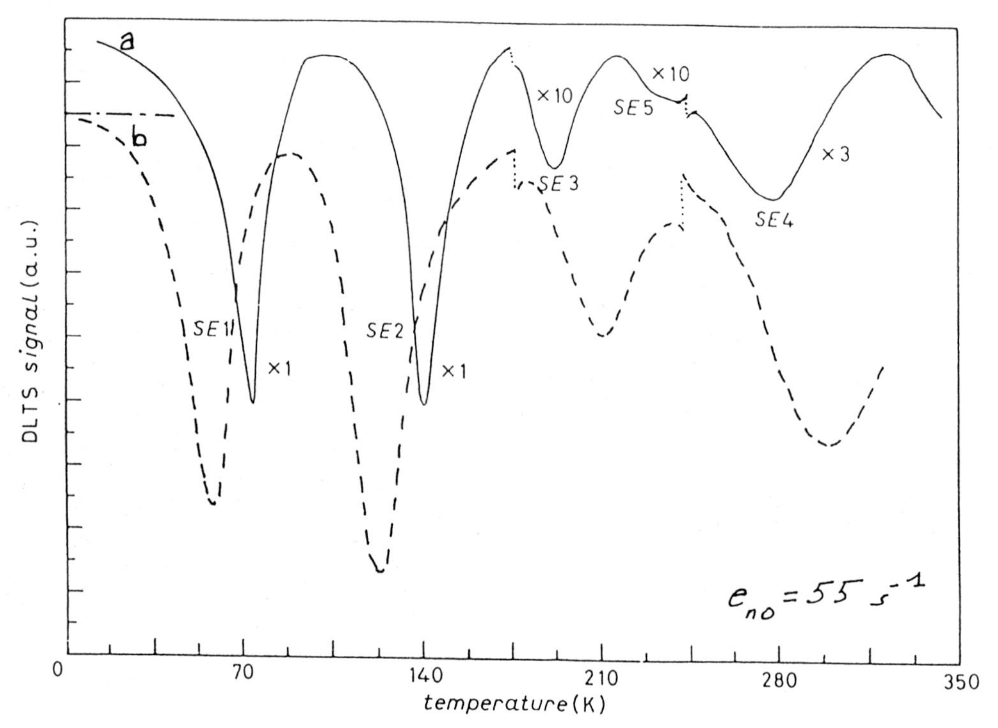

Fig. 3. Typical DLTS spectrum obtained in an irradiated 20-20 A SL (curve a) (from Ref. 4), and in an irradiated 40-40 Å SL (b).

Typical results are shown on Fig. 3 for defects introduced by electron irradiation in 20-20 Å and 40-40 Å GaAs-GaAlAs SL (irradiation at 1 MeV with a dose of $3x10^{15} cm^{-2}$). Prior the irradiation, the structure does not exhibit any DLTS peak in the same temperature range (4K to 350K). Five electron traps (labelled SE1 to SE5) are found. The values of the energy levels are obtained from an Arrhenius anlysis of the emission rate. For this, we record the temperature dependence of the peak for different emission rate windows. According to Eq. (2), the plot of $Ln(T^2/e_n)$ versus 1/T provides the associated energy level. However, one must take into account the temperature dependence of the capture cross-section due to a MPE process, and we have to insert the following equation[17] in Eq. (2)

$$\sigma_n = \sigma_\infty \exp\left(- \frac{E^a}{kT}\right) \tag{4}$$

Thus, the slop of the Arrhenius plot gives an apparent activation energy of the defect $E= E^t+E^a$. The apparent capture-cross section σ_∞ is then deduced from the extrapolation to $T^{-1}=0$ of the Arrhenius plot. By this mean, we have determined the apparent activation energy of irradiation defects in the SL, except for SE5 which appears as a shoulder on the SE4 peak, with a too low concentration to allow a correct measure. The values of these energies are listed and compared with the values in both GaAs and GaAlAs bulk materials in Table 1.

Table 1. Characteristics of the defects (ionization energy E^t and capture cross-section activation energy E^a) introduced by electron irradiation in GaAs, GaAlAs and SL.

	Bulk material			superlattice		
					20-20 Å	40-40 Å
	Level	E^t (eV)	E^a (eV)	Level	$E^t + E^a$ (eV)	$E^t + E^a$ (eV)
GaAs	E1	0.045	-	SE1	0.140	0.088
	E2	0.140	0	SE2	0.185	0.208
	E3	0.300	0.1	SE3	0.340	0.366
				SE4	0.550	0.461
GaAlAs	E'1	0.180	-			
	E'2	0.270	-			
	E'3	0.690	0.14			

Preliminary measurements have been performed on a 40-40 Å SL in order to probe the influence of the layer depth on the defects properties[16].The DLTS spectrum exhibits a similar shape, but the energy level of the SE_1 to SE_4 defects are changed (see Table 1). These first results will be briefly discussed latter (see section 5) in terms of modification of the SL band structure.

In SL, we have obtained by DLTS similar results as in bulk materials. We have demonstrated that the same behavior of the emission rate occurs. Then, we have been able to determine irradiation-induced defects in the SL. From these results, we shall see how to determine the SL band offset by an original method, and how to obtain a characterization of the native defects in the SL.

CAPTURE CROSS-SECTION OF DEFECTS IN SUPERLATTICES

In order to obtain the activation energy (E^a) of the capture cross-section when the capture of free electrons occurs via a MPE process, the variation of σ versus T is required. This done by a measurement of the filling kinetics of the trap which is obtained by the filling pulse width dependence of the the DLTS peak amplitude[21]. To overcome the difficulties involved by the application of very short pulses (the capture process is usually fast) on large capacitances, which are usually used, Stiévenard et al. [22] have developed a method based on a simplified analysis of the kinetics of the filling of the trap in the slow regime of filling. The study of the capture kinetics was performed using filling pulse width in the range 50 ns to 20 ms. Fig. 4 gives the kinetics of traps SE_1,SE_2 and SE_4 in a 20-20 Å irradiated SL. The SE_3 level was not precisely analysed because it just appears as a shoulder between SE_2 and SE_4 with a too low DLTS signal amplitude. We have shown[12] that SE_1 and SE_2 are already filled for a pulse width of 50 ns. This means that the activation energies of their capture cross-sections are expected to be very low. From the variation of the capture kinetics versus temperature, we have deduced the temperature dependence of the capture cross-section of SE_4, and by Eq. (4) we have computed [12] its activation energy E^a_{SE4}=0.145 eV ± 0.055 eV. This results will be discussed in the following by taking into account the SL band offset and the SL band structure in a simple model of configuration coordinate diagram as theoretically explained elsewhere[8].

Preliminary results [16] on a 40-40 Å irradiated SL show that the activation energies of the capture cross-sections of SE_1 and SE_2 are significantly increased compared to the low values in the 20-20 Å SL. This is briefly demonstrated in Fig. 4. These results indicate that the layer step of the SL strongly influence the capture processes.

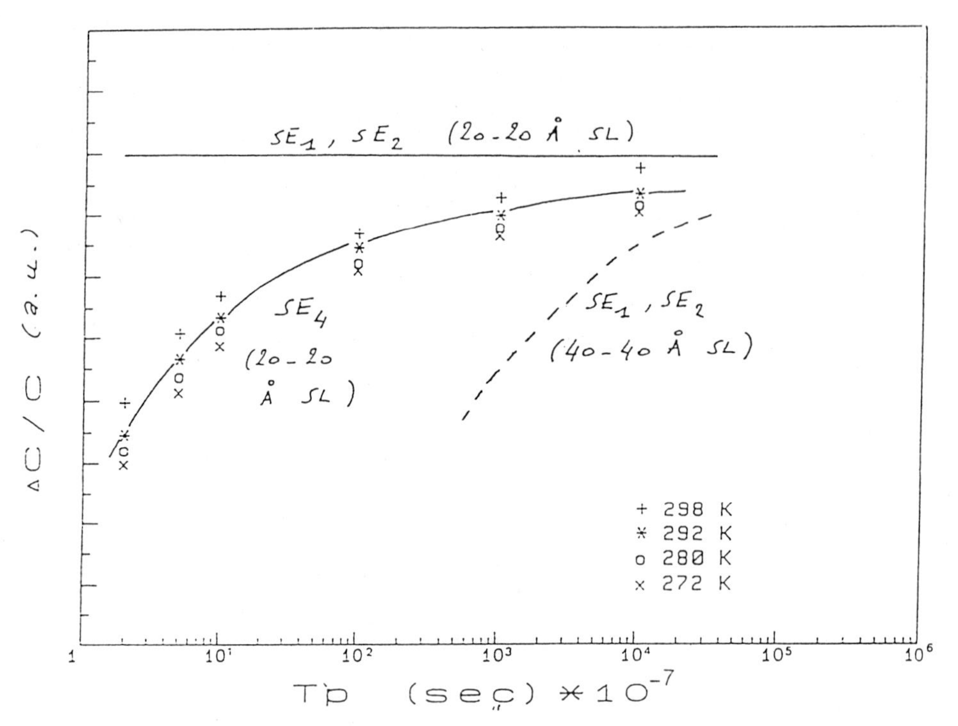

Fig. 4. Amplitude of the capacitance transient versus the filling pulse width for the defects SE1,SE2 and SE4 in the 20-20 Å SL (full lines), and for SE1 and SE2 in the 40-40 Å SL (dashed lines).

SUPERLATTICE BAND DISCONTINUITY DETERMINATION

The observation of the electron emissions from a localized state situated in the one of layers into a common band associated to the SL allows the determination of the band offset (Δ) between the conduction bands and the position of the bottom of the first conduction band of the SL (δ). Indeed, as explained in other lecture [8], deep levels are localized enough so that their wave functions are not influenced by the layered structure and their ionization energies remain identical to the ones they have in the bulk materials. This was proved by Deveaud et al.[23] in their low temperature photoluminescence studies of manganese in SL. Thus, we expect to observe the same defect levels in the SL but with energy position rigidly shifted in both layers. The general relations between the observed levels are (see Ref. 8):

$$SE_i = E_i + \delta \qquad \text{(in the GaAs layer)} \qquad (5a)$$
$$SE_i = E'_i - \Delta + \delta \qquad \text{(in the GaAlAs layer)} \qquad (5b)$$

When E_i and E'_i are known, the new energy levels in the SL obtained through DLTS experiments allow the deduction of both quantities Δ and δ. The central issue is the determination of the layer to which belong the defect which generates a DLTS peak in the SL. This can be done unambiguously if we have not only one single level but a series of levels, since the energy difference between two levels of a serie is different in each bulk constituants and remain unchanged in the SL. Moreover, since the introduction rate of a given defect in GaAs and GaAlAs are similar, we also expect that the relative concentration of the different defects will be the same in the SL, because the volumes of the two materials involved are equal (in our case, the width of the well and the barrier are the same in the SL studied).

The starting point to determine Δ and δ from our DLTS experiments is to know if the E'_1 level in the GaAlAs layers is above or below the conduction miniband in the SL. Because $E'_1 = 0.18$ eV is most probably smaller than the expected value for the band offset Δ, let us examine carefully only the first possibility: E'_1 is a resonant level in the SL conduction band and it is not observed by DLTS. Then the following correspondance of levels is deduced[4] (Fig. 5): $SE_1 = E'_2$, $SE_2 = E_1$, $SE_3 = E_2$, $SE_4 = E_3$ or E'_3 or a combination of both. The ambiguity for SE_4 will be clarify after a carefull study of the capture cross-section of defects in the SL . The level SE_5 is not considered in a first approch because it cannot be easily decorelated from SE_4. With the energy levels summarized in Table 1, the previous levels correspondance leads to the resolution of the following system:

$$\left(0.14 - E^a_{SE_1}\right) = 0.27 - \Delta + \delta \qquad (6a)$$
$$\left(0.185 - E^a_{SE_2}\right) = 0.045 + \delta \qquad (6b)$$
$$\left(0.34 - E^a_{SE_3}\right) = 0.14 + \delta \qquad (6c)$$
$$\left(0.55 - E^a_{SE_4}\right) = 0.3 + \delta \ (\text{if } SE_4 = E_3) \qquad (6d)$$
$$\text{or} = 0.69 - \Delta + \delta \ (\text{if } SE_4 = E'_3) \qquad (6e)$$

If we assume, in a first approach, that the activation energies of the capture cross-sections of SE_1 and SE_2 are negligible as demonstrated in previous work[12] , and because we have an experimental observation in good agreement between the energy levels of (SE_1 , SE_2) and (E'_2 ,E_1) respectively, the solution of Eqs. (6a) and (6b) leads to Δ=0.27 eV and δ=0.14 eV . This determination of Δ gives a band offset repartition[24] of 64%-36% in good agreement with the most recent results[25,26,27] which suggest that the ratio $\Delta E_c/\Delta E_g$ is between 0.6 and 0.7[28]. This would indicate that the recent value Δ=0.36 eV used to fit most recent photoluminescence data on identical structures[24] has been over-estimated.

DISCUSSION

In spite of this accurate determination of the SL band offset, the position of the SL conduction miniband (δ=0.14 eV) is not consistent with the starting hypotesis namely that E'_1 is a resonant level. In fact, to satisfy this hypothesis, we have to verify $\Delta-\delta>E'_1$ as shown by Fig. 5. This

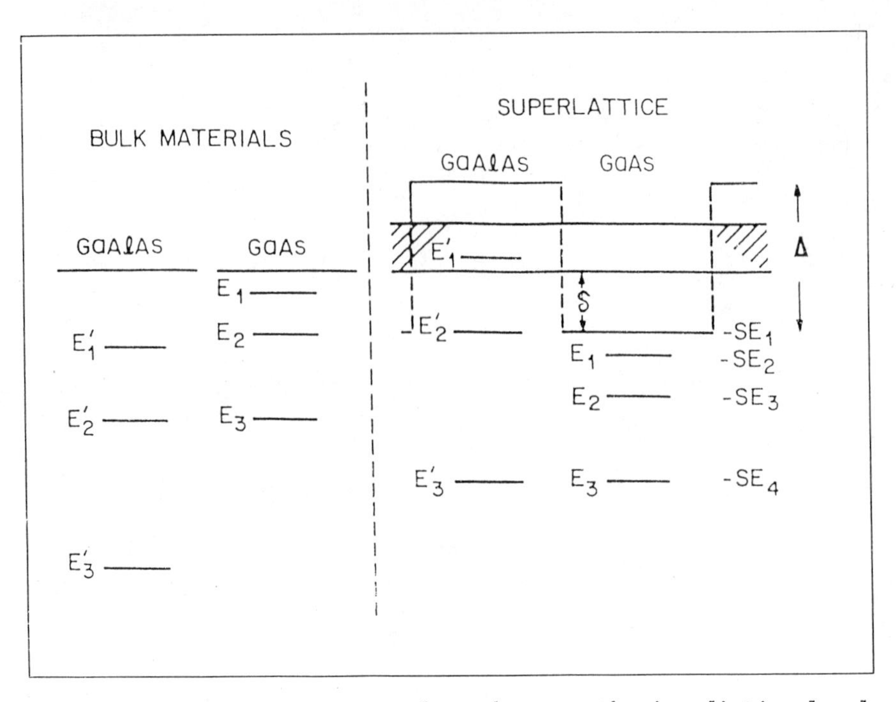

Fig. 5. Scheme of the correspondance between the irradiation levels in bulk GaAs, GaAlAs and in the 20-20 Å SL.

fact may be due to the inaccuracy of the determination of the energy level E'_2 in GaAlAs, especially because the activation energy of its capture cross-section is unknown (Table 1). Also, the correspondance between the levels SE_3 , SE_4 and E_2,E_3(or E'_3) respectively is poorly verified in that case (Fig. 5). But, from the determination of the capture cross-section activation energy of SE_4 at 0.145 ± 0.055 eV[12], we can used Eq. 6d (assuming that SE_4=E_3) to obtain a more accurate value of δ. We found δ=0.1 eV and then, the E'_1 level is too close to the SL conduction miniband to be observed by DLTS measurements.

The assumption $SE_4 = E_3$ is validated by the fact that the capture cross-section activation energy of this defect is 0.145 eV in the SL which is larger than its value in the bulk GaAs (0.1 eV) as theoretically predicted elsewhere[8]. However, the other possibility, namely that SE_4 would be the E'_3 level in GaAlAs layer, do not satisfy the theoretically predicted decrease of the activation energy measured in the SL when the defect is located in the barrier layer. Moreover, we can calculate the activation energy of SE_4 from the know characteristics for E_3 in GaAs (E^t =0.3 eV, E^a =0.1 eV and the Franck-Condon shift d_{fc}=0.075 eV) using Eq. 14 in Ref. 8. This calculation leads to E^a_{SE4}=0.21 ± 0.05 eV, close to the experimental determination (0.145 ± 0.055 eV) taking into account the uncertainty of about 0.05 eV.

Finally, let us briefly comment the preliminary DLTS measurements of irradiation-induced defects in 40-40 Å SL. We have shown (Fig. 3 and Table 1) a modification of the energy levels SE_1 to SE_4. Accordingly with the fact that the energy levels are unchanged in the SL layers and in the bulk materials, that is an obvious experimental evidence of the layer step dependence of the SL band structure parameters Δ and δ.

NATIVE DEFECTS IN GaAs-GaAlAs SL

Once the SL band structure is known (Δ,δ), it is possible to characterize the native defects in the SL. We have previously noticed that the C-V characteristics exhibit a hysteresis with the bias sweep[5,6]. This indicate a non-equilibrium situation induced by the presence of deep traps. However, no DLTS peak has been detected in the temperature range 4-400 K, with a rate window analysis varying between 2×10^{-3} and 1 s^{-1}. This indicates that the traps have no time to respond at the scale of 1 second. Thus, the C-V hysteresis shown at temperature below 300 K are due to native traps in the SL which exhibit a weak emission rate in this temperature range. To study such defects (invisible by DLTS because it is a fast analyzer), we have performed capacitance transient study at several temperature (C-t,T). The measurement procedure is as follows. The sample is biased at 0 V, untill all the defects are filled. Then a sudden bias variation is applied (-3 V), and we recorde the time release of the capacitance induced by the emptying of the defects (time from 0 to 1000 s). The bias is reset to 0 V before any change of temperature. From a carefull analysis of the transient, we can determine the emission kinetics which allow us to distinguish bulk traps from interface states[5].

At temperatures higher than 20 °C, the capacitance transients (Fig. 6) are characterized by exponential time decay. In the range 90-130 °C, the capacitance variations $\Delta C(t)$ are small, and they may be accuratly fitted by electron emission from a majority carriers trap, according to Eq. (1). The temperature dependance of this emission rate gives the energy level of the defect involved in this transient : $SL_1 = 1.02 \pm 0.05$ eV. The concentration deduced by Eq. (1) is about $10^{15} cm^{-3}$. In the temperature range 20-90 °C, the transient is fitted by two exponential components (Fig. 6). An other defect emits its electrons in the range of the smallest time constant, with a greater emission rate. The temperature dependance allows to decorelate the two defects , and we obtain a second level $SL_2 = 0.06 \pm 0.02$ eV. Its concentration is about $5 \times 10^{15} cm^{-3}$.Using the level correspondance given by Eq. (5), and the previously dermination of $\Delta = 0.27$ eV and $\delta = 0.1$ eV, we have identified[5] the SL_1 level as the MBE related E center in GaAlAs (1.22 eV) [29], and the SL_2 level as one of the known trap levels(0.24 eV) in the bulk MBE GaAlAs(30% Al)[29] with the same concentration ($5 \times 10^{15} cm^{-3}$).

In the temperature range 50-100 K a very slow capacitance transients have been observed[5,6](Fig. 6). The thermodynamic equilibrium of the sample is not yet reached at $t = 2 \times 10^3$ s after the bias excitation. In order to avoid any spurious effect which can be observed in thermodynamic systems out of the equilibrium, the sample is reset to 0 V and heated to the room temperature before any change of temperature and then, it is cooled from 300 K to the measurement temperature, maintaining the zero voltage. Fig. 6 gives a typical capacitance transient thus obtained at 77 K. This transient cannot be fitted by a finite set of exponential decays, but the capacitance relaxes according to a logarithmic law. In the SL context, this behavior is just a particular example of the T Ln t relaxation rate observed in any thermodynamic system where the equilibrium is reached through a thermally activated process of a broad distribution of relaxation times (as in semiconductor-insulator interfaces, grain boundaries, dislocations). For our case, a defect density was estimated of about $10^{11} cm^{-2}$ at the GaAs-GaAlAs interfaces. This value is in agreement with other DLTS results at a single GaAs-GaAsAl heterojunction[30]. A more accurate characterization of the native GaAs-GaAlAs interface states is to day under study in order to improve the modelization of the interface states behavior in SL.

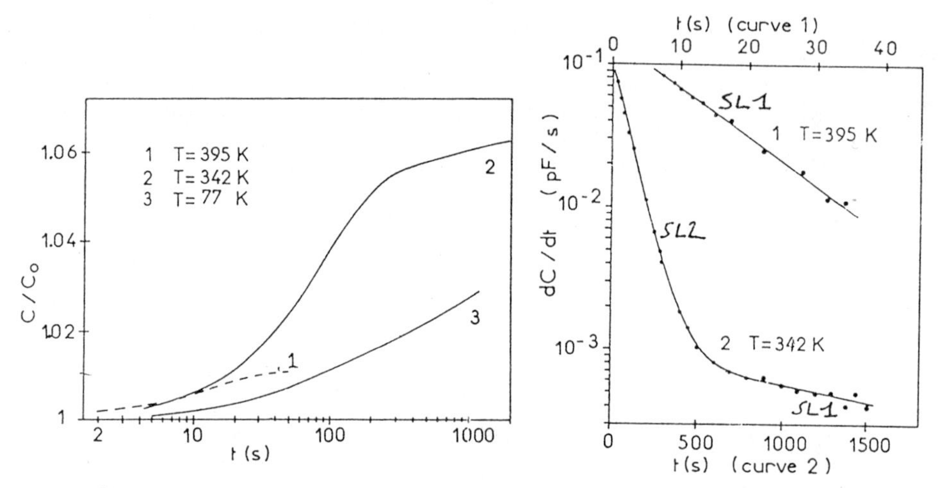

Fig. 6. Typical capacitance transients for the virgin 20-20 Å SL in the range 4-400 K (left side). Plot of Ln dC/dt vs t used to determine the emission rates: curve 1 corresponds to the SL1 level (1.02 eV), and curve 2 to a mixing of SL1 and SL2 (0.06 eV) (right side). Dots are experimental data and full lines theoretical (from Ref. 5)

CONCLUSION

We have demonstrated using capacitance-voltage measurement, the behavior of the electronic transport in SL, where conduction electrons tunnel through the barrier and then propagate along the direction perpendicular to the layers, in a SL conduction miniband. We then showed that the electrical characterization techniques of defects by capacitance transients can be succesfully applied to probe the defects in SL[4,5,6,7].

In addition, we have shown that DLTS experiments on SL[4,7] provides a new powerfull method to determine the band offset at the interface of two semiconductors. Any method requires a common reference level in the both materials. In previous attempts using heterojunctions, these levels were chosen arbitrarily, as the TM impurity levels [13] for example, which may induce inaccuracy in the determination of the band line-up. The basic advantage of our method, inherent to the choice of SL, is the intrinsic existence of a reference level, directly accesible to experiment, namely the edge of the SL conduction miniband. Moreover, the choice of a set of several levels (electron irradiation-induced defects) well-known in each bulk semiconductors constituting the SL, instead of a single impurity level, allows an unambigous determination of the band offset(Δ) and of the location of the SL miniband(δ). We have illustrate this method on 20-20 Å and 40-40 Å GaAs-GaAlAs(30% Al) SL.Therefore, our method, contrary to previous ones, does not involve any ad hoc hypothesis concerning a correlation between deep levels and a common reference level, and the method is universal in essence.

When the band offset parameters are known ,we have shown how the analysis of the capacitance relaxation time decay leads to the determination of the native defects (bulk and interface) in the SL[5,6]. Two bulk traps in the GaAlAs layer have been detected, presumably related to defects usually detected in the MBE grown GaAlAs layer. The logarithmic time decay, observed at low temperature, satisfying the universal behavior of disordered systems has been attributed to the relaxion towards equilibrium through a distribution of states at (or near) the GaAs-GaAlAs interfaces. Thus, such dynamic properties have proved sensitive enough to give information on electron traps at (or close to) the GaAs-GaAlAs interfaces.

AKNOWLEDGEMENTS

The authors whish to thank J.C. Bourgoin, who has iniated this study, for fruitful discussions, and M. Lannoo for his interest and encouragement. We are indepted to D. Deresmes, A. Mauger, S.L. Feng, and F. Sillon for their contributions in the electrical characterizations, and to B. Deveaud and A. Regreny for suplying the samples. This work is supported by Centre National d'Etude des Telecommunications under contract n° 86 6 B 00 790 92 45.

REFERENCES

1. C. Weisbuch, R.C. Miller, R. Dingle, A.C. Gossard and W. Wiegmann, Solid Stat. Commun. 38:709 (1981).
2. B. Deveaud, J.Y. Emery, A. Chomette, B. Lambert and M. Baudet, Appl. Phys. Lett. 45:1078 (1984).
3. H.L. Störmer, J.P. Eisenstein, A.C. Gossard, W. Wegmann and K. Baldwin,Phys. Rev. Lett. 56:85 (1986).
4. D. Stievenard, D. Vuillaume, J.C. Bourgoin, B. Deveaud and A. Regreny, Europhysics Letters 2:331 (1986).
5. F. Sillon, A. Mauger, J.C. Bourgoin,B. Deveaud, A. regreny and D. Stievenard, " Defects in semiconductors," H.J. Von Bardeleben, ed., Trans Tech Publications, Aedermannsdorf (1986).
6. A. Mauger, S.L. Feng and J.C. Bourgoin, Appl. Phys. Lett. 51:27 (1987).
7. J.C. Bourgoin, A. Mauger, D. Stievenard, B. Deveaud and A. Regreny, Solid State Commun. 62:757 (1987)
8. J.C. Bourgoin and M. Lannoo, this conference
9. D. Pons and J.C. Bourgoin, Phys. C: Solid State Phys. 18:3839 (1985).
10 S. Loualiche, G. Guillot, A. Nouailhat and J.C. Bourgoin Phys. Rev. B 26:7090 (1982).
11. D.V. Lang, R.A. Logan and L.C. Kimmerling, Phys. Rev. B 15:4874 (1977).

12. D. Stievenard, D. Vuillaume, S.L. Feng and J.C. Bourgoin, "Proceedings of MSS3" to be published
13. J.M. Langer and H. Heinrich, Phys. Rev. Lett. 55:1414 (1985).
14. B. Plot, B. Deveaud, B. Lambert, A. Chomette and A. Regreny, J. Phys. C: Solid State Phys. 19:4279 (1986).
15. H. Kroemer, W.Y. Chien, J.S. Harris and D.D. Edwall Appl. Phys. Lett. 36:295 (1980).
16. S.L. Feng, J.C. Bourgoin, D. Stievenard and D. Vuillaume, unpublished.
17. J.C. Bourgoin and M. Lannoo, "Point defects in semiconductors", Springer Verlag, Berlin (1983).
18. The reader can found more detailled information in:
D. Pons, Appl. Phys. Lett. 37:413 (1980).
D. Stievenard, M. Lannoo and J.C. Bourgoin, Solid State Electron. 28:485 (1985).
D. Stievenard and D. Vuillaume, J. Appl. Phys. 60:973 (1986).
and in Ref. 17.
19. D.V. Lang, J. Appl. Phys. 45:3023 (1974).
20. Two other filtering method are currently used involving a one-phase or two-phases lock-in amplifier. Details can be found in:
A. Mircea, A. Mitonneau, J. Allais and M. Jaros, Phys. Rev. B 16:3665 (1977).
D. Pons, Thesis, Paris (1979).
D. Pons, P.M. Mooney and J.C. Bourgoin, J. Appl Phys. 51:2038 (1980).
21. D. Pons, Appl. Phys. Lett. 37:413 (1980).
22. D. Stievenard, J.C. Bourgoin and M. Lannoo, J. Appl. Phys. 55:1477 (1984).
23. B. Deveaud, B. Lambert, B. Plot, A. Chomette, A. Regreny, J.C. Bourgoin and D. Stievenard, J. Appl. Phys. to be publieshed
24. B. Deveaud, A. Regreny, D. Stievenard, D. Vuillaume, J.C. Bourgoin and A. Mauger, "Proceeding of the ICPS 18" to be published.
25. W.I.Wang, E.E. Mendez and F. Stern, Appl. Phys. Lett. 45:639 (1984).
26. M.O. Watanabe, J. Yoshida, M. Mashita, T. Nakanisi and H. Hojo, J. Appl. Phys. 57:5340 (1985).
27. G.B. Noris, D.C. Look, W. Kopp, J. Klem and H. Morkoc, Appl. Phys. Lett. 47:423 (1985).
28. H.Z. Chen, H. Wang, A. Ghaffari, H. Morkoc and A. Yariv, Appl. Phys. Lett. 51:990 (1987)
29. P.M. Mooney, R. Fisher and H. Morkoc, J. Appl. Phys. 57:1928 (1985)
30. A. Okuma, S. Misawa and S. Yoshida, Surf. Sci. 174:331 (1986)

STUDIES OF THE DX CENTRE IN HEAVILY DOPED n^+GaAs

L. Eaves*, J. C. Portal[+], D. K. Maude* and T. J. Foster*

*Department of Physics, University of Nottingham,
Nottingham NG7 2RD, U.K.
[+]Dept. de Genie Physique, INSA, 31077 Toulouse and
SNCI-CNRS, Avenue des Martyrs, 38042 Grenoble, France

ABSTRACT

The effect of hydrostatic pressure on the electrical properties of MBE-grown GaAs layers heavily doped in Si with Sn is described. It is shown how these measurements provide fundamental information about the DX centre in n^+GaAs. Shubnikov-de Haas measurements show that as increasing pressure is applied, electrons are trapped out from the Γ conduction band minimum of GaAs into the localised DX states, with an accompanying increase in carrier mobility. Optical illumination at $T \lesssim 100K$ causes persistent photoconductivity in which free carriers are released back into the conduction band with an accompanying fall in mobility. The results show that the DX centre produces a resonant donor level between the Γ- and L- conduction band minima at a concentration comparable with the doping level. The energy and occupancy of the DX level are calculated using Fermi-Dirac statistics. For the Si-doped samples, comparison with local vibrational mode measurements indicate that the DX level can be identified with the simple substitional donor, Si_{Ga}. In the heaviest doped samples ($n \gtrsim 1 \times 10^{19}$ cm^{-3}) we conclude that the DX level is partially occupied at 300 K and atmospheric pressure, thus acting to limit the free carrier concentration. Our results are discussed in terms of other recent work on the DX centre in heavily doped GaAs.

INTRODUCTION

The properties of the DX centre in (AlGa)As have been the subject of considerable interest for more than a decade[1,2]. The centre, with the character of a deep donor level, is present in all n-type (AlGa)As at concentrations comparable to the doping level. The characteristic feature of the DX centre is its thermally activated capture cross section which gives rise to metastability and associated persistent photoconductivity (PPC) at low temperatures. The variation of the capture and emission processes as a function of Al-mole fraction ($0.2 < x < 0.8$) has been reported recently[3]. The DX centre has not been observed previously in GaAs at atmospheric pressure. However, Mizuta et al[4] have recently shown in DLTS measurements that the DX centre becomes occupied in n-GaAs and in n-(AlGa)As with low Al mole fraction (x) at pressures above a critical value which decreases with increasing x. The DX level appears to follow closely the L-band edge in the direct gap alloy composition range[5,6,7] and becomes resonant with the Γ conduction band for $x \lesssim 0.22$. Since the application of hydrostatic pressure has a similar effect on the band structure to increasing x, it is expected that the DX centre in GaAs will move with a similar pressure coefficient to that of the L-minima.

Several alternative models have been proposed for the nature of the defect giving rise to the PPC[1,8-10] in (AlGa)As. In particular, it has been suggested recently that PPC is associated with states arising from substitutional donor impurities[4,9] rather than with defect-impurity complexes. The question of whether the PPC is due to large lattice relaxation, as originally proposed by Lang and Logan[1] has also been the subject of recent debate[9,10].

EXPERIMENT

In this paper we investigate as a function of hydrostatic pressure up to 15 kbar the free carrier concentration, mobility and the persistent photoconductivity, of a series of six n-GaAs layers heavily doped ($0.2-1.8 \times 10^{19}$ cm^{-3}) with either Si or Sn. The layers (labelled a to f) were prepared by molecular beam epitaxy. Samples a, d, e and f were grown at Philips and samples b and c at IBM. Enhanced doping was achieved by lowering the substrate temperature during growth[11]. For samples a and d, the substrate temperature was 450°C. The Si- or

Sn-doped n^+ layers varied in thickness between 0.15 and 10 μm. The carrier concentration was measured at low temperature and with magnetic fields up to 18 T using the de Haas-Shubnikov effect (dHS). This measures the radius of the Fermi sphere in k-space and yields directly the number (n) of conduction electrons in the Γ-minimum. Pressure was applied by means of a liquid clamp cell manufactured by Unipress (Warsaw). An important feature of these cells is that the pressure can be applied at room temperature only. When measurements are subsequently made at low temperatures the applied pressure is "frozen in". Cooling to low temperatures is accompanied by a small decrease of pressure (typically < 10%). The pressure on the sample is continuously monitored in situ with a calibrated n-InSb resistance manometer. The cooling procedure to the helium temperatures required for the dHS measurements involved a precool to 77 K of ~20 minutes' duration. The subsequent cooling to 4 K took place over ~5 minutes. The occupation of the DX centre is fixed at some intermediate temperature (probably about 120 K) at which the capture and emission rates become so slow that the electron population of the trap is no longer in equilibrium with the free electron density. In order to study PPC, samples could be illuminated under pressure at low temperatures by means of a visible light-emitting diode mounted inside the cell.

Typical dHS measurements on a Si-doped sample with initial (P = 0) carrier concentration of 1.1×10^{19} cm^{-3} are shown in Figure 1 for a range of applied pressures. The pressures quoted are those measured at 4.2 K. As P is increased, the period of the dHS oscillations increases, corresponding to a decrease in carrier concentration. Note also that the amplitude of the oscillations increases with pressure and that the oscillatory structure becomes observable down to lower magnetic fields. This is a clear indication that the mean collision time τ and mobility μ increase with increasing pressure since the amplitude of dHS structure is controlled by the factor $\omega_c\tau$, where $\omega_c = eB/m^*$ is the cyclotron frequency. Since the period $\Delta(1/B)$ of the dHS structure and the zero field resistance give, respectively, the electron carrier concentration n and the conductivity σ, the variation of μ with P can be determined. Typical results for some of the samples investigated are shown in Figure 2. For the most heavily doped sample, n falls immediately on increasing the pressure above atmospheric (i.e. Figure 2(d)). This indicates that for this doping the DX level is very close to the Fermi energy and partially occupied even at atmospheric pressure. In the more lightly

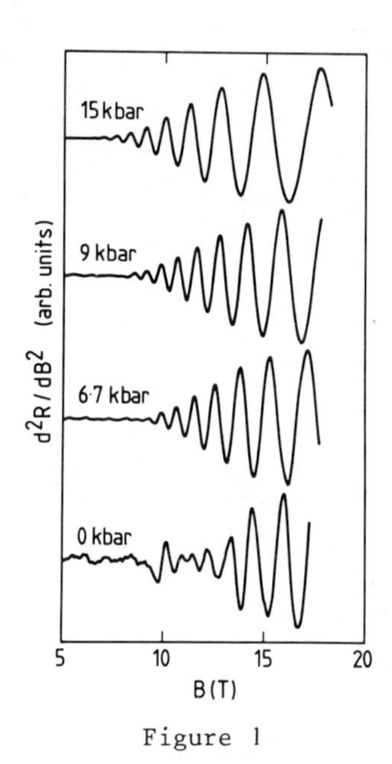

Figure 1

Typical dHS oscillations in the second derivative of the transverse magnetoresistance of an Si-doped sample with initial (P = 0) carrier concentration of 1.1×10^{19} cm^{-3}, T = 4 K, $\underline{J}\perp\underline{B}$.

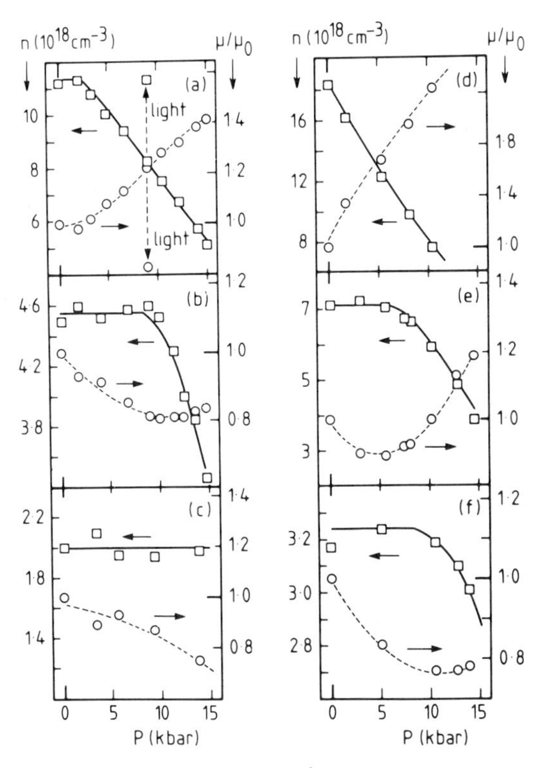

Figure 2

Carrier concentration n (□) and mobility μ (o) versus pressure P for a range of Si (a,b,c) and Sn (d,e,f) doped samples. The mobility is normalised to its atmospheric pressure value. The solid lines show the expected variation of n with P calculated using Fermi-Dirac statistics as described in the text. The effect of illumination on n and μ at 9 kbars is indicated in Fig. 2(a) for the most heavily Si-doped sample. The broken curves through the mobility data are guides to the eye only.

doped samples, e.g. Figure 2(b), (e) and (f), the carrier concentration remains roughly constant up to a critical value P_c and then falls sharply. The sample doped at 2×10^{18} cm^{-3} (Figure 2(c)) shows no significant change of n up to 15 kbar. It can be seen that in all samples a decrease in n with P is accompanied by an increase in μ. In lighter-doped samples, over the range of pressure for which n remains constant, μ is found to decrease slowly with pressure. This effect is associated with the increase of the electron effective mass of the Γ-conduction band.

These results indicate that the loss of free carriers with increasing pressure is associated with capture at donor-like localised resonance levels located between the Γ- and L-conduction band minima. This capture process is described by the relation $D^+ + e^- \rightarrow D^\circ$. The neutralisation of the ionised donors by electron capture and the corresponding decrease in ionised impurity scattering explains the increases in mobility and in amplitude of the dHS structure.

To test if the deep donor level showed the PPC effects which characterise the DX centre in (AlGa)As, we investigated the effect of pulses of light on the carrier concentration and mobility with the sample maintained under pressure at low temperature. Here it is important to note that the pressure in these experiments was initially applied at room temperature and the sample was cooled slowly. Therefore, if the pressure is above the threshold value for trapping out onto the donor-like resonance levels, a fraction of the DX donor states corresponding to thermal equilibrium ($T \simeq 120$ K) should be filled with electrons after cooling in the dark to low temperatures. By illuminating the sample with light from an LED under pressure at low temperatures, the free electron concentration, measured by the dHS effect, was restored to its zero pressure value. The changes in n and μ following illumination at 9 kbar are indicated by broken arrows in Figure 2(a) for the most heavily Si-doped sample. This increase in n persisted long after switching off the light. It was accompanied by a persistent decrease of μ and of the amplitude of the dHS structure. Such behaviour is fully consistent with photo-ionisation of a donor level. The PPC indicates that the centre involved has the same character as the DX centre in n-(AlGa)As. Note that on illuminating under pressure, the mobility does not fall to its zero pressure value but to a somewhat lower value due to the pressure-dependent effective mass.

DISCUSSION

The role of the hydrostatic pressure in producing the observed trapping-out of free electrons from the Γ-conduction band minimum must be related to its effect on the energy of the deep donor level and on the conduction band structure. It is clear from the measurements that the deep donor responsible for the trapping is present at a concentration comparable to that of the Si or Sn doping level.

We have previously shown[12] that, for samples doped at $n \leq 5 \times 10^{18}$ cm^{-3}, the variation of n with P can be fitted by assuming that electrons capture at a donor level whose energy relative to the Γ-minimum decreases at a rate of 4.8 meV/kbar. This value is close to the pressure coefficient of the energy difference between the Γ and L minima[13], which suggests that the level may have some of the character of these minima. As the pressure is increased above a critical value the energy of the DX level moves towards the Fermi energy ε_F of the electrons in the Γ-conduction band, which then remains pinned at the energy of the deep level. The decrease of n with P is governed by the above pressure coefficient and by the density of states of the Γ minimum at ε_F. It is important to note that although the carrier concentration is measured at 4.2 K the actual occupation of the DX centre and hence the carrier concentration is fixed at some intermediate temperature (~120 K) at which the capture and emission rates become so slow that the electron population in the DX level is no longer in equilibrium with the free electron density. Using Fermi-Dirac statistics it is possible to calculate the position of the DX centre relative to the Γ conduction band at 120 K. We assume that the number of DX centres N_{DX} is equal to the donor doping N_D. Using the usual Fermi-Dirac expression together with the charge neutrality equation, it is easy to show that, in the absence of compensation, the energy of DX above the Γ minimum is given by

$$\varepsilon_d = \varepsilon_F - kT \, \ln\left[\tfrac{1}{2}\left(\frac{N_{DX}}{n} - 1\right)\right]. \qquad (1)$$

As will be discussed later in the paper, we have evidence (LVM measurements) that the compensation of the Philips-grown samples is small. (In any case, a compensation ratio $N_A/N_D \lesssim 1/3$ has very little effect on the calculated value of ε_d). The Fermi energy ε_F in the conduction minimum is calculated taking into account non-parabolicity and the pressure dependence of the effective mass[14,15]. A value of N_D slightly greater than the zero pressure carrier concentration is used.

The calculated values of the energy of the DX level together with the pressure dependence of the Γ, X and L minima[13] are shown in Figure 3 for two of the samples. The positions of the X and L minima illustrated in this diagram are the positions for undoped GaAs. Band gap renormalisation[16] at the high doping levels used here probably shifts the X and L minima to higher energies relative to the Γ conduction minimum and may also affect the pressure coefficients, even at constant carrier concentration.

As can be seen from Figure 3, for a particular sample, the variation of the position of DX as a function of hydrostatic pressure is approximately linear. From the slope, we can estimate the pressure coefficient for the energy of the DX level relative to the Γ-minimum. With increasing doping the DX level is shifted to higher energies. For a particular dopant the pressure coefficient of DX also increases with increasing doping. These results are tabulated in Table 1. At these high doping levels it is possible that the observed reduction of the binding energy of the electron on DX arises from Coulomb and screening effects. These effects are known to occur for the much more extended shallow donor states of the Γ-minimum at low doping levels ($\sim 10^{16}$ cm^{-3})[17]. Above the critical pressure the electrons become trapped and neutralise ionised donors, thereby reducing the strength of the coulomb interactions. Hence the level would become deeper relative to the L-minima, giving rise to an increased rate of shift of the donor level relative to the Γ-minimum. The zero pressure values of ε_d as a function of doping shown in Figure 4 are calculated by assuming that for pressures below the critical pressure the energy difference between DX and L remains fixed. Values of ε_d reported by Theis et al[18] using DLTS are also plotted for comparison. The solid lines in Figure 2 show the expected variation of n with P, calculated using an iterative method since equation (1) cannot be solved analytically for n. The variation of ε_d with pressure was calculated using the pressure coefficient in Table 1. Non-parabolicity and pressure dependence of effective mass were taken into account when calculating the Fermi energy ε_F.

It can be seen that this procedure, taking full account of the Fermi statistics,provides a good fit to the experimental data shown in Figure 2. Note that the fit is a significant improvement over that given in our earlier paper[12] where, for simplicity, it was assumed that the Fermi level was precisely pinned at the DX level for pressures above the critical pressure at which the level starts to fill.

Table 1

Zero Pressure Carrier Conc. ($x10^{18}$ cm^{-3})	$E_F(P=0)$ (meV)	P_c (kbar)	N_D ($x10^{18}$ cm^{-3})	ε_d (meV)	$d\varepsilon_d/dP$ (meV/kbar)
(a) [Si] 11.3	226	2	12.00	270	- 9.4
(b) [Si] 4.6	126	10	4.60	223	- 9.1
(c) [Si] 2.0	73		pressure has no effect on n.		
(d) [Sn] 18.0	287	0	21.00	311	-13.3
(e) [Sn] 7.1	175	7	7.15	251	-10.7
(f) [Sn] 3.2	109	12	3.25	206	- 6.0

The measured carrier concentration n, at P = 0, for the samples studied are indicated in column 1. The Fermi energy ε_F is calculated taking into account non-parabolicity. P_c is the critical pressure at which n starts to fall. N_D is the donor doping level used in the Fermi-Dirac calculation. The energy of the deep level above the Γ minimum (ε_d) is calculated using the Fermi-Dirac statistics as described in the text. The final column is the pressure coefficient of the energy of the deep level relative to the Γ minimum used to obtain the fitting curves in Figure 2.

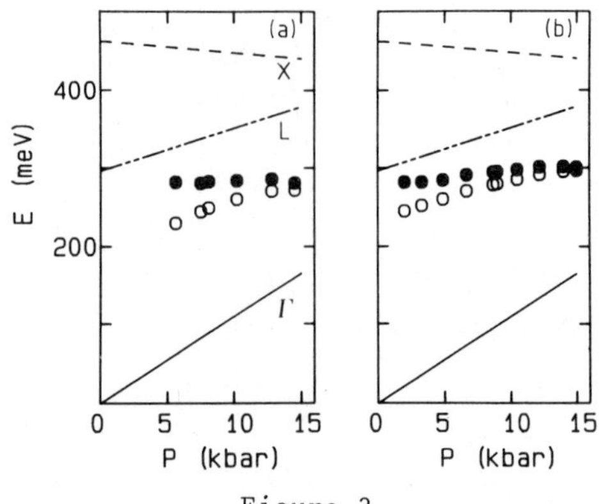

Figure 3

Energy of the DX level (•) and the Fermi energy (o) versus pressure for (a) a 7.1×10^{18} cm^{-3} Sn-doped sample and (b) a 1.1×10^{19} cm^{-3} Si-doped sample. Note that the positions of the X and L minima illustrated are for undoped GaAs. Band gap renormalisation at these high doping levels probably shifts the X and L minima to higher energies relative to the Γ conduction band.

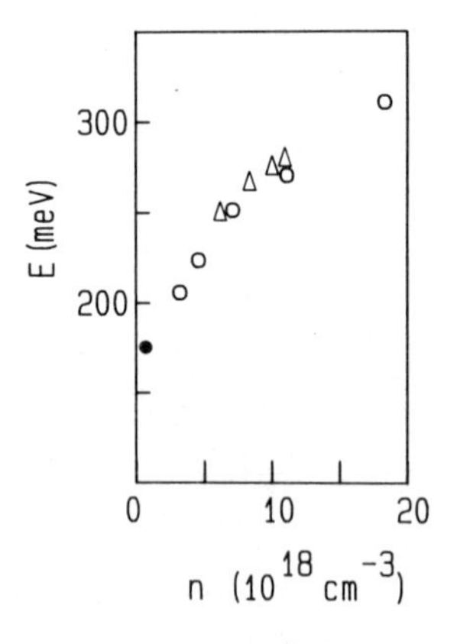

Figure 4

Energy of the DX level relative to the Γ band (ε_d) (o) versus carrier concentration calculated using the procedure described in the text. Values reported by Theis et al., for ε_d (Δ) are shown for comparison. The energy of the DX level deduced by Mizuta and coworkers [4,6] is also indicated (•).

Our analysis of the occupancy of the Γ-conduction band and the DX level indicates that at room temperature and atmospheric pressure the DX level has a significant occupancy (~10%) in the heaviest doped samples ($n \gtrsim 1 \times 10^{19}$ cm^{-3}). Thus the DX level is acting to limit the free carrier concentration. This interpretation explains why the carrier concentration deduced from Hall measurements at room temperature for sample(a)is significantly lower than that obtained from the low temperature dHS measurements at atmospheric pressure: as the sample is cooled from room temperature, the occupancy of the DX level decreases as electrons thermalise into unoccupied states of the Γ-conduction band.

In the above analysis, we have followed other authors by neglecting the so-called band gap renormalisation. At high doping levels ($\gtrsim 5 \times 10^{18}$ cm^{-3}) this effect has been estimated to decrease the direct band gap by up to $\Delta E_G \simeq 40$ meV[16]. However, over the range of dopings of interest here, the change in direct bandgap is small, ~10 meV. The uncertainties about the size of the renormalisation do not affect our estimates of the energy ε_d of the donor level relative to the Γ-minimum. The pressure dependence of the carrier concentration in the lighter-doped samples indicates that the level has the same pressure coefficient as the L-minimum. The critical pressures at which we observe n to decrease suggest that the energy of the deep level relative to the Γ-minimum increases with increasing doping. Of course, such linear behaviour assumed in Figure 2 would not be expected to continue over a very much wider range of pressure than employed in the present work. This is because the energy of the DX level and its pressure coefficient apparently themselves depend on the carrier concentration, which itself decreases with increasing pressure.

THE MICROSCOPIC NATURE OF THE DX CENTRE IN GaAs

It is clear from the data presented above that the DX level is present at concentrations comparable to the doping level. One is therefore led naturally to consider the possibility that the level is associated with the substitutional donor impurity atom itself[4] rather than a complex[1].

Far infrared local vibration[19] mode (LVM) measurements made by

Murray and Newman[20] on sample(a)show that our measured carrier concentration at atmospheric pressure of 1.1×10^{19} cm^{-3} agrees closely with the concentration of Si atoms occupying Ga-sites (1×10^{19} cm^{-3}). It is also very close to the calibrated Si-flux employed in the growth (1.3×10^{19} cm^{-3}) and the concentration of Si given by secondary ion mass spectrometry measurements (1.4×10^{19} cm^{-3}). The LVM measurements also show that for sample a the concentration of Si-Si pairs, Si-X complexes and Si_{As} are at least one order of magnitude lower than Si_{Ga}. It is particulary noteworthy that it is only in the situation when the free electron concentration is much less than the calibrated Si flux that LVM measurements reveal compensation due to Si on As-sites and to complexing of Si with an unknown defect X[19,20]. These observations, coupled with the electrical measurements, lead us to support the view[4,9,10] that the level causing the pressure-induced trapping and the PPC is associated with the substitutional donor, rather than a donor-related complex. Such a view would also explain recent results of passivation on Si donors and DX centres in (AlGa)As[21]. In this work, it was found that both the Si-shallow donor states and DX states in hydrogen passivated n-(AlGa)As recovered on annealling with very similar activation energies. In contrast with this, deep levels in GaAs usually recover at much higher annealling temperature than do shallow donor levels. The apparently anomalous result is explained if the DX- and shallow donor levels are associated with the same centre, namely the substitutional Si_{Ga} impurity.

CONCLUSIONS

We have shown that pressure-induced trapping of electrons and associated PPC in n^{+}GaAs is due to DX levels which occur in the energy range between the Γ- and L-conduction band minima. The precise location of the DX level relative to the Γ-minimum depends on the free carrier concentration, the DX level moving to higher energy with increasing n. We have argued that the electrical data, coupled with LVM measurements, indicate that the DX level is associated with the substitutional donor, a conclusion which has also been reached recently by other workers.

A controversial topic is whether or not the thermally activated capture cross section that gives rise to the PPC is due to a large lattice relaxation as was originally proposed by Lang and Logan[1]. Some authors have suggested that the lattice relaxation is rather small[9,10] (see also [8]). However, very recent pressure-dependent DLTS measurements[22] give a small value for the pressure coefficient dE_B/dP = -2.1 meV/kbar of the capture barrier height. This value is more consistent with the large lattice relaxation model. Clearly, a good theoretical model is required to explain how a simple substitutional donor can give rise to states with quite strong lattice relaxation. Such a model would also need to consider the relative importance of the L- and X-conduction band minima in determining the character of the DX state.

Acknowledgement - This work is supported by SERC and CNRS. The work described was done in collaboration with Drs. R. B. Beall, L. Dmowski, J. J. Harris, M. Heiblum, M. Nathan and P. Simmonds. We are grateful to Dr. T. N. Morgan for many useful discussions, to Dr. R. Murray and Prof. R. C. Newman for communicating to us their unpublished far-infrared absorption measurements and for invaluable discussions and to Dr. J. B. Clegg for performing the SIMS measurements.

REFERENCES

1. D. V. Lang and R. A. Logan, Phys. Rev. Lett. 39:635 (1977). See also D. V. Lang, R. A. Logan and M. Jaros, Phys. Rev. B 19:349 (1987).
2. R. J. Nelson, Appl. Phys. Lett. 31:351 (1977).
3. P. M. Mooney, E. Calleja, S. L. Wright and M. Heiblum, Proceedings of the International Conference on Defects in Semiconductors, ed. H. J. von Bardeleben, Material Science Forum 10-12:417-422 (1986).
4. M. Mizuta, M. Tachikawa, H. Kukimoto and S. Minomura, Japan. J. Appl. Phys. 24:L143 (1985).
5. T. N. Theis, Proceedings of the International Conference on Defects in Semiconductors, ed. H. J. von Bardeleben, Material Science Forum 10-12:393-398 (1986).
6. M. Tachikawa, M. Mizuta, H. Kukimoto and S. Minomura, Japan. J. Appl. Phys. 24:L821 (1985).
7. T. N. Theis and S. L. Wright, Appl. Phys. Lett. 48:1374 (1986).
8. A. K. Saxena; Solid State Electronics 25:127 (1982).
9. H. P. Hjalmarson and T. J. Drummond, Appl. Phys. Lett. 48:656 (1986).
10. J. C. M. Henning and J. P. M. Ansems, Semicond. Sci. and Technol.

2:1 (1987). See also Proceedings of the International Conference on Defects in Semiconductors, ed. H. J. von Bardeleben, Material Science Forum 10-12:429-434.

11. J. H. Neave, P. J. Dobson, J. J. Harris, P. Dawson and B. A. Joyce, Appl. Phys. A 32:195 (1983).
12. D. K. Maude, J. C. Portal, L. Dmowski, T. Foster, L. Eaves, M. Nathan, M. Heiblum, J. J. Harris and R. B. Beall, Phys. Rev. Lett. 59:815 (1987).
13. J. S. Blakemore, J. Appl. Phys. 53 (10):R123 (1982).
14. L. G. Shantharama, A. R. Adams, C. N. Ahmad and R. J. Nicholas, J. Phys. C: Sol. State Phys. 17:4429 (1984).
15. A. Raymond, J. L. Robert and C. Bernard, J. Phys. C 12:2289 (1979).
16. R. A. Abram, G. J. Rees and B. L. H. Wilson, Advances in Physics 27: 799 (1978).
17. J. Leloupe, H. Djerassi, J. H. Albany and J. B. Mullin, J. Appl. Phys. 49 (6):3359, June 1978.
18. T. N. Theis, P. M. Mooney and S. L. Wright, preprint.
19. J. Maguire, R. Murray, R. C. Newman, R. B. Beall and J. J. Harris, Appl. Phys. Lett. 50 (9):516 (1987).
20. R. Murray and R. C. Newman; unpublished, private communication.
21. J. C. Nabity, M. Stavola, J. Lopata, W. C. Dautremont-Smith, C. W. Tu and S. J. Pearton, Appl. Phys. Lett. 50:921 (1987).
22. M. F. Li, P. Y. Yu, E. R. Weber and W. Hansen, Appl. Phys. Lett. 51:349 (1987).

SHALLOW AND DEEP IMPURITY INVESTIGATIONS: THE IMPORTANT STEP TOWARDS A A MICROWAVE FIELD-EFFECT TRANSISTOR WORKING AT CRYOGENIC TEMPERATURES

W.Prost, W.Brockerhoff, M.Heuken, S.Kugler, and K.Heime
Universität-GH-Duisburg, SFB 254, D-4100 Duisburg

W.Schlapp and G.Weimann
Forschungsinstitut der DBP, D-6100 Darmstadt

I Introduction

Current, amplification, output power and frequency limits of field-effect transistors (FETs) are determined by both the carrier concentration and the carrier transport properties (mobility and velocity) in the channel. In GaAs MESFETs the current conducting channel itself is doped with shallow impurities up to levels of 10^{17} or $10^{18}cm^{-3}$. These high concentrations degrade the mobility.
In the heterostructure FET (HFET, also called HEMT, MODFET, SDHT, TEGFET) the carrier supplying layer and the quasi two dimensional electron gas (2DEG) channel are spatially separated, thus enabling an almost independent optimization of both carrier concentration and mobility. Moreover when cooling down both types of devices to cryogenic temperatures dramatically different transport properties can be observed. The mobility of highly doped MESFETs remains constant due to the temperature independent Coulomb scattering, whereas the HFET exhibit at 77K a mobility in the 2DEG which is typically 10 times higher than the room temperature value and about 20 times higher than the mobility of carriers in a MESFET channel.

The HFET provides comparable performance at room temperature /1/, but excellent device properties at cryogenic temperatures are reported up to now (digital /2/, analog /3/ and low noise /4/) under white light illumination, only. In the dark however deep level effects degrade the device performance dramatically /5,6/. Deep level analysis by means of PhotoFET /7/, PhotoCAP /8/ and Low Frequency (LF-) NOISE /9/ in our laboratory has correlated this effect mainly with a level of 0.40eV to 0.44eV below the conduction band egde (DX-center) /10/. Various attempts started to eliminate the degradation of HFET performance at cryogenic temperatures /11,12,13/.

This paper will report about the improvement achievable by the use of both doped and undoped superlattices (SL) in HFETs. Methods of investigation of deep and shallow impurities will be briefly introduced. Results on conventional as well as on SL HFETs will be shown and the consequences for the design of a FET for microwave amplification at cryogenic temperatures which operates degradation-free in the dark will be illustrated.

II Methods of investigation applied to AlGaAs/GaAs HFETs

A conventional AlGaAs/GaAs HFET (cf. fig.1) consists of a GaAs caplayer, a highly doped $Al_xGa_{1-x}As$ carrier supplying layer ($N_D \approx 1 \cdot 10^{18}cm^{-3}$, an undoped $Al_xGa_{1-x}As$ spacer layer, and an undoped GaAs buffer with the 2DEG at the heterointerface towards the spacer. All samples reported here were grown by MBE /14/ on GaAs substrate.

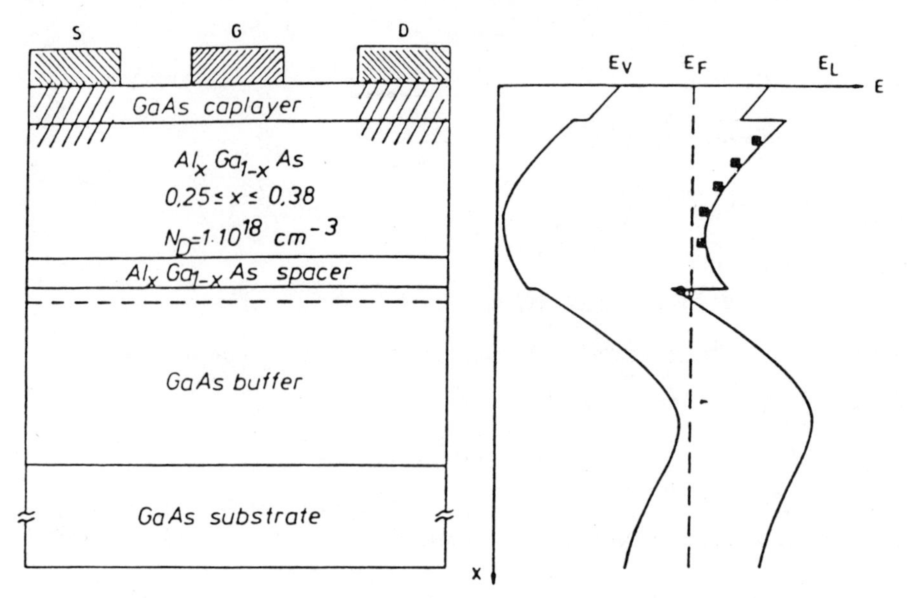

Fig.1. Structure and band diagram of a conventional n-AlGaAs/GaAs HFETs

Transconductance, mobility and gate-modulated carrier concentration profiling

In 1981 Jay and Wallis published the magnetotransconductance and mobility profiling method for the application to homogeneously doped GaAs-MESFET channels /15/. From the decrease of transconductance in a magnetic field a real and direct depth resolved mobility profile can be obtained. In this chapter the magnetotransconductance method and its extension /16/ to gate-modulated carrier concentration profiling will be described. For a better understanding the method will be applied at first to MESFETs and then special problems of HFETs are taken into account. The transconductance g_m is an important device parameter which determines gain and frequency limits. It is defined as:

$$(1) \quad g_m = \frac{\Delta I_D}{\Delta V_{GS}} \approx \frac{i_d}{v_{gs}}$$

(I_D = drain current, V_{GS} = gate-source bias voltage,
i_d = ac drain current in the modulated part of the channel,
v_{gs} = applied modulating ac gate voltage)

The current i_d is determined by

$$(2) \quad i_d = g(v_{gs}) \cdot V_{DS}$$

$$(3) \quad g(v_{gs}) = q \cdot \mu \cdot \frac{W}{L} \cdot n_S^*; \; n_S^* = n \cdot d^*$$

where L and W are the channel length and width, respectively, and n is the carrier-concentration per cubic centimeter, d^* is the thickness of the modulated part of the channel.

Substituting (3) into (2) and then into (1), setting $v_{gs}=dV_{GS}$ and solving for n_s^* results in

$$(4) \quad n_S^*(V_{GS}) = \frac{g_m(V_{GS})}{\mu(V_{GS})} \cdot \frac{L}{g \cdot V_{DS} \cdot W} \cdot v_{gs}$$

It has to be pointed out that n_s* is only a fraction of the total sheet concentration n_s. n_s* represents those carriers which contribute to the small-signal ac current i_d, while n_s carriers support the much larger dc current I_D. Thus $\mu(V_{GS})$ is the mobility of the carriers n_s* and therefore may be different from the mean (Hall-)value of the mobility. The total sheet carrier concentration n_s may be obtained by integration of eq.(4) from V_{Bi} (built-in voltage) to V_T (threshold voltage). Note that $g_m \propto V_{DS}$ in the low-field regime; n_s* is therefore independent of V_{DS}. The transconductance "profile" $g_m(V_{GS})$ is simply measured by varying the bias voltage V_{GS}. The modulation of the channel thickness by the ac voltage v_{gs} results in the ac current i_d which is proportional to the conductivity of the modulated part of the total channel thickness. In a magnetic field B this conductivity varies due to the geometric magnetoresistance effect (the contact width W is much larger than the contact separation; W/L≈100 in the devices investigated here). Hence at any given V_{GS}

$$(5) \quad \mu(V_{GS}) = \frac{1}{B}\sqrt{\frac{g_m(V_{GS},0)}{g_m(V_{GS},B)} - 1}$$

with $g_m(V_{GS},0)$ transconductance without and $g_m(V_{GS},B)$ transconductance with an applied magnetic field B. By applying this method to HFETs it is clear that the dependence of g_m, μ and n_s* on gate voltage can be determined in a way identical to MESFETs. However the gate bias does not correspond to a certain depth d and thickness d* in the channel and thus no "depth" profiling is possible. Recent publications indicate that the wave function is modulated by the gate bias /27/. The gate-voltage profiles do however represent the carrier concentrations and mobilities which are available for the ac operation of the device.

Optical activation of deep levels

Charging a trap in the space-charge region under the gate by illumination with monochromatic light varies the gate capacitance C_G (PhotoCAP /17/). If the gate voltage is held constant the depth of the space-charge region changes and hence the conducting channel is varied. This finally results in a variation of the drain current (PhotoFET /18/). From the sign of the current or capacitance variation, respectively it is possible to discriminate electron and hole traps. Activation energies of deep levels can be deduced from the threshold in the variation ΔI_D or ΔC_G with photon energy. The threshold is defined as the energy at which ΔI_D or ΔC begins to change, because this is the minimum energy necessary for exciting a carrier from a deep level. An increase of ΔI_D or ΔC_G indicates donor-like levels (D), a decrease indicates acceptor-like levels (A). The PhotoCAP method is similar to the PhotoFET method. But in addition this method allows a gate-voltage profiling (and in correlation with the simultaneously obtained capacitance-voltage characteristic of the gate) an estimate of trap concentration versus gate bias /8/.
Both methods are straightforward and simple. In HFETs traps both in the AlGaAs and the GaAs buffer layer contribute to the PhotoFET effect /7/. Figure 2a,b shows the band diagram of a conventional HFET (cf. fig.1) with a trap level in the AlGaAs which is occupied in the

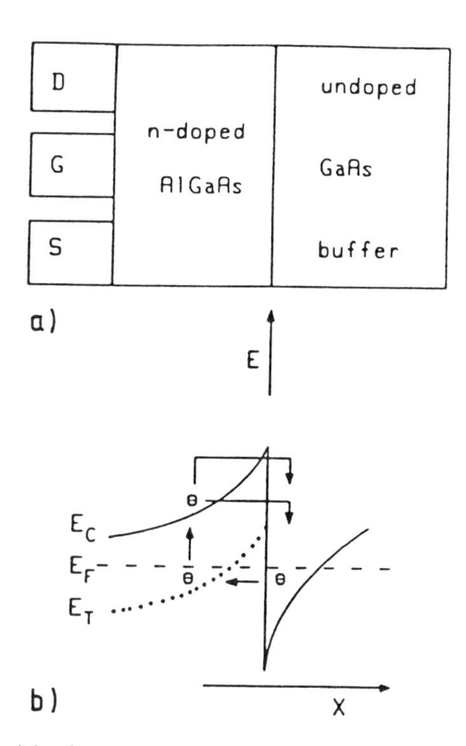

Fig. 2.
a) Cross-section of a conventional HFET
b) Band diagram showing the generation of a carrier from a traps in the AlGaAs, its transfer into the 2DEG and its recombination

dark. Monochromatic light with energy $h \cdot \nu > E_C - E_T$ is able to ionize the trap level and an electron is emitted into the conduction band. The emitted electron is transferred into the 2DEG by thermionic emission or by thermionic field emission. Consequently the current increases. The recombination of electrons from the 2DEG into the trap level is determined by tunneling processes. Deep levels in the GaAs buffer layer may also contribute to variations in the 2DEG concentration if they are ionized by incident light (fig.3). An electron from a donor-like level will be transferred into the 2DEG by the influence of the electric field and increase the drain current, too.
PhotoFET spectra of a conventional HFET (cf. fig.1) for two photon energy ranges are shown in fig. 4a,b. In the lower energy range we have found deep donor-like levels at 0.44 and 0.47eV (DX-center) and a strong acceptor like level of 0.6eV. These three levels were present in all investigated HFET structures. In the high energy range (cf. fig. 4b) there are donor and acceptor like levels which relate to the GaAs buffer or GaAs substrate /17,18/.

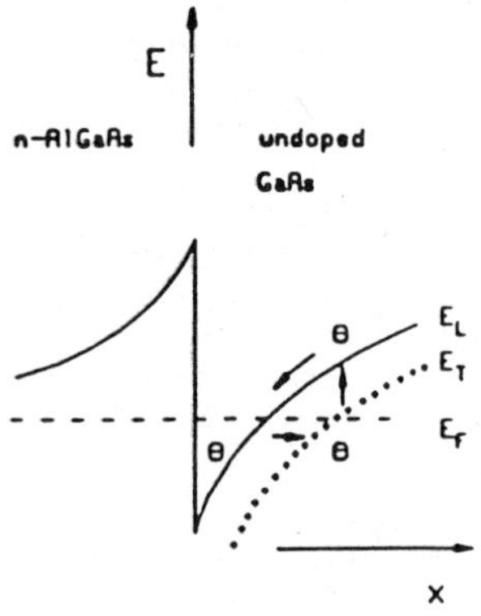

Fig. 3.
Same as fig. 2, but now with recombination in the GaAs buffer

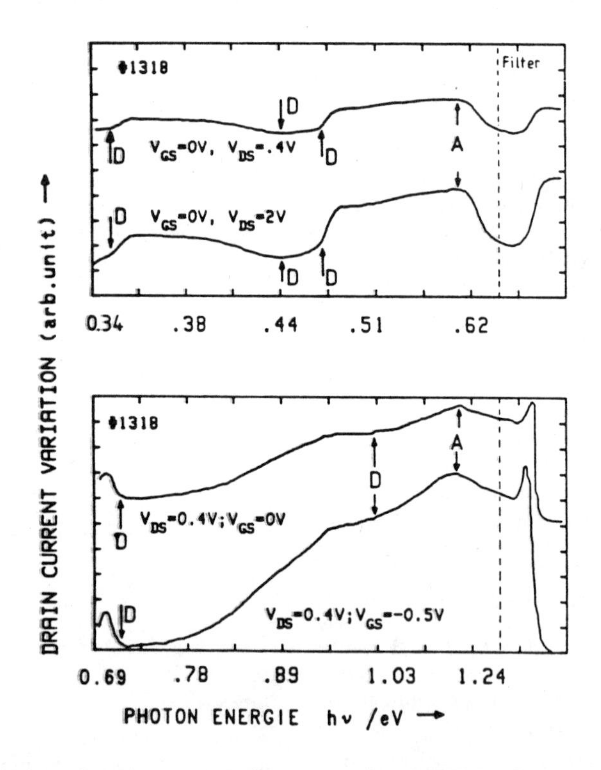

Fig. 4a,b: Variation of drain current of a conventional HFET with energy of incident photons (PhotoFET spectrum for two different energy ranges)

We have investigated samples with an Al-content from 25% to 38% and a doping level of about $1 \cdot 10^{18} cm^{-3}$. For this data no significant Al content dependence is expected /19,20,21/. In fig.5 the trap concentration of the two main donor like levels in a conventional HFET (fig.1, x=0.25) are shown. With increasing forward bias the trap concentration increases from about $10^{13} cm^{-3}$ to about $10^{17} cm^{-3}$.

Both levels are present in the AlGaAs layer. With positive increasing bias the conduction band in the AlGaAs layer is lowered,more and more levels fall below the FERMI level and trap electrons. This effect is responsible for a degradation free I-V-characteristic at low temperatures if the cool down cycle has been done at negative bias when all traps are empty /11/. The time-constant τ for the recombination process is very high (some minutes up to hours) at low temperatures and allows a degradation-free operation for times $t<\tau$.

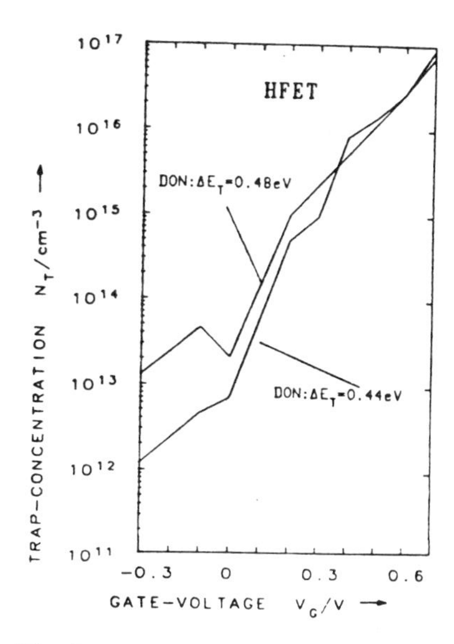

Fig. 5.
Trap concentration of the donor like levels 0.44eV and 0.48eV versus gate bias in a n-AlGaAs/GaAs HFET (x=0.25).

Thermal activation of deep levels

Deep level transient spectroscopy (DLTS) /22/ and LF-NOISE /23/ measurements use the temperature dependence of gate capacitance and noise spectrum, respectively. Both methods were applied to AlGaAs/GaAs heterostructures successfully /24,25,9/ and provide results which are in very good agreement. In this paper the LF-NOISE method will be briefly introduced. LF-NOISE measurements on HFETs were performed by the author's group for the first time. Low frequency noise is a major drawback in FETs since in non-linear circuits like mixers and oscillators it is up-converted in the high-frequency range /28,29,30/.

It is reported that low frequency noise in HFETs is higher than in MESFETs with otherwise comparable parameters /28,31/. It is therefore important for HFET improvement to discover the origin of the low frequency noise. LF-NOISE measurements indicated spectra (fig.6) which can be explained by a superposition of several Lorentzian-type spectra. These kind of spectra is characteristic for generation-recombination /9/.

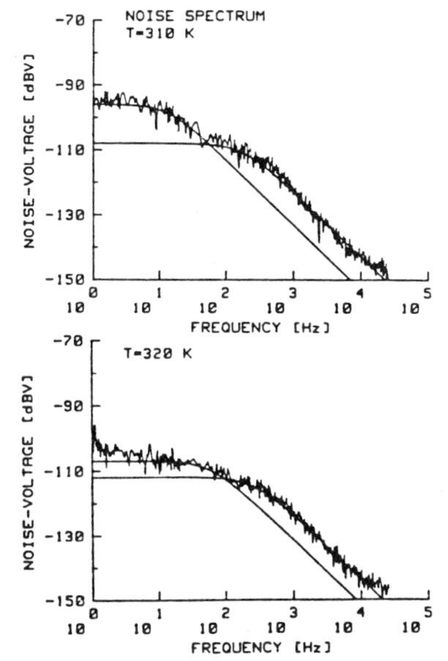

Fig. 6.
Low frequency noise spectra of a conventional HFET at low gate and drain bias at two different temperatures

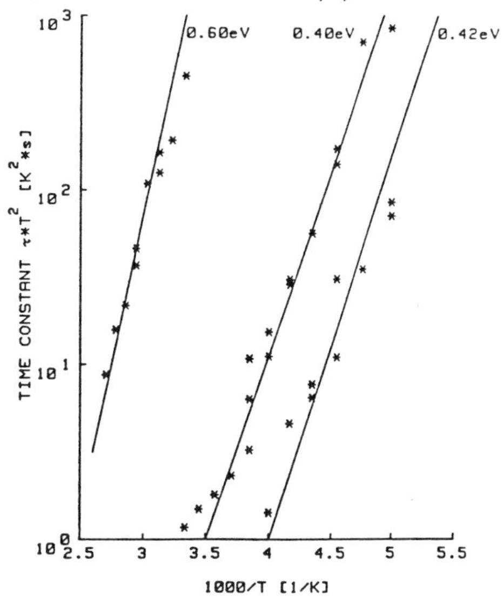

Fig. 7.
Arrhenius plot of the time constants $\tau \cdot T^2 = 1/2\pi f_0$ versus 1/T. Straight lines from /24/, points: noise measurements

The time constant τ (emission time constant) is characteristic for a given trap and depends on temperature:

$$(6) \quad \frac{1}{\tau} \propto T^2 \cdot e^{-E_T/kT}$$

The emission time constant τ is identical to that deduced from DLTS measurements. Therefore a close correlation between DLTS and LF-NOISE is expected. The statistic time distribution of current pulses due to the trapping and release of a carrier leads - if transformed into the frequency domain - to the Lorentzian-type spectrum /9/,

$$(7) \quad W(f) = \frac{A}{1 + (2\pi f \tau)^2}$$

(W= frequency spectrum of current or voltage; A= amplitude). The corner frequency f_i is determined by the emission time constant:

$$(8) \quad f_i = \frac{1}{2\pi\tau}$$

From a plot of $\tau \cdot T^2$ versus 1/T the activation energy of the trap responsible for the noise contribution can be evaluated. An experimental result of our noise measurements is shown in fig.7. The points are obtained from noise measurements, the straight lines are results (best fits) from DLTS measurements obtained by other authors on similar AlGaAs layers /24,25/. The excellent correlation is evident.

III Undoped AlGaAs/GaAs SL buffer layer

Using a conventional HFET structure (cf. fig.1) two disadvantageous features have been demonstrated:

- carrier deconfinement near pinch-off
- influence of deep levels both from the AlGaAs doping layer and the GaAs substrate (cf. fig.4b)

These effects can be strongly reduced by the incorporation of superlattices both below and above the 2DEG. In this chapter the improvement due to the use of an undoped AlGaAs/GaAs superlattice between the GaAs substrate and the GaAs buffer will be discussed.

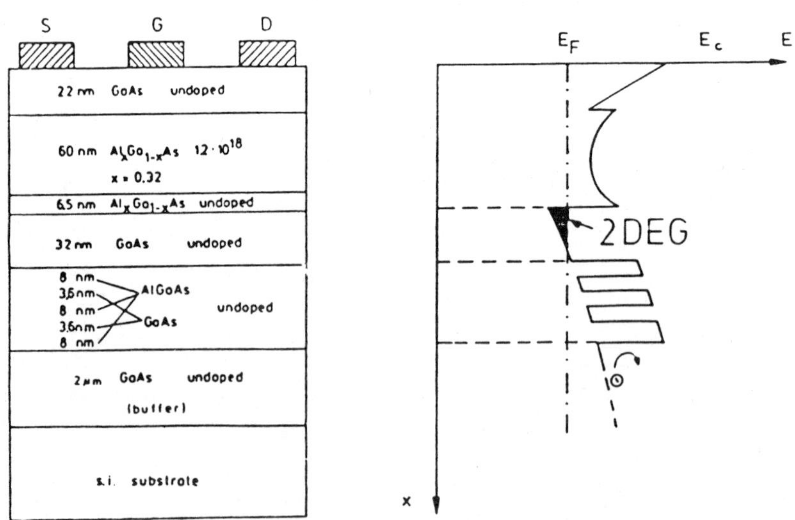

Fig. 8. Cross-section and band diagram of a HFET with an undoped superlattice barrier layer

The carrier confinement was enhanced by using a 32nm GaAs buffer, only (instead of 1000nm in the conventional HFET). This is indicated by the sharp decrease of n_s* towards pinch-off (fig.9). At forward bias the mobility is slightly reduced due to low-mobility carriers in the AlGaAs /16/. A further advantage of this superlattice is the elimination of most of the substrate effects on the 2DEG: carriers generated from deep levels in the substrate cannot be transferred into the 2DEG due to the AlGaAs/GaAs barriers. The PhotoFET spectra of this sample are shown in fig.10. In the low energy range (cf. fig.10a) the donor (0.43eV, 0.47eV) and acceptor (0.62eV) like levels are still present (cf. fig.4a). They all relate to the unchanged n-AlGaAs layer. In the high energy range the diagram is rather different from the conventional structure. All effects due to the substrate-related traps are eliminated. The shape of the PhotoFET spectrum is identical to the lamp spectrum. We assume that photoemission of electrons over the gate (barrier height≈0.7eV) is responsible for this effect.

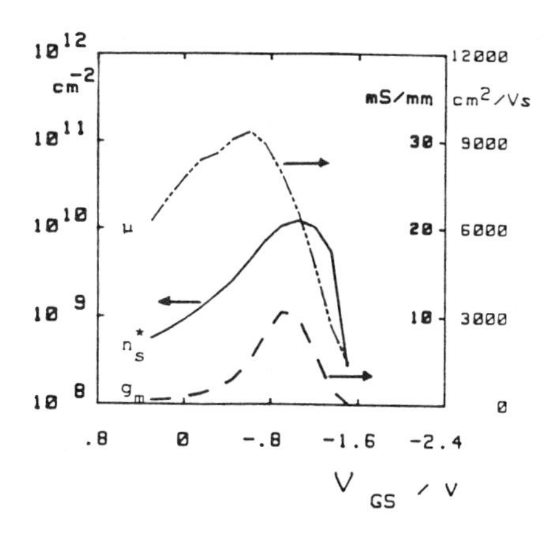

Fig. 9.
Transconductance, mobility and sheet carrier concentration profile in a quantum well structure HFET with an undoped superlattice barrier layer obtained from magnetotransconductance measurements: T=300K, V_{DS}=50mV, V_{gs}=50mV

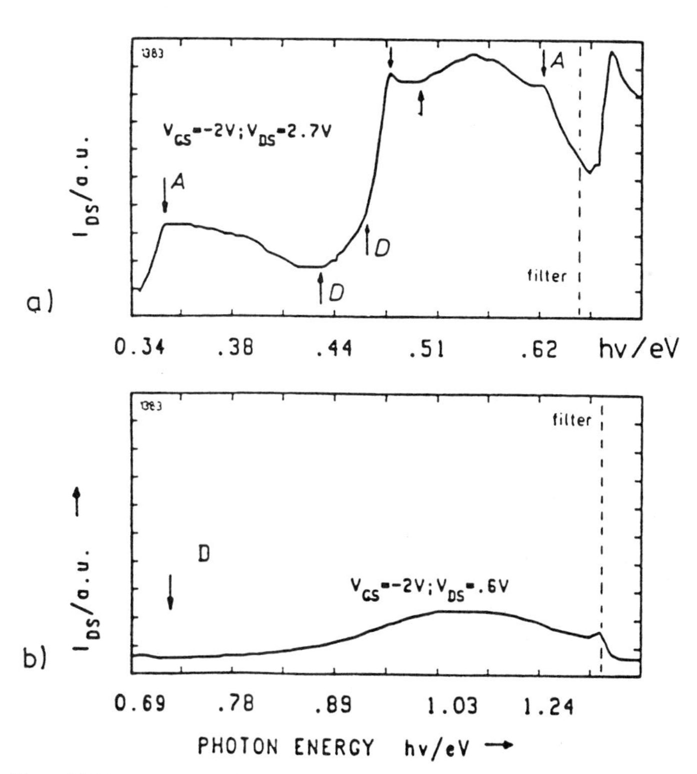

Fig. 10. PhotoFET spectrum of HFET with an undoped AlGaAs/GaAs buffer layer a) low photon energy range, b) high photon energy range

IV Doped n-GaAs/AlAs SL

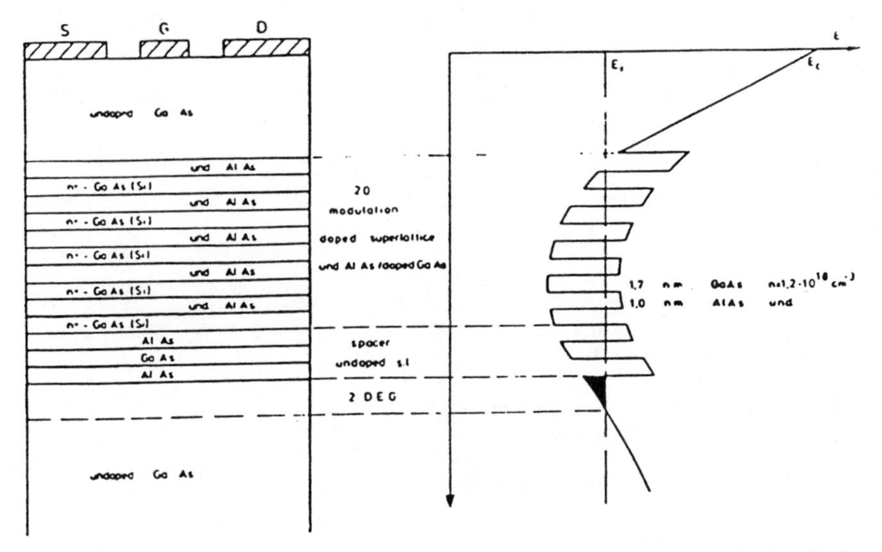

Fig. 11. Configuration (left) and band diagram (right) of a HFET with superlattice doping layer (SL HFET)

Chapter III has demonstrated the potential of an undoped GaAs/AlGaAs SL barrier layer in order to suppress the influence of the substrate and to increase the confinement of carriers. Nevertheless the n-AlGaAs related traps are still present (cf. fig.10a). Based on the hypothesis that the DX-center is due to the neighbourhood of Al, Ga and a dopant atom /32/, the n-AlGaAs dopant layer which contains this three components was replaced by an undoped AlAs/doped GaAs SL /32,13,33,34/ or an undoped AlGaAs/n-GaAs SL /35,36/. The SL provides a spatially separation of the dopant atom from Al (cf. fig.11). The following impurity investigations were performed on a n-GaAs/AlAs SL HFET with homogeneously /34/ or δ-doped /33,14/ GaAs wells. In fig.12a,b the sheet carrier concentration n_s obtained from VAN DER PAUW/HALL measurements of a conventional structure (cf. fig.1) and a SL donor layer structure (cf. fig.11) are shown.

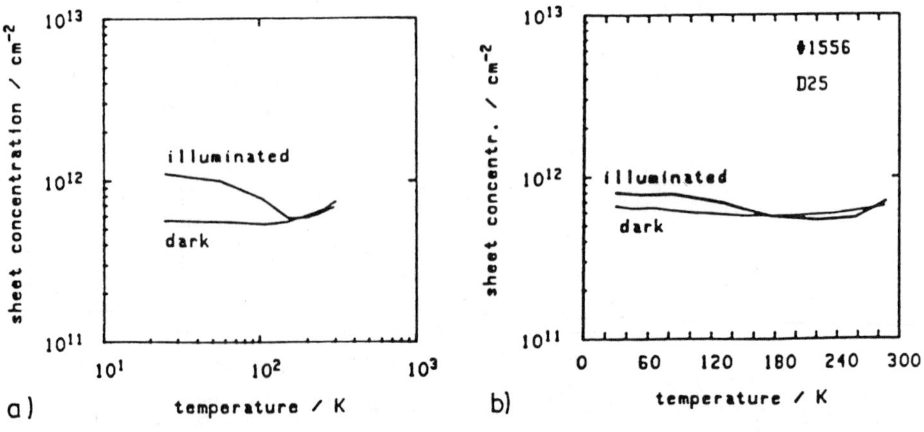

Fig. 12. Hall sheet carrier concentration of HFET structures versus temperature a) conventional HFET (x=0.38) b) n-GaAs/AlAs SL HFET (δ-doped)

The measurements were performed in absolute darkness or under red LED illumination, respectively. During the cool down cycle no bias was applied. The difference between illuminated and dark n_s values represents the deep level concentration of the structure. The ratio

(9) $\eta = n_{s,shallow}/n_{s,trap}$,

can be calculated from these measurements by

(10) $\eta = n_{s,dark}/(n_{s,ill}-n_{s,dark})$

At 30K we obtained for the conventional structure η=1.04 and for the SL HFET η=5.1 indicating the 5 times higher suppression of deep levels in the structure with the SL donor layer.
The deep level concentration can be estimated from PhotoCAP measurements, too. In fig.13 the remaining concentration in a SL structure of a 0.47eV deep level is plotted vs. gate bias. The maximum value is less than $10^{15}cm^{-3}$ in comparison to about $10^{17}cm^{-3}$ in a conventional structure (cf. fig.5) indicating an improvement by a factor of 100 for this special level. In addition the 0.44eV level is very weak in the SL structure.

LF-NOISE measurements are not able to detect any deep level in the range of 0.4eV to 0.5eV. The voltage noise of the SL HFET is lower than in a conventional structure.

Table 1. Summary of deep level analysis applied to conventional AlGaAs/GaAs HFETs and n-GaAs/AlAs SL HFETs

n-AlGaAs/GaAs HFETs

	Activation energy E_T /eV			
Type	PhotoFET	PhotoCAP	LF-NOISE	DLTS [1]
DON	0.35	0.37		
	0.44	0.44	0.40	0.40
	0.47	0.48	0.42	0.42
	0.70	–	0.60	0.60
	1.01	–		
ACC	0.36	0.38	acceptor like levels not detectable	
	0.60	0.59		
	0.73			
	0.93			

[1] after HIKOSAKA and KÜNZEL

n-GaAs/AlAs SL-HFETs

	Activation energy E_T /eV			
Type	PhotoFET	PhotoCAP	LF-NOISE	DLTS [1]
DON	0.37	0.37	0.36	
	(0.43) [2]	(0.44) [2]		
	0.47	0.47	0.59	0.5–0.6
	0.70	–		
ACC	0.39	0.38	acceptor like levels not detectable	
	0.60	0.59		

[1] after BABA (1986)
[2] weak

Table 1 summarizes the results of the deep level analysis applied to SL HFET. The DX-center levels were strongly suppressed and can be detected by PhotoCAP or PhotoFET method, only. DLTS measurements were performed by other groups and support our results /33/.

Influence of center or δ-doping in the GaAs wells

The basic idea of the suppression of the DX-center is the spatial separation of dopant atom from Al. If the n-AlGaAs dopant layer is replaced by a SL with homogeneously doped GaAs wells it is clear that at each AlAs-GaAs (or AlGaAs/GaAs) interface the separation is not given and it is shown that in these structures the DX-center is suppressed but still present /34/. Three groups independently confirmed that center doping in GaAs wells as thin as 2nm improves spatial separation over that in homogenously doped wells /33,36,38,39/. But using the PhotoCAP method we still found DX-center like levels in a δ-doped n-GaAs/AlAs SL. In the work on δ-doping of bulk GaAs a minimum thickness of the d-layer of about 3nm /40/ is reported. The result of this consideration is that a total elimination of the DX-center by use of such narrow GaAs wells it not possible. But the fact that best results are obtained with a δ-doping scheme clearly proves that a significant doping profile exists in a GaAs well which is as narrow as 1.7nm.

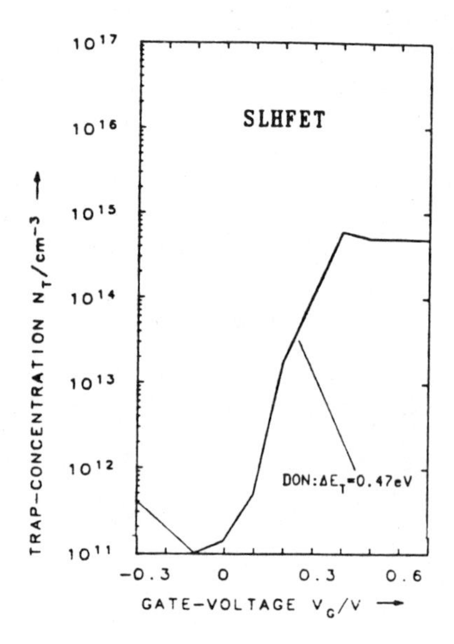

Fig. 13.
Trap concentration of the donor like level E_T=0.47eV versus gate bias in a n-GaAs/AlAs Sl

V Conclusion

In homogeneously doped AlGaAs layers the DX-center consists of two donor like levels separated by 0.02-0.04eV with almost equal concentrations. The doped SL donor layer strongly suppress both levels. Especially the lower one is hardly detectable anymore. Accordingly the main acceptor like level at 0.6eV is reduced. But a total elimination of the DX-center even by use of a center doping scheme seems not to be possible.

Using n-GaAs/AlAs /33,34/ or n-GaAs/$Al_{0.6}Ga_{0.4}As$ SL /35,36/ several groups have successfully fabricated SL HFETs. The best transconductance value's reported so far for a 1μm gate length are 269mS/mm and 313mS/mm at room temperature and 77K, respectively /36/. From RF measurements room temperature values are reported /38/: f_c (MAG=1) = 18GHz. This result was obtained with a high source resistance due to an undoped cap layer. At low temperature this resistance is strongly reduced due to the higher mobility in the 2DEG and a maximum oszillation frequency is calculated of 29GHz and 110GHz at room temperature and at 77K, respectively. This excellent RF performance can be achieved in packaged devices in absolute darkness.

Acknowledgements

The authors are gratefully indebted to G.Howahl for device fabrication, U.Doerk C.Heedt, H.Gurski and J.Krauss for sample preparation and measurements and to G.Oberheide for typing the manuscript.
This work was in part supported by Deutsche Forschungsgemeinschaft and Stiftung Volkswagenwerk.

References

/1/ H.Dämbkes, W.Brockerhoff, K.Heime, K.Ploog, G.Weimann, W.Schlapp
Electronics Letters, Vol.20, No.15, 1984, 615-618

/2/ C.P.Lee, D.Hou, S.J.Lee, D.L.Miller, R.J.Anderson
GaAs IC Symp.Techn.Dig., Oct, 1983

/3/ P.C.Chao, S.C.Palmateer, P.M.Smith, U.K.Miskra, D.H.G.Duk, J.C.M.Hwang
IEEE Electron Dev.Lett., Vol.EDL-6, No.10, 1985

/4/ M.W.Pospialski, S.Weinreb, P.C.Chao, C.K.Mishra, S.C.Palmateer, P.M.Smith, J.C.M.Hwang
IEEE Transactions on Electron Devices, Vol. ED-33, No.2, February 1986

/5/ A.Kastalsky, R.A.Kiehl
Proc.GaAs and Related Compounds 1985

/6/ T.J.Drummond, R.J.Fischer, W.F.Kopp, K.Morkoc, K.Lee, M.S.Shur
IEEE Electron Dev., Vol.ED-30, No.12, 1983

/7/ M.Heuken, L.Loreck, K.Heime, K.Ploog, W.Schlapp, G.Weimann
IEEE Electron Devices, ED-33, 693, May 1986

/8/ M.Heuken, J.Kraus, K.Heime, W.Schlapp, G.Weimann
Proc. E-MRS, Straßbourg, Spring 1987

/9/ L.Loreck, H.Dämbkes, K.Heime, K.Ploog, G.Weimann
IEEE Electron Device Lett., Vol. EDL-5, No.1, 1984

/10/ D.V.Lang, R.A.Logan, M.Jaros
Phys. Rev.B., Vol.19, No.2, 15.January 1979

/11/ J.Y.Chi, R.P.Holmstrom, J.P.Salerno
IEEE Electron Device Lett., Vol.EDL5, No.9, 1984

/12/ R.Bhat, W.K.Chan, A.Kastalsky, M.A.Koza, P.S.Davisson
Appl.Phys.Lett. 47(12), 15.Dec. 1985

/13/ T.Baba, T.Mizutani, M.Ogawa, K.Ohata
Jap. Jornal of Appl.Phys., Vol.23, No.8, 1984

/14/ G.Weimann
Festkörperprobleme XXVI,(1986)pp 231-250

/15/ P.R.Jay, R.H.Wallis
IEEE Electron Device Lett., Vol.EDL-2. No.10, Oct.1981

/16/ W.Prost, W.Brockerhoff, K.Heime, K.Ploog, W.Schlapp, G.Weimann, H.Morkoc
IEEE Electron Devices, ED-33, May 1986

/17/ F.J.Tegude, J.Baston, N.Arnold, K.Heime
Proc. 2nd Conf. on Semi-Insulating III-V Materials
Evian, Frankreich 1982

/18/ F.J.Tegude, K.Heime
IEEE Electronics Device Lett., Vol.16, p.22, 1980

/19/ K.Yamanaka, S.Naritsuka, M,Mannoh, T.Yussa, M.Mihara, M.Ishii
J.Vac.Science & Techn. Vol.2, No.2, 1984

/20/ M.O.Watanabe, K.Morizuku, M.Mashita, Y.Ashizawa, Y.Zohta
Jap. Journal of Applied Physics, Vol.23, No.2, 1984

/21/ T.Ishikawa, T.Yamamoto, K.Kondo, J.Komeno , A.Shibatomi
Inst. Phys. Conf. Ser. No.83

/22/ D.V.Lang
Journal of Applied Physics, Vol.45, No.7, Juli 1974

/23/ L.Loreck
Dissertation Universität-GH-Duisburg, 1985

/24/ K.Hikosaka, T.Mimura, S.Kiyamizu
Inst.Phys.Conf. Ser.No. 63, 233-238, 1981

/25/ H.Künzel, K.Ploog, K.Wünstel, B.L.Zhon
J.Electron.Mat., Vol.13, No.2(1984), 281-308

/26/ L.Loreck, H.Dämbkes, K.Heime, K.Ploog, G.Weimann
Proc.Noise in Physical Systems, North-Holland Pub.Co., Amsterdam 1983, 261

/27/ L.Loreck, H.Dämbkes, K.Ploog, K.Heime
IEEE Intern. Electron Devices Metting, 107, 1983

/28/ J.M.Dieudonne, M.Pouysegur, J.Graffeuil, J.H.Cazaux,
IEEE Trans.Electron Devices, ED-33, 572, 1986

/29/ R.J.Trew, M.A.Khatibzadek, N.A.Masnari,
IEEE Trans.Electron Devices, ED-32, 1985
/30/ H.J.Siweris, B.Schiek, IEEE Trans.Microwave Theory
Techniques, MTT-33, 233, 1985
/31/ R.J.Trew, private communications, 1985
/32/ T.Baba, T.Mizutani,M.Ogawa
Japan J.Appl. Phys. 22 (1983) L627-629
/33/ T.Baba
Microelectronic Engineering 4 (1986) 195-206
/34/ M.Heuken, W.Prost, S.Kugler, K.Heime, W.Schlapp, G.Weimann
Inst. Phys. Conf. Ser. No.83, Chapter 8
/35/ R.Fischer, W.T.Masselink, J.Klem, T.Henderson, H.Morkoc
Electronics Letters, 30th August 1984, Vol.20, No.18
/36/ C.W.Tu, W.L.Jones, R.F.Kopf, L.D.Urbanek, S.S.Pei
IEEE Electron Device Lett.,Vol.7, No.9, 1986
/37/ E.F.Schubert and K.Ploog
Japanese Journal of Applied Physics, Vol.24, No.8, August 1985
/38/ W.Prost, W.Brockerhoff, K.Heime, W.Schlapp, G.Weimann
11th Workshop on Compound Semiconductor Devices and
Integrated Circuits, Grainau, FRG, 1987
/39/ T.Baba private communication, Oktober 1986
/40/ K.Ploog
Journal of Crystal Growth 81 (1987) 304-313
/41/ G.Bosman, R.J.J.Zijlstra
Sol.State Electronics, 25, (1982), 273,

ELECTRONIC STATES IN HEAVILY AND ORDERED DOPED SUPERLATTICE SEMICONDUCTORS

Inder P. Batra

IBM Research Division
Almaden Research Center
650 Harry Road
San Jose, California 95120-6099

C. Y. Fong

Department of Physics
University of California
Davis, California 95616

ABSTRACT

We have calculated the electronic band structure of two heavily and ordered doped superlattice semiconductors: (I) GaAs/AlAs doped with Si and (II) Ge/Si doped with Al and As. For case (I), the properties of the donor states in 2- and 3-layers of GaAs and AlAs have been examined. The donor state charge is found to be non-hydrogenic. The polarization of the charge affects the band alignment. For case (II), superlattice grown on Si substrate with 2 atomic layers of Ge and 10 atomic layers of Si is considered. We have doped Al and As in the Si region but near the interfaces. The properties of the acceptor and the donor states in the presence of the thin Ge-region have been analyzed. No impurity related states emerge in the band gap region.

1. INTRODUCTION

One of the fundamental properties of semiconductor materials is the ability to drastically alter their electronic and optical properties with doping. As the trend towards smaller electronic devices continues, the need to understand the quantitative difference between doping these heterostructures and their bulk counterparts is important. Previous theoretical work on impurity states in quantum well (QW) structures focused on the description of shallow hydrogenic impurities within the effective-mass approximation (EMA).[1-4] A detailed description of shallow hydrogenic impurities in QW structures was recently given by Csavinszky and Elabsy[5] and elegantly discussed by Srivastava and Rössler[6] at this workshop. The validity of the EMA to describe deep and shallow impurity states in semiconductors was investigated by Pantelides and Sah,[7,8] and Bernholc and Pantelides.[9] Their results showed that for isocoric (same core) substitutional impurities, the EMA is applicable for both deep and shallow levels. Failure of the EMA occurs for non-isocoric substitutional impurities. For example, the non-isocoric substitution of a Si atom for a Ga atom would not be properly described within the EMA. On the experimental side, an important observation is the change in the interface potential barrier heights with doping. Recently, Capasso et al.[10] showed that the interface band discontinuities can be tuned by the use of doping interface dipoles.

Here, we present the electronic properties of two heavily and ordered doped superlattice semiconductors: (I) GaAs/AlAs doped with Si and (II) Ge/Si doped with Al and As. For case (I), the properties of the donor states in 2- and 3-layers of GaAs and AlAs have been examined. For case (II), superlattice grown on Si substrate with 2 atomic layers of Ge and 10 atomic layers of Si is considered. These are discussed in separate sections below.

2. DOPED GaAs-AlAs SUPERLATTICES

In this section, we report the electronic properties of Si donor states located at and near the interface of superlattice GaAs/AlAs (001). We make no assumptions about the screening or nature of the donor states. The main focus is to present the charge distributions of the Si donor states and to discuss how the heavy doping affects the interface dipoles and hence the band edge discontinuities. For a detailed and an in depth analysis, the reader may wish to consult the work by Nelson et al.[11]

2.1 Superlattice Geometry and the Method

The model of the interface we employ consists of thin slabs of GaAs and AlAs repeated periodically. The undoped superlattices consist of two (16-atoms) and three (24-atoms) layers, each of GaAs and AlAs arranged in the ideal [001] geometry. The GaAs lattice constant, a (= 10.69 a.u.), is used for the entire superlattice since the lattice match between GaAs and AlAs is nearly perfect. The interface is n-type doped by the substitution of a Si atom for a Al or Ga atom. The unit cells for the undoped three layer superlattice and with one doping situation is shown in Fig. 1. For future reference, we designate Fig. 1(a) as $(GaAs)_3-(AlAs)_3$ and Fig. 1(b) as ${}_{Si}(GaAs)_3-(AlAs)_3$. The Si subscript is used to indicate the region of doping. The basal plane area (a^2) normal to the superlattice growth direction is twice the primitive unit cell area. In the plane

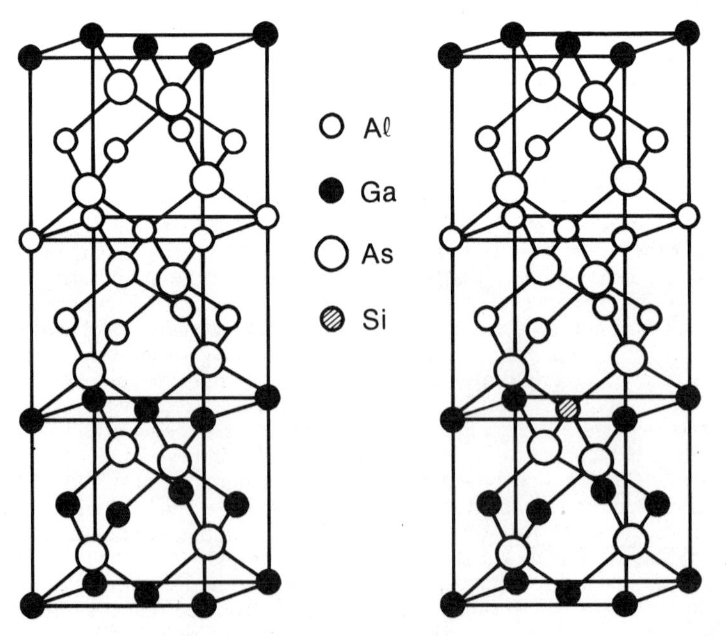

Fig. 1. The undoped and doped three layer superlattices, $(GaAs)_3$-$(AlAs)_3$ and ${}_{Si}(GaAs)_3$-$(AlAs)_3$.

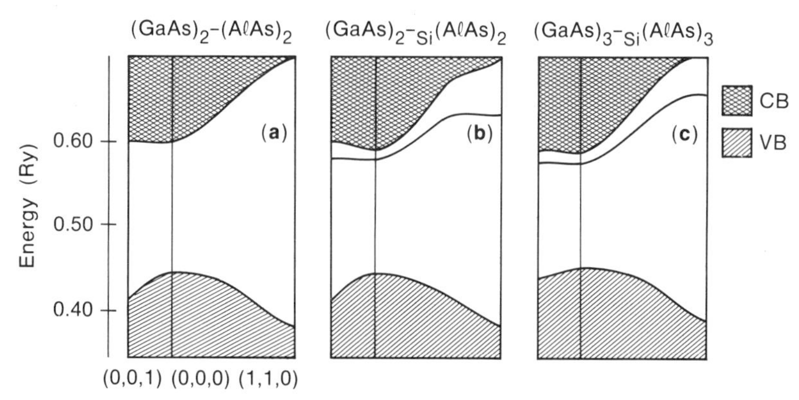

Fig. 2. Energy dispersion curves for $(GaAs)_2$-$(AlAs)_2$ (a), $(GaAs)_2$-$_{Si}(AlAs)_2$ (b), and $(GaAs)_3$-$_{Si}(AlAs)_3$ (c) in directions parallel ([001]) and perpendicular ([110]) to the superlattice growth direction. The occupied VB band and unoccupied CB regions are shaded in. The Si related impurity band in (b) and (c) is located about 0.1 eV below the bottom of the CB.

perpendicular to the superlattice growth direction, the Si impurities are second-nearest neighbors, at distance a, and in the plane parallel to the superlattice growth direction, the Si impurities are separated by n × a. The tetragonal unit cell used in our calculation has a volume of na^3 for $(GaAs)_n$-$(AlAs)_n$ superlattice; for the substitutional doping of selective regions, the volume was kept fixed. The calculations were done using the the well-known self-consistent pseudopotential method.[12] The norm-conserved ionic pseudopotentials of Bachelet, Hamann, and Schluter[13] were used here to describe the valence electron-atomic core interaction.

2.2 Results and Discussion

When a Si atom is substituted for a group III Ga or Al atom, one expects very little change in the valence band (VB) states. These states are predominantly associated with As anion related states, whereas the conduction band (CB) states are more associated with the non-bonding cation states. In Fig. 2, the band structure of the undoped and doped superlattices is shown. To better isolate the changes in the band structure due to the addition of a Si atom, we have shaded in the occupied VB and unoccupied CB regions. In Fig. 2(b), where a Si atom has substituted an Al atom at the interface, $(GaAs)_2$-$_{Si}(AlAs)_2$, we see the appearance of an extra band about 0.1 eV below the bottom of the host CB. This extra band, which we identify as the Si impurity band, is formed by the Si non-bonding states. In the [001] direction, the impurity band is dispersionless, since in this direction, the Si impurity wavefunctions do not overlap. In Fig. 2(c), where we have substituted a Si atom for an Al atom in the middle of the AlAs region, $(GaAs)_3$-$_{Si}(AlAs)_3$, we see similar features. The formation of the Si impurity band is not restricted to doping in the AlAs region. Doping in the GaAs region, $_{Si}(GaAs)_3$-$(AlAs)_3$, gave similar features in the electronic structure. This is not a surprising result, since, as we shall see below, the impurity band charge density is fairly localized in all cases. The impurity band in the [001] direction is completely occupied with a large effective mass, while in the [110] direction the impurity band is only partially occupied with a much smaller effective mass. The extra electron of the Si impurity has formed a two-dimensional metal parallel to the interface. The two dimensional metallic property of the Si impurity band is a consequence of the high doping concentration.

The impurity band charge densities for the two doped superlattices, in the (100) plane parallel to the superlattice growth direction, are given in Fig. 3. The charge density is localized around the Si impurity. The impurity charge density also shows a substantial polarization towards the GaAs region when the impurity is placed at the interface (Fig. 3(b)). The charge density distributions clearly show that the Si impurity band is predominantly associated with the non-bonding p_z states of Si. This is unlike the classic picture[1] of a hydrogenic donor state. We can understand the p-like character of the Si donor state with the following argument. The cation site has an atomic configuration of

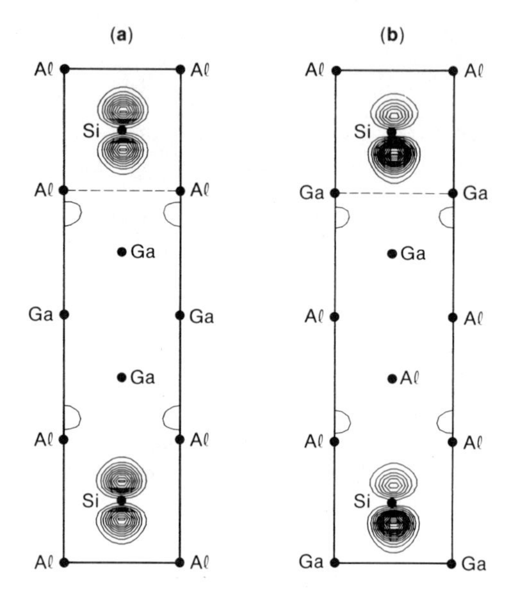

Fig. 3. The charge density of the Si impurity band in the (100) plane parallel to the superlattice growth direction, for $(GaAs)_3$-$_{Si}(AlAs)_3$ (a), and $_{Si}(GaAs)_3$-$(AlAs)_3$ (b). The dotted lines mark the periodicity of the unit cell.

ns^2np^1, where n = 3 for Al and n = 4 for Ga. This is the necessary atomic configuration to form the sp^3 bond with As. When a Si atom with atomic configuration $3s^23p^2$ is substituted at the cation site, it possesses an extra p electron. The As atom which is bonded to the Si atom is also bonded to three group III atoms. For these bonds, the As atom provides the extra p electron to form the sp^3 bond. Thus, the As atom feels an overall tendency, since it has already formed three sp^3 directional bonds to do the same with Si. Consequently, the extra donor electron of Si can occupy the non-bonding p-state.

Figure 4 suggests an explanation for the asymmetry found in Fig. 3(b). Since the AlAs bond is more ionic relative to the GaAs bond, the charge transfer to the As atom

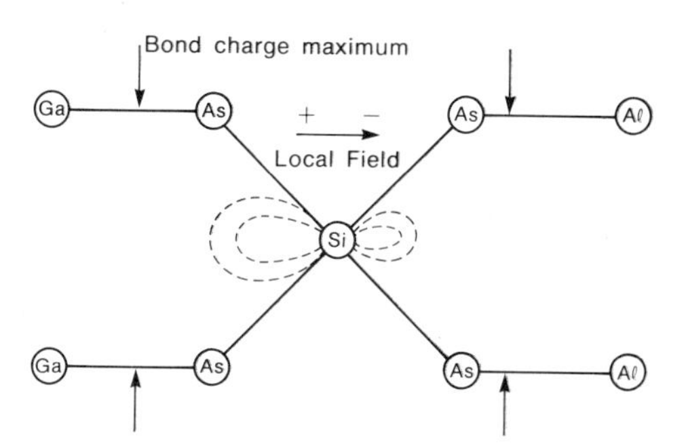

Fig. 4. Explanation of the asymmetry in the impurity charge distribution found in Fig. 3(b). Due to the difference in ionicity of GaAs and AlAs, a local field is set up around the Si impurity causing the charge to shift towards the GaAs region.

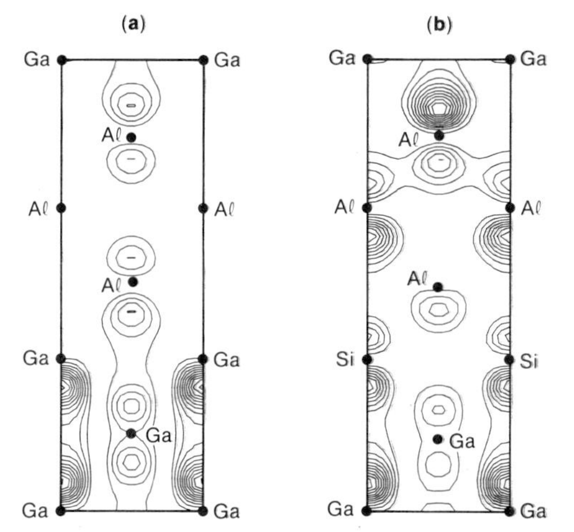

Fig. 5. Charge density contour plots, in the (100) plane parallel to the superlattice growth direction, of the first GaAs related CB state for $(GaAs)_3$-$(AlAs)_3$ (a), $_{Si}(GaAs)_3$-$(AlAs)_3$ (b).

will be larger. This sets up a local field which acts on the impurity charge, favoring a polarization towards the GaAs region. The polarization changes the interface dipole and hence the band alignment.

Let us now briefly discuss with an illustrative example as to how the Si doping of the interface affect the band edge states. The states for the undoped superlattices have been discussed by several authors and can be conveniently found in Ref. 14. Figures 5(a) and 5(b) give the charge density of the first GaAs related CB state (the fourth CB state) in the (100) plane parallel to the superlattice growth direction for $(GaAs)_3$-$(AlAs)_3$ and $_{Si}(GaAs)_3$-$(AlAs)_3$, respectively. Comparing Fig. 5(a) (undoped three layer superlattice) and Fig. 5(b), we see substantial change in the localization of this state when a Si atom is substituted in the GaAs region. The impurity band strongly hybridizes with this (nearest) conduction band state which is manifested in the state charge density.

In summary, for all cases studied, the Si donor formed an impurity band about 0.1 eV below the bottom of the host CB. If the impurity was placed at the interface region, there was a slight tendency for the donor charge density to shift towards the GaAs region. The presence of donor impurities at and near the interface affects the band edge discontinuity. While some of these features are consistent with results obtained from EMA theory, the electronic structure we find for the Si donor state is quite different. We saw no essential difference in the electronic structure of the Si donor states for the isocoric substitution of a Si atom for an Al atom as compared to the non-isocoric substitution of a Si atom for a Ga atom. Whether or not our results would be significantly altered by increasing the unit cell size, especially in the lateral direction remains to be seen.

3. DOPED Si/Ge SUPERLATTICES

Because of the lattice constant mismatch (~4%), Si/Ge is one of the strained superlattice semiconductor. It is known to grow in thin layers of Ge (~with maximum of 6 atomic layers in each region) on Si without introducing dislocations.[15] The band alignment has been studied both experimentally[16] and theoretically.[17] With 6 atomic layers each, the sample exhibits staggered band alignment with the maximum of the VB in the Ge-region and the bottom of the CB in the Si-region. Such band alignment offers several interesting ways to dope the sample. Here, we report the theoretical findings of one configuration of doping in Si/Ge. This configuration is similar to the n-i-p-i (n doped-intrinsic-p doped-intrinsic) structure in GaAs suggested first by Döhler.[18]

3.1 Model and Calculations

We model the Si/Ge superlattice as grown on Si-substrate. It consists of 10 layers of Si atoms and 2 layers of Ge atoms. The tetragonal distortion is in the Ge-region. The lattice constant c, along the $\hat{z}$ direction, is determined by preserving the atomic volume of the Ge as in pure crystal which leads to a(Ge) = 6.14Å. The unit cell is shown in Fig. 6. The lattice constant $a_{\parallel}$ is 5.43Å, whereas c is 16.645Å. There are a total of 24 atoms/unit cell. The Al and As atoms are treated as the impurity atoms. We substitute a Si atom at the left-hand side of the Ge-region by the As atom and another Si atom at the right side of the Ge-region by the Al atom. We do not relax the lattice after the impurity atoms are substituted. The impurity atoms are separated by the Ge-region which is serves as an intrinsic material. The Si-region at the left-hand side of As atom is treated as another intrinsic region. The two impurity atoms are at the third nearest neighbor to each other. Of course, there are many configurations similar to the n-i-p-i structure. The configurations discussed above are the simplest one.

We used the ionic pseudopotential of Ge, Al, As, and Si determined by Pickett et al.[19] and Ho et al.[20] The form of this ionic pseudopotential is given by

$$V_{ion}(G) = \frac{a_1}{G^2}\{\cos(a_2 G) + a_3\}e^{a_4 G^4} \tag{1}$$

where G is the magnitude of the reciprocal lattice vector in a.u. The values of a's in Eq. (1) for the four elements are given in Table 1. The self-consistent procedure was

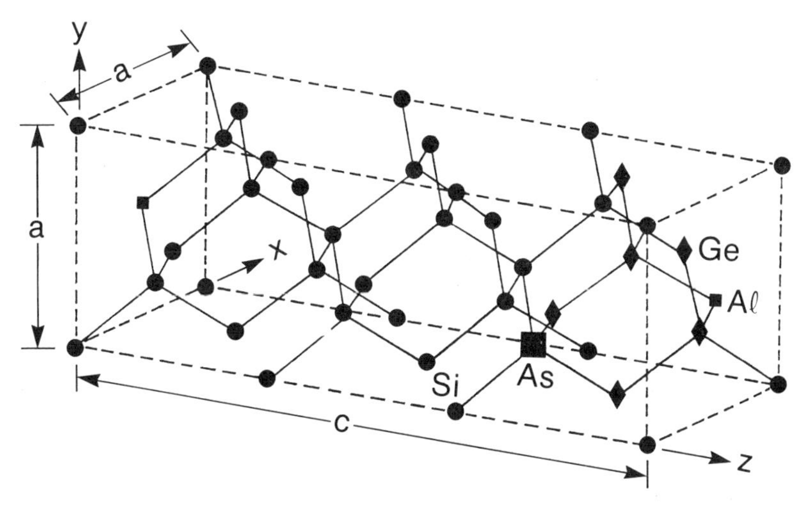

Fig. 6. The unit cell of doped Si/Ge. Si atom is indicated by a dot and Ge atom is denoted by a diamond. The Al atom is indicated by a small square while the As atom is denoted by a large square. The undoped structure is obtained by replacing impurity atoms with Si atoms.

carried out at only k = 0. We anticipated that the donor state of the As atom should be partially occupied. Therefore, in determining the self-consistent electron-electron interaction, we weigh the states at the gap by 1/2.

Table 1. The Parameters of the Ionic Pseudopotentials

Element	$a_1(Ry/a_0^2)$	$a_2(a_0^{-1})$	a_3	$a_4(a_0^{-4})$
Si	− 1.214	0.791	− 0.352	− 0.018
Ge	− 1.058	0.803	− 0.312	− 0.019
As	− 0.78	1.045	− 0.166	− 0.015
Al	− 0.63	1.047	− 0.134	− 0.029

3.2 Results and Discussion

In order to facilitate the discussions of the impurity states in doped Si/Ge, we shall present first the results of the undoped Si/Ge, especially the electronic states near the band edge. The band structure of the strained Si/Ge superlattice is plotted in Fig. 7 along two different symmetry directions. The gap is 0.19 eV. It is an indirect gap in k-space. Because of the local density functional approximation, this value is expected to be smaller than the experimental value in samples with the same atomic layers. At this moment, experimental information is lacking. However, it should be emphasized that the band ordering of the present calculations are likely to be correct.

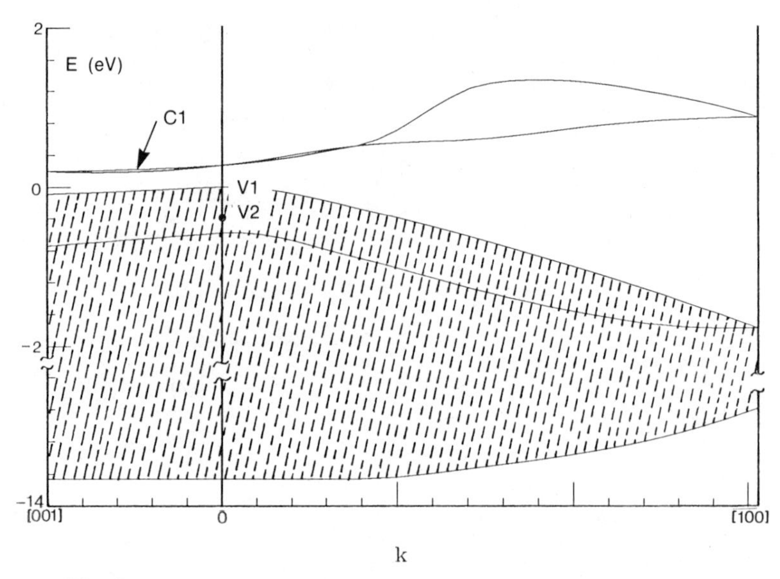

Fig. 7. Band structure of undoped Si/Ge along [100] and [001] directions. The bands C1, V1, and V2 are identified.

To determine the nature of the states at the gap, we plot in Fig. 8, the charge density of the band edge states (band 49 (C1) and band 48 (V1)) in a section of plane (100) passing through the bonds. Also shown is the charge density of another VB state, V2 (the 47-th band), in panel 8(c). The projected positions of Si and Ge atoms which form the bond in this plane are shown by the dots and the diamonds, respectively. It is clear that most of the charge is concentrated in the bond between the two Ge atoms. The charge density of the lowest CB, C1 state is mostly located in the middle of the Si-region. Therefore, the Si(10)/Ge(2) superlattice is also an indirect gap semiconductor in real space. This is consistent with the 6 layers each Si/Ge showing a staggered band alignment with the VB in the Ge-region and CB in the Si-region.[21] The energy of the V2 state is 0.54 eV below the top of the VB (V1). From the charge distributions, this state is an interface state. There are charges at the Si-Ge bonds. As we will see later, this state is mostly influenced by doping.

Before discussing the results of detailed calculations, let us discuss first the physical situation. Naively, the presence of the As atom at the left-hand side of the Ge-region will contribute a partially occupied donor state. Its energy should be below the bottom of the CB. On the other hand, the replacement of a Si atom at the right-hand side of Ge-region by the Al atom will create a hole state. Its energy should be slightly higher than one of the highest energy states of the Si-region. However, because the highest occupied state is in the Ge-region, it is unlikely to have a hole state in the middle of the VB for the ground state of the system. Furthermore, the Al atom is placed near the interface region. Its proximity to the Ge-region may cause charge transfer from the Ge-Ge bond to fill the hole state. Therefore, the naive picture should be modified. The possible modifications can be posed in the form of the following questions:

(i) Is it possible to transfer charge from the Ge-Ge bond to fill the hole state, so that the resultant hole state is Ge-related? In this case, the naive picture for the donor state may be valid.

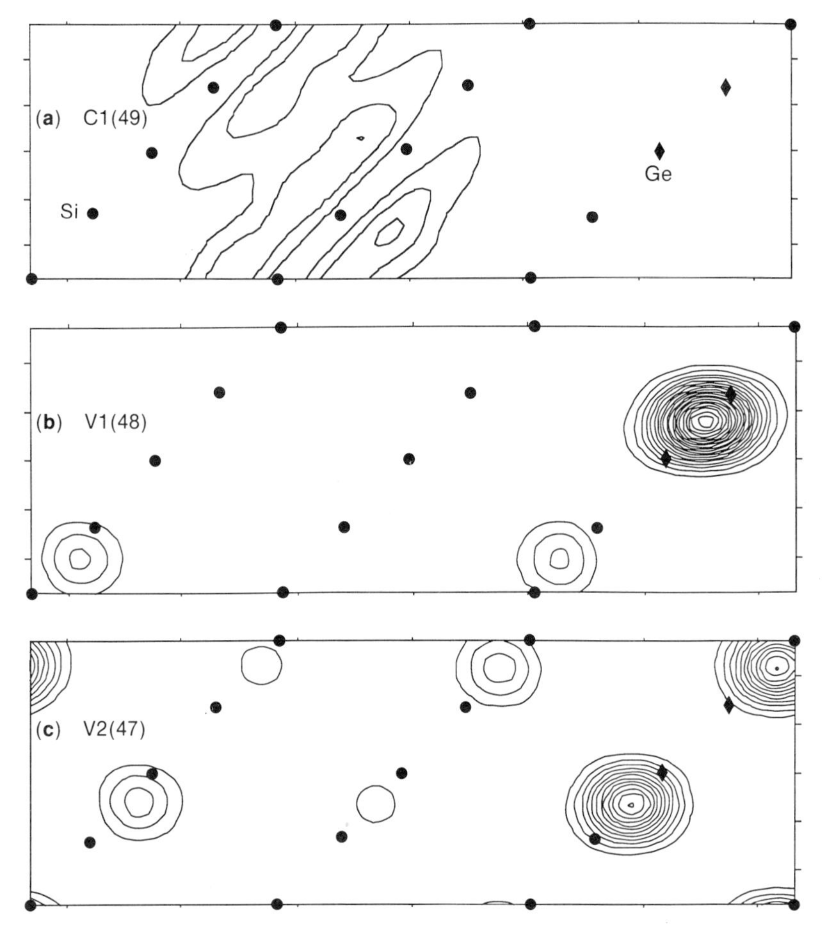

Fig. 8. (a) Charge density of the bottom of the CB, C1 (band 49) for the undoped case at k = (0,0,0.25) in a section of (100) plane with x = 0.125a. Atoms forming the bonds are projected on the plane. (b) Charge density of the top of the VB, V1 (band 48) at Γ. (c) Charge density of V2 (band 47) at Γ.

(ii) Even though the As and the Al atoms are separated apart as the third nearest neighbor, is it possible to partially transfer charge from the donor state of As to the acceptor state of Al? In this case, the Ge-Ge bonding state may not be disturbed significantly. The acceptor state is expected to be filled and above the Ge-related state for the ground state of the system. Because of the partial transfer of charges, the donor state energy may be increased. Consequently, is it possible for the donor state and the lowest conduction of the undoped sample to mix?

To discriminate the two possibilities, we performed the numerical calculations.

In Fig. 9, we show the charge distribution of the 47-th band $V2_d$. As compared to the undoped case, the results are very similar to the one given in Fig. 8(b) which corresponds to the top of the VB V1 for the undoped case. There is a slight depletion of charge in the doped case as seen from the contour between the Ge-Ge bond now being

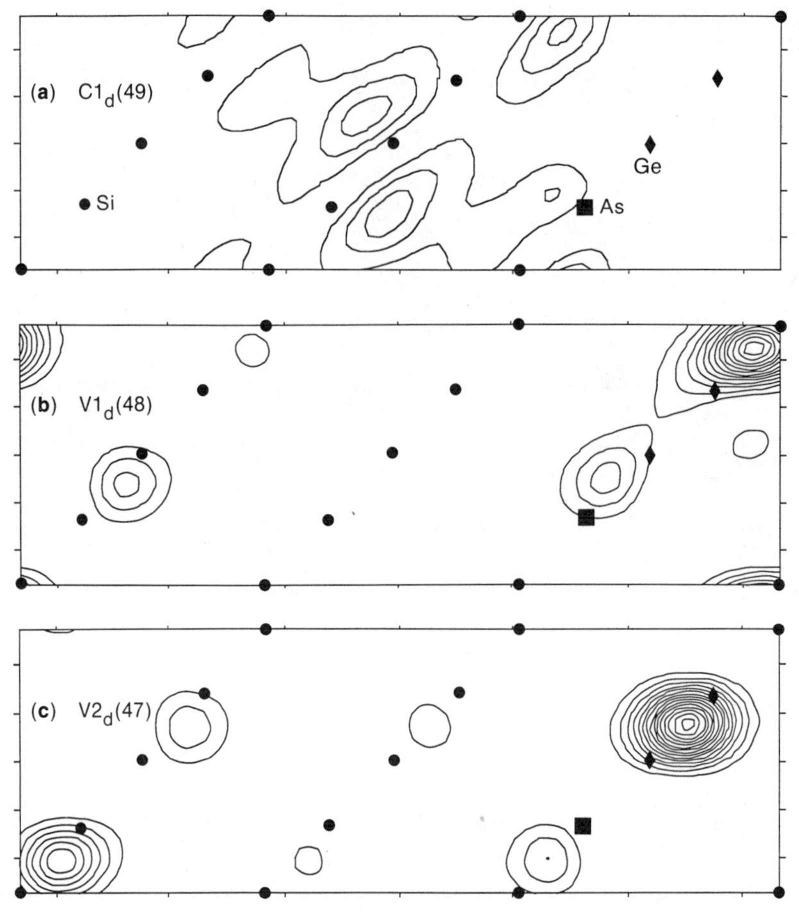

Fig. 9. (a) Charge density of the bottom of the CB, $C1_d$ (band 49) for the doped case at k = (0,0,0.25) in a section of (100) plane with x = 0.125a. Atoms forming the bonds are projected on the plane. (b) Charge density of the top of the VB, $V1_d$ (band 48) at Γ. (c) Charge density of $V2_d$ (band 47) at Γ.

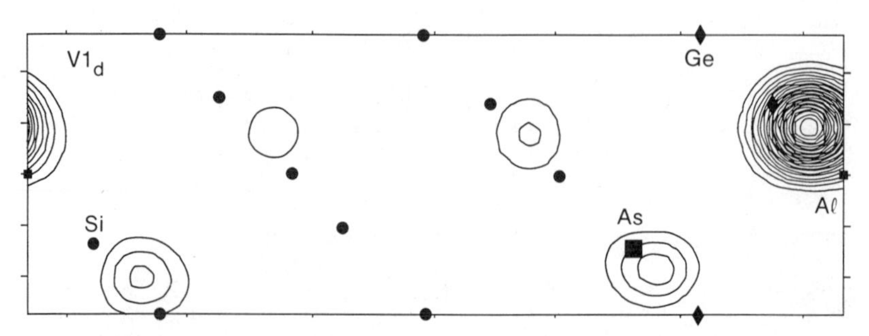

Fig. 10. Charge density of the top of the VB, $V1_d$ (band 48) for the doped case with x = 0.375a.

extended less to the right-hand side edge as compared to the one shown in Fig. 8(b). However, the character of Ge-related state is manifested clearly in Fig. 9. The charge density of the top of the VB, $V1_d$, for the doped sample is shown in Fig. 9(b). The contribution to this state comes from the As, Ge, and Si atoms at the interface. Compared to the results in Fig. 8(c), the charge distribution is not symmetric about the middle of Ge-Ge bond. This is caused by the presence of As atom. Since we anticipate this state to be associated with Al hole state, but Fig. 9 does not cut Al plane, we plot the charge density in another section in Fig. 10. In this plot, the plane is passing through the Ge-Al bond, i.e., displaced along $\hat{x}$-direction by a/8, with respect to the plane in Fig. 9. There is a bond charge between Ge and Al. The maximum of the charge is closer to the Ge atom. This state is associated with the Al atom. The picture derived from these charge distributions suggests that the whole state takes charge from the nearby Ge-Ge and Ge-Si bonds. These bonds, in turn, make up the loss of charge from the extra charge of the As atom. The Coulomb interaction of the doubly occupied state pushes up the energy of the Al related state. On the other hand, the depletion of some of the charge from the Ge-Ge bond lowers the energy of the corresponding state. The energies of the 47-th and 48-th bands are nearly degenerate. Therefore, the second possibility mentioned above is the modification after doping the sample.

The lowest conduction band of the doped sample is shown in Fig. 9(a). As compared to the results in Fig. 8(a), this case shows clearly that there are charge contributions from the As atom. Because the charge of the donor state contributed by the As atom has been partly transferred to the neighboring Ge-Ge and Ge-Si bonds, its binding energy is less than the case of a single donor state. Therefore, there is a strong mixing of the charge from the bottom of CB of the undoped case with the less bound donor state. Consequently, the gap energy of the doped sample is larger than the undoped one. The doped sample is still a semiconductor instead of a metal due to the charge transfer.

In summary, the staggered band alignment of strained Si/Ge superlattice provides many interesting ways of doping. We have studied the simplest n-i-p-i type doping. Because the donor and acceptor atoms are separated only by the third nearest neighbor distance and the acceptor atom is in the Si-region, there is charge transfer from the donor state to the acceptor state. Comparing to the undoped case, this acceptor state is completely filled and becomes the top of the VB. On the other hand, the donor state increases its energy and mixes strongly with the bottom of the CB for the undoped case.

4. CONCLUDING REMARKS

The model doped calculations based on first principles point to some shortcomings of the EMA. For GaAs-AlAs, the possibility of tuning the band edge discontinuity by dopants is verified. For Si/Ge doped with n- and a p-type donor simultaneously leads to no new impurity related states in the gap region. These results suggest further studies such as to increase the separation and the concentration of impurity atoms will be valuable.

ACKNOWLEDGMENTS

We would like to thank Mr. L. H. Yang and J. S. Nelson for performing some of the calculations. The work at Davis was supported by San Diego Supercomputer Center.

REFERENCES

1. G. Bastard, Phys. Rev. B24:4714 (1981).
2. C. Mailhoit, Yia-Chung Chang, and T. C. McGill, Phys. Rev. B26:4449 (1982).
3. R. L. Greene and K. K. Bajaj, Solid State Commun. 45:825 (1983).
4. C. Priester, G. Bastard, G. Allan, and M. Lannoo, Phys. Rev. B30:6029 (1984).
5. P. Csavinszky and A. M. Elabsy, Phys. Rev. B32, 6498 (1985).
6. G. P. Srivastava, These Proc.; U. Rössler, These Proc.
7. S. T. Pantelides and C. T. Sah, Phys. Rev. B10:621 (1974).
8. S. T. Pantelides and C. T. Sah, Phys. Rev. B10:638 (1974).

9. J. Bernholc and S. T. Pantilides, Phys. Rev. B15:4935 (1977).
10. F. Capasso, K. Mohammed, and A. Y. Cho J. Vac. Sci. Technol. B3:1245 (1985).
11. J. S. Nelson, C. Y. Fong, I. P. Batra B. M. Klein, and W. E. Pickett, Phys. Rev. B (to be published).
12. M. Schlüter, J. R. Chelikowsky, S. G. Louie, and M. L. Cohen, Phys. Rev. B12:4200 (1975); J. Ihm, A. Zunger, and M. L. Cohen, J. Phys. C12:4409 (1979); M. T. Yin and M. L. Cohen, Phys. Rev. Lett. 45:1004 (1980); K. C. Pandey, ibid. 49:223 (1982); I. P. Batra and S. Ciraci, Phys. Rev. B33:4313 (1986).
13. G. B. Bachelet, D. R. Hamann, and M. Schlüter, Phys. Rev. B26:4199 (1982).
14. I. P. Batra, S. Ciraci, and J. S. Nelson, J. Vac. Sci. Technol. B5:1300 (1987).
15. F. Ciedeira, A. Pinczuk, J. C. Bean, B. Batlogg, and B. A. Wilson, App. Phys. Lett. 45:1138 (1984).
16. G. Abstreiter, H. Brugger, T. Wolf, R. Zachai, and Ch. Zeller, in: "Two-Dimensional Systems: Physics and New Devices," G. Bauer, F. Kuchar, and H. Heinrich, eds., Springer-Verlag, Berlin, 130 (1986).
17. C. G. Van de Walle and R. M. Martin, Phys. Rev. B34:79 (1972); B52:553 (1972).
18. G. H. Döhler, Phys. Status Solidi B52:79 (1972); B52, 553 (1972).
19. Warren E. Pickett, Steven G. Louie, and Marvin L. Cohen, Phys. Rev. B17:815 (1978).
20. K. M. Ho, Marvin L. Cohen, and M. Schlüter, Phys. Rev. B15:3888 (1977); also see Ref. 12.
21. J. S. Nelson, private communication.

PROPERTIES OF IMPURITY STATES IN n-i-p-i SUPERLATTICE STRUCTURES

Gottfried H. Döhler

Institut für Technische Physik
Universität Erlangen
Erwin-Rommel-Str. 1, 8520 Erlangen, FRG

Abstract

The primary effect of the static potential of the impurity atoms in doping superlattices is the formation of space charge induced quantum wells. In most of our previous studies of n-i-p-i doping superlattices we have focussed our interest to the remarkable features which result from the spatial separation between electrons and holes, such as tunability of the electronic structure, electron-hole recombination lifetimes, increased by many orders of magnitude compared with those of bulk semiconductors, or huge optical non-linearities. In this lecture we will concentrate on the point defect aspects of impurities in doping superlattices. Topics to be discussed will include the impurity band formation by shallow (donor) and less shallow (acceptor) impurities over the whole concentration range from very low to high concentrations. We propose various n-i-p-i and hetero n-i-p-i structures which should be idealy suite for optical studies of impurity- and Hubbard bands, and for electrical investigations of the density of states of the conductivity in these bands as a function of (tunable) carrier and dopant density. In particular it is expected that these structures represent unique systems for investigations of the Mott-Hubbard transition in two dimensions.

I. Introduction

Since the first proposal of n-i-p-i doping superlattices as crystals with tunable electrical properties and their first theoretical investigation[1,2] a considerable amount of experimental work has been performed, which confirmed the theoretical predictions[3,4]. These results have stimulated further theoretical investigations and proposals for new experimental studies[5-13]. Although all the remarkable features of the n-i-p-i doping superlattices derive from impurities, almost no attention had been paid to the impurity **states** in these structures. We have almost all the time eliminated them from our discussion by replacing the random distribution of the point charges within the two- or three-dimensional doping layers by a uniform two- or three-dimensional charge density. In this way a one-dimensional mesoscopic periodic potential is obtained and the well-known tunable electronic structure of n-i-p-i superlattices with tunable carrier concentration, band-gap, subband spacing, and lifetimes is obtained for this system with purely space-charge induced quantum wells. In fact, also some of the properties which are related to the impurity states and to their random distribution have at least qualitatively been discussed in the past[1, 5].

In this paper we first review some of the theoretical and experimental problems associated with the impurity atoms and their random distribution. In the following section we discuss possibilities to overcome these problems. In the last section, finally, we propose new hetero - n-i-p-i - structures, which appear ideally suited for experimental and theoretical investigations of impurity and Hubbard band formation and, in particular, of the Mott-Hubbard transition in two dimensions.

2. Problems associated with the point-charge character and the random distribution of impurities in doping superlatices

Fig. 1 shows an example of self-consistently calculated band-gap, subband energies, and electron and hole quasi Fermi-levels as a function of two-dimensional carrier density $n^{(2)}$ in a n-i-p-i crystal. As mentioned before, the space charge of the impurities has been assumed to be uniformly distributed within the respective layers. The quasi Fermi-level of the holes Φ_p has been taken as a constant reference energy. Formally, the impurity state associated with an **additional** impurity can be calculated. It is interesting to note, that for the present case of a three-dimensional doping layer, the energy of impurity states as a function of their position with the layer would approximately increase in parallel with the potential itself. Therefore, they would become resonant states of the subbands at relatively small distances from the center of the doping layer. This problem does not arise in the case of δ-doped layers. For this special case the energy of impurity levels in a n-i-p-i crystal and the increase in binding energy with increasing internal space charge fields had been discussed first in our original paper[1]. More detailed studies have been given by Crowne, Reinecke and Shanabrook[14].

In order to estimate the significance of these calculated impurity energies, one has to compare them with the potential fluctuations which are present in the layers as a result of the random distribution of impurity atoms. The fluctuations of this random potential, given by the expression

$$v_{rand}(r) = - \sum_{\substack{\text{all ionized}\\ \text{donors}}} e^2/(\kappa_0 |r-r_i|) + \sum_{\substack{\text{all ionized}\\ \text{acceptors}}} e^2/(\kappa_0 |r-r_j|) \qquad (1)$$

can become very large in the case of a compensated n-i-p-i

crystal in the ground state. With increasing population of the layers by electrons and holes, respectively, the amplitude of these fluctuations decreases rapidly.

It turns out, that the effect of increasing free carrier concentration is very strongly dependent on the effective mass of the charge carriers. The relatively large effective mass of holes in the valence bands of most crystals results in relatively large acceptor ionization energies and a relatively small effective Bohr radius. Thus, a treatment of the effect of increasing population in terms of step by step neutralization of the deepest lying states in this random potential is appropriate. From computer simulations[15] it follows that the potential fluctuations become smaller than the acceptor binding energy at a relatively low ratio of the (2-dimensional) hole to acceptor concentration $p^{(2)}/n_A^{(2)}$ even at rather high values of $n_A^{(2)}$. For the much lower effective masses of electrons in the conduction band of direct gap semiconductors the situation is substantially different. A considerably larger ratio of electron to

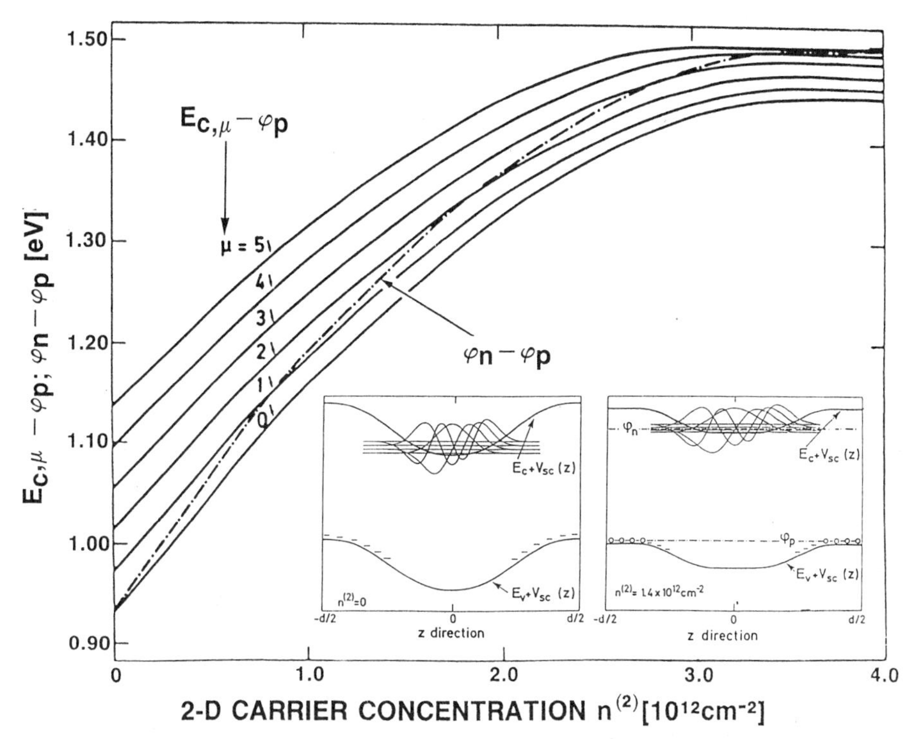

Fig. 1 Relation between (two-dimensional) carrier concentration in the layers and effective band gap and quasi Fermi level difference, calculated self-consistently for a GaAs doping superlattice with $n_D = n_A = 10^{18} cm^{-3}$ and $d_n = d_p = 40nm$. The inset shows the envelope wave functions $\zeta_{c\mu}(z)$ for the lowest subbands for the ground state and for an excited state with $n^{(2)} = 1.4 \times 10^{12} cm^{-2}$.

acceptor concentration $n^{(2)}/n_D^{(2)}$ is required in order to obtain an amplitude of the random potential (Eq. (1)) less than the donor binding energy at the corresponding donorconcentration $n_D^{(2)}$. It is found that at the typical donor concentrations used in n-i-p-i crystals the kinetic energy effects become already very important. The Fermi-energy of a two-dimensional electron gas

$$\varepsilon_F = (\hbar^2/2m_c)(2\pi n^{(2)}) \tag{2}$$

becomes larger than the donor binding energy at these concentrations. This makes qualitatively plausible that the donor impurity potentials will no longer be strong enough for "condensation" of the electrons in an impurity band. Thus, impurity bands will no longer play a significant role in this case. The appropriate picture is a free electron gas which is screening the random potential of ionized donor potentials[5].

In contrast, the kinetic energy of heavy holes in the valence band becomes comparable to the acceptor ionization energy only at fairly high concentrations. It turns out, that, up to sheet carrier concentrations of $p^{(2)}$ of the order of about $10^{19}cm^{-3}$ the description in terms of an impurity band remains appropriate for acceptors.

From our previous discussion it becomes obvious, that all the energy levels observed in transitions between conduction subbands and the impurity band or some valence subbands exhibit significant broadening due to the imperfect screening of the space charge potential fluctuations. Also, it becomes understandable, that a correct treatment of these problems is quite complicated.

2.2 Acceptor impurity band wave functions

Transitions between conduction and valence subbands can be calculated straight forward by evaluation of the corresponding dipole matrix elements[16]. From our previous discussion, however, it follows that holes will populate not valence subbands, but rather the acceptor impurity band. A calculation of these transitions poses new problems. First of all, the wave function of an impurity band is not known. But even, if it is approximately replaced by a suitable superposition of acceptor wave functions, the following problem still remains unsolved. The maximum contribution to the dipole matrix element for valence to conduction band transitions comes from a rather narrow region somewhere between the centers of two adjacent doping layers. The exact position depends on the shape of the self-consitent potential and the effective masses of the electrons, light and heavy holes. Calculations of the **bulk** acceptor states have been performed by a variational method involving light and heavy hole states[17]. Although these calculations yield good values for the acceptor **energies**, it is questionable whether the wave **functions** obtained by this method are in a satisfactory manner describing their amplitude far away from the acceptor. This, however, is just the relevant range of distances, for a calculation of the transitions between the acceptor and the conduction subbands.

3. Possibilities to avoid impurity related problems in n-i-p-i doping superlattices

3.1 Weakly compensated δ-doped n-i-p-i - structure with low doping concentrations

In order to be specific, we consider a structure as shown in Fig. 2. This n-i-p-i design resembles the one considered in our original work[1,2]. We assume an acceptor concentration much lower than the donor concentration. All acceptors will be ionized in the ground state. The ionized donors will be those which are closest to the negatively charged acceptors.

In the following we will discuss this structure for the case of much lower doping levels than previously considered. In terms of tunable properties this structure becomes quite uninteresting. The intriguing aspects result from this structure representing an ordered version of the well-known donor acceptor pair luminescence[18]. The photon-energy emitted in a recombination between an electron on a donor and a hole on an acceptor is given by[18]

$$\hbar\omega = E_g{}^0 - E_D - E_A + e/\kappa_0 r_{DA} \tag{3}$$

(this expression can be easily understood by considering the total energy difference between the initial state, characterized by a neutral acceptor and the electron sitting on a neutral donor, and the final state, characterized by the electron sitting on the acceptor and experiencing the negative electrostatic potential of the donor in the distance r_{DA}, given by $e^2/\kappa_0 r_{DA}$. The transition probability for this process is determined by the overlap between the donor and acceptor wave function.Therefore, it depends exponentially on the distance r_{DA},[18]

$$W(r_{DA}) = W_0 \exp-(r_{DA}/a^*). \tag{4}$$

a^* is not much different from the effective Bohr radius of the less localized impurity involved.

In a uniform bulk crystal with a random distribution of donors and acceptors the photo-luminescence is determined by the capture and recombination of photo-excited electrons and holes. The Coulomb term in Eq. (3), of course, leads to a considerable broadening of the donor-acceptor pair luminescence spectra. The contribution of the pairs with low distance r_{DA} is, of course, particularly strong. These transitions also cannot easily be saturated because of the high transition probabilities at low distances (see Eq. (4)). The width of the luminescence line associated with donor-acceptor transitions can be estimated from the distribution function of donors neighbouring the acceptors (also for the bulk case we assume n_D much larger than n_A). The probability of finding the next nearest donor within a distance interval (r, r+dr) is given by

$$p(r)dr = 3(r^2/r_0{}^3) \exp-(r/r_0)^3 dr \tag{5}$$

Two third of all nearest neighbours are found within the distance interval between r_- and r_+, with r_-, r_+ and r_0 given by

$$r_- = 0.567\, r_0 \tag{6a}$$
$$r_+ = 1.215\, r_0 \tag{6b}$$
$$r_0 = (4\pi n_D/3)^{-1/3} \tag{6c}$$

r_0 corresponds to the average nearest neighbour distance.

Inserting the values for r_- and r_+ into Eq. (3), we find that two thirds of all transitions are found within the energy interval

$$\Delta\hbar\omega^{(3)} = e^2/\kappa_0 r_- - e^2/\kappa_0 r_+ \tag{7}$$

For GaAs with a donor concentration of 10^{16} cm^3 we find

$$\Delta\hbar\omega = 7.1 - 3.3\text{meV} = 3.8\text{meV} \tag{8}$$

From Fig. 2 it is obvious that we suppress all the transitions at distances less than d_i by introducing our one-dimensional ordering. We can think of our n-i-p-i - structure as being obtained by a projection of all the impurities within a superlattice period $d = 2d_i$. This results onto planes separated by the distance d_i in sheet impurity concentrations of

$$n_D^{(2)} = n_D d \tag{9}$$

$$n_A^{(2)} = n_A d. \tag{10}$$

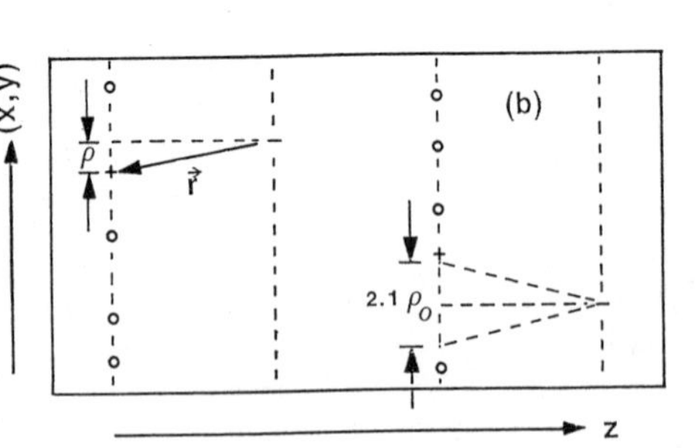

Fig. 2 n-i-p-i structure, as considered in our original work[1,2]. In the upper part the doping profile along the direction of growth, z, is shown. In the lower part a real space picture is shown schematically. The length of the vector r pointing from a (minority) acceptor state to its nearest (majority) donor neighbour is $r = (d_i^2+\varrho^2)^{1/2}$. Approximately 2/3 of all neighbouring donors are found within a circle of diameter $2.1\varrho_0$ centered at the normal projection of the acceptor position onto the neighbouring n-layers. In the ground state (depicted here) all acceptors are negatively charged, A^-, whereas the nearest donors are positive, D^+. In the fully excited state all the impurities are neutral.

Now the probability of finding the nearest donor within a

distance r (see Fig. 2) becomes strongly peaked near d_i. For the probability of finding the next donor within a distance interval (ϱ, $\varrho+d_\varrho$) within the two-dimensional doping layer away from the nearest distance point is now

$$p(\varrho)d\varrho = 2(\varrho/\varrho_0{}^2)\exp-(\varrho/\varrho_0)^2 d\varrho \tag{11}$$

with

$$\varrho_0 = (2\pi n_D{}^{(2)})^{-1/2} \tag{12}$$

(the factor 2 in Eq. (12) results from the fact that each acceptor is facing **two** neighbouring donor layers). Quite analogous to our previous three-dimensional considerations we now define quantities $r_- = d_i$ and $r_+ = [d_i{}^2+(\varrho_{1/3})^2]^{1/2}$ such, that two thirds of all the nearest donors are found within this interval. Thus, we obtain $\varrho_{1/3} = 1.05\varrho_0$. For the luminescence linewidth we find therefore

$$\Delta\hbar\omega^{(2)} = e^2/\kappa_0 - e^2/\kappa_0[d_i{}^2 + (\varrho_{1/3})^2]^{1/2} \tag{13}$$

For the n-i-p-i version of our previous example we find therefore

$$\Delta\hbar\omega^{(2)} = 2.34 - 2.26\text{meV} = 0.08\text{meV} \tag{14}$$

Comparing the values from Eqs. (14) and (8), we observe a dramatic reduction in impurity pair luminescence linewidths. We believe that this strong narrowing of the linewidth, indeed, should be observable. The only energy in a luminescence experiment which could interfere with the donor acceptor pair luminescence could be due to excitons bound to neutral acceptors. This contribution to the luminescence would have a much faster decay after excitation compared with the donor acceptor pair luminescence, which is given by Eq.(4) with $r_{DA} \simeq d_i$.

So far, we have neglected the Coulomb broadening of these lines due to other non-neutral donor-acceptor pairs. This broadening will be negligible if the donor acceptor transitions are saturated or almost saturated. Because of the uniform distribution of transition probabilities, which, at the same time are relatively small (for our present example they can be estimated to be of the order of 10^4 s^{-1}) saturation can be achieved with moderate excitation intensities quite easily.

Apart from the narrowing of the luminescence spectra, which is interesting by itself, this new system offers the exciting possibility to study the formation (of 2-dimensional) impurity bands. If we study a set of samples in which only the donor concentration in the n-layers differs, we expect a broadening of the luminescence spectra when the donor energy level distribution broadens due to the formation of an impurity band. The observed shape of the luminescence spectrum should reflect directly the density of state distribution $N(\varepsilon)$ of the impurity band. Note, that the acceptors are randomly distributed in the p-layers and, therefore, probe the impurity band independently on the special donor distribution within the n-layer.

A more elegant way of doing this study is probably to use one single samle with an impurity concentration higher than required for the impurity band formation. In this case, the impurity band formation can now be studied as a function of a magnetic field B_z applied perpendicular to the doping layers. The formation of the impurity band becomes now a function of the ratio ϱ_0/a_D with the effectice donor Bohr radius a_D now decreasing with increasing magnetic field[20,21].

We have discussed here only the case of donor band formation. Note, that similarly also the formation of an acceptor band can be studied. Instead of the GaAs system, which was discussed here as an example it seems attractive, or even advantageous, to study other systems like GaP, the classical example for donor-acceptor pair luminescence.

3.2. IV-VI n-i-p-i's.

The IV-VI compound semiconductors have a very large dielectric constant and very low (average) effective masses for both, electrons and holes. As a consequence, no bound impurity states exist in these crystals. Therefore, the properties of IV-VI compound n-i-p-i crystals are not unfavorably affected by impurity states. The intriguing properties of this class of n-i-p-i structures has been discussed in detail in Ref. 6.

3.3. Hetero - n-i-p-i - structures

A third possibility **to get rid** of the impurity states in doping superlattices is based on the hetero n-i-p-i concept.

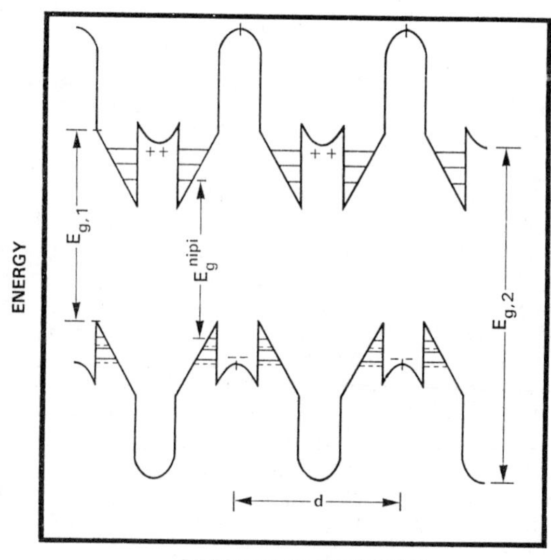

Fig. 3 Real space band diagram of a type-II hetero n-i-p-i. n- and p-doping is confined to the central part of the larger band gap material with band gap $E_{g,2}$. Electrons and holes are spatially separated from each other but also from their parent donors or acceptors, respectively.

A special version, a type-II hetero - n-i-p-i, is shown in Fig. 3. It is obvious that electrons and holes are now spatially separated, not only from each other but also from their parent donors and acceptors, respectively. The advantage is two-fold. First, the influence of potential fluctuations as a broadening mechanism for the subbands is now strongly reduced. Second, holes no longer populate acceptor impurity bands. Instead, they reside in hole subbands whose properties are readily accessible by the theory[22]. Recent photoluminescence and time-resolved luminescence measurements have, in fact, demonstrated much narrower luminescence lines[23]. Also distinct lines due to recombination from excited subbands could clearly be distinguished. In the next section we will discuss some specific properties of hetero - n-i-p-i - structures with very low doping concentrations which have not been considered previously.

4. Study of (2-dimensional) impurity- and Hubbard-bands in hetero - n-i-p-i's

In section 3.1 we discussed a new possibility to observe the density of states of impurity bands, $N(\varepsilon)$, for the special case of a completely or nearly completely filled impurity band. In this section we outline that hetero - n-i-p-i's provide a unique possibility to study impurity bands **as a function of their population**. This is, of course, a consequence of the tunability of the carrier concentration in n-i-p-i - structures. We will find that the situation in hetero - n-i-p-i's of appropriate design differs from that in the conventional n-i-p-i's by the fact that the tunability of the carrier density is not limited to the range between 0 and the density which just neutralizes all the impurities. We can increase the carrier density even beyond this limit. This implies the possibility to populate the "Hubbard-band"[lit z], formed by double occupancy of the impurity states. The most interesting aspect of this system is perhaps the possiblity to have a unique system for studying the Mott-Hubbard-transition[24] as a function of ϱ_0/a_D or ϱ_0/a_A, similar to the case of the simple impurity band formation in section 3.1. In this respect, our present system differs from any other system we are aware of. We do not see any possibility to design a similar bulk system in which the conductivity and the density of states could be investigated in the range of population higher than one carrier per impurity (we exclude the case of impurities which exist in a singly and doubly ionized state and which are separated by a large energy difference. The Mott-Hubbard transition can probably never be studied in such systems).

4.1. Hetero - n-i-p-i for the study of (2-dimensional) conduction in impurity and Hubbard-bands

Fig. 4 shows the band diagram for the conduction and valence bands of a hetero - n-i-p-i in the ground state (b), the excited state with neutral impurities (c), and with the Hubbard-band populated (d). The doping profile is indicated in part (a). The doping concentration in the p-layers is assumed to be large enough to prevent carrier freeze-out at low temperatures. The width of the layer with the n-type δ-doping at its center is choosen such, that the quantized energy levels are well separated. The binding energy E_D^1 of the donors is enhanced compared to the pure bulk material by a factor between one and two, depending on the well width.

In the neutral donor state (c) the difference between the electron and hole quasi Fermi levels, Φ_n and Φ_p, at zero temperature is just

$$\Phi_{np}^0 \sim E_g^1 + E_{c,0} - E_D^1 - E_A. \tag{15}$$

This state can be reached by applying an external potential

$$eU_{np}^0 = \Phi_{np}^0 \tag{16}$$

between selective n- and p-type contacts to the hetero n-i-p-i[4,26,27]. We can avoid significant leakage currents between the n- and p-layer if we choose sufficiently large values of the superlattice period d and the barrier heights ΔE_c and ΔE_v, respectively. If U_{np} differs from the value determined by Eqs. (15) and (16), the sheet electron density in the layer varies according to

$$n^{(2)} = n^{(2)}_D + (U_{np} - U^0_{np})\,\kappa_0/\pi ed. \tag{17}$$

For $n_D^{(2)} = 10^{11}$ cm^{-2} and d = 100nm, e.g., $n^{(2)}$ vanishes in AlGaAs/GaAs hetero n-i-p-i for

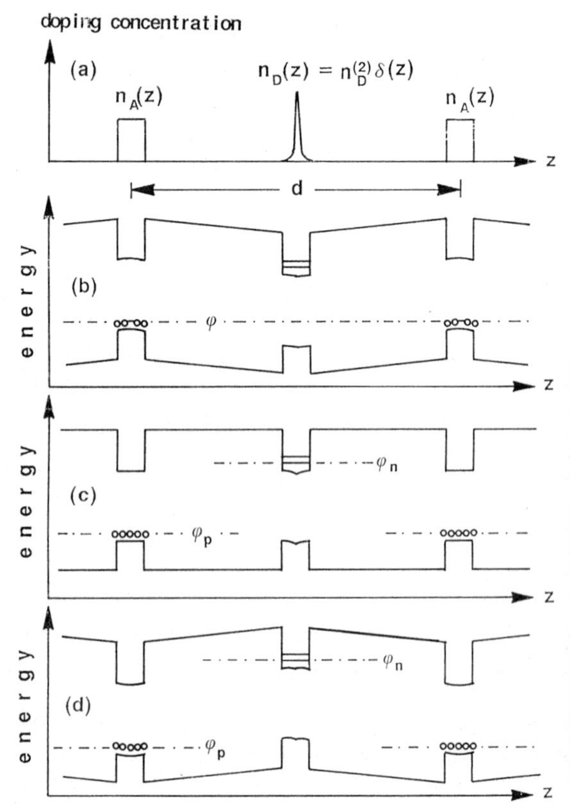

Fig. 4 Hetero n-i-p-i for the observation of (2-dimensional) impurity conduction in the neutral-impurity - and in the impurity - Hubbard-band as a function of sheet electron concentration $n^{(2)}$ in the interval $0 \lesssim n^{(2)} \lesssim 2n_D^{(2)}$. Part (a): doping profile. Parts (b) - (d): Schematic band profiles for different sheet electron concentration. (b): Ground state ($n^{(2)} = 0$); (c): Neutral-donor state ($n^{(2)} = n_D^{(2)}$): (d): Hubbard band completely filled ($n^{(2)} = 2n_D^{(2)}$). $n^{(2)}$ can be tuned by a bias U_{np} applied between selective n- and p-type contacts which inject cariers until $\Phi_n - \Phi_p = eU_{np}$ is satisfied. The conductivity in the n-layers will be measured by a small bias between two n-type contacts.

$U_{np} - U^0_{np} < -35mV$

At $U_{np} - U^0_{np} > 0$, on the other hand, D^- centers will be formed in the n-layers by the excess electrons, provided that $n_D^{(2)}$ is less than the critical density $n^{(2)}_{D,MH}$ for the Mott-Hubbard transition. These D^- centers form the Hubbard impurity band. The center of this band is shifted to higher energies, compared to the center of the D^0 impurity band by the repulsive correlation energy U_H.

For our example we find, that at a voltage

$$U_{np} = U^0_{np} + 35mev \qquad (19)$$

the number of electrons in the GaAs layers has increased up to $n^{(2)} = 2n_D^{(2)}$, which corresponds to a completely filled Hubbard band. This means, that we can shift the position of the electron quasi Fermi level from the lower edge of the D^0 impurity band up to the upper edge of the D^- Hubbard band by varying the voltage U_{np} between selective n- and p-contacts by a rather small amount (~ 70mV in our example). Over the whole range the leakage currents through the $Al_xGa_{1-x}As$ barriers are expected to be very low at very low temperatures.

In order to study the conduction processes in the D^0 and the D^- bands one has only to measure the current I_{nn} due to a small bias U_{nn} applied between two selective n-type contacts with the voltage U_{np} as a parameter[26,27]. A schematic picture of the density of states and the position of the electron quasi Fermi level Φ_n as a function of sheet electron density $n^{(2)}$ is shown in Fig. 5.

At low temperatures we expect to observe the Mott-Anderson metal-insulator transition at low $n^{(2)}$ provided that $n_D^{(2)}$ is not too low. The most interesting tuning range, however, is $n^{(2)} \simeq n_D^{(2)}$, i.e., from slightly below to slightly above this value. Depending on the doping level $n_D^{(2)}$ and/or the strength of a magnetic field B_z perpendicular to the layers, the Mott-Hubbard metal-insulator transition due to the breakdown of the Hubbard gap at sufficiently strong interaction between neighbouring donor states, can be studied by conductivity measurements as a function of the position of Φ_n.

4.2 Hetero n-i-p-i for the optical investigation of the density of states in the impurity- and the Hubbard band

The structure shown schematically in Fig. 6 consists of weakly n-doped quantum wells with built-in electric space charge fields due to the space charge in the comparatively high doped n- and p-layers (see part (a) of Fig. 6 for the doping profile). In the ground state there are no carriers in the tilted quantum wells. Excitation with photons with an energy above the effective band gap, E_g^{eff} creates electron-hole pairs within the tilted quantum wells.

$$\hbar\omega > E_g^{eff,0} = E_g^1 + E_{c,0} - E_D^{1,} - eFd^1 + E_{vh,0} \qquad (20)$$

During the relaxation process the electron-hole pairs become spatially separated. The electrons fill up the donor impurity bands, first the D^0- impurity and subsequently the D^--Hubbard band until a steady state between generation and recombination processes is reached. The recombination processes of interest are the luminescent tunneling recombinations between the electrons in the donor bands and the holes in the uppermost heavy-hole subband. The shape of the luminescence lines reflects the density-of-states distribution $N(\varepsilon)$ of the D^0 and D^- donor bands, according to

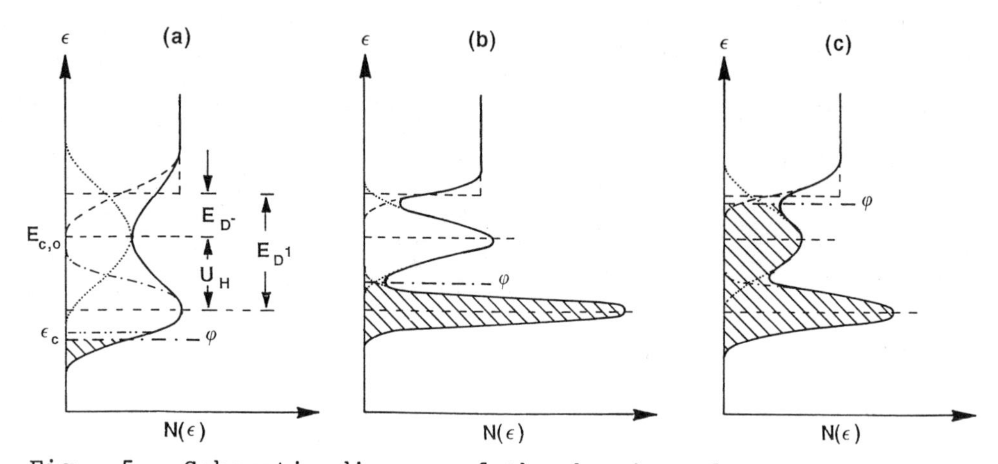

Fig. 5 Schematic diagram of the density of states of the lowest 2-dimensional subband with impurity bands due to neutral, singly occupied, and doubly occupied centers. Although the centers may be acceptors as well, the situation is shown for donors. Also, excited levels of the neutral and the doubly occupied donors are omitted. Part (a): Low sheet electron density ($n^{(2)} << n_D^{(2)}$). Only a small fraction of the donors is neutralized by electrons. Therefore, the conduction subband (dashed line), the neutral-donor impurity band (dash-dotted line), centered at about E_D^1 (the renormalized donor binding energy in the potential well) below the conduction subband, and the Hubbard band (dotted line), centered at the D^--energy, i.e. at the Hubbard correlation energy U_H above the D^1 level, they all are broadened due to the random potential of the ionized D^+- centers. The full line depicts the total density of states curve $N(\varepsilon)$. The Fermi level, Φ, is assumed to lie below the mobility edge, ε_c (which, in 2-dimensional systems, separates weakly localized states from strongly localized ones). The position of E_c depends on the product of effective Bohr radius, a_D, and sheet doping concentration, $n_D^{(2)}$. Part (b): Same, however, for $n^{(2)} = n_D^{(2)}$. Almost all donors are neutral. The bands are no longer broadened by the random Coulomb potential of D^+-centers. At high values of $a_D n_D^{(2)}$ the Mott-Hubbard insulator metal transition is expected to occur due to a collapse of the Hubbard gap[24]. Part (c): Same, however, for $n^{(2)} = 2n_D^{(2)}$. There is, again, Coulomb broadening present, due to the charged D^- centers. At intermediate sheet electron concentrations, $n^{(2)}$, upon tuning the Fermi level moves back and forth between regions of strong and weak localization, and, therefore, between regions of zero and finite conductivity at zero temperature[28,31].

their population, if the dispersion of holes populating the upper heavy-hole subband can be neglected. This requires a low hole concentration in the tilted quantum well even for the case that the D^0 and the D^- donor bands are filled with $2n_D^{(2)}$ electrons. To meet this goal the height of the barriers between the tilted quantum wells and the neighbouring n- and p-type quantum wells has to be chosen appropriately.

The probabilities for escaping from the tilted quantum wells by tunneling are proportional to exp $\{-4/3[m(\Delta E-E_0]^{3/2}/\hbar eF$, where the appropriate values for the effective mass, m, band edge of-set, ΔE and lowest bound state energy, E_0, have to be chosen for electrons and heavy holes, respectively. With the values for the system $Al_xGa_{1-x}As/GaAs$, $\Delta E_c/\Delta E_v = 65/35$, $m_c/m_{hh} = 0.168$ and $E_0 \propto [(eF)^2/m]^{\frac{1}{3}}$ they turn out to be nearly identical for electrons and heavy holes, if the Al-content is the same in both types of barriers. It should be noted that the dependence on the ratio $\Delta E_c/\Delta E_v$ is very sensitive. It might even be useful for an accurate determination of Hetero junction band off-sets.

The design of the structure as shown in Fig. 6 allows for an external recombination of electrons and holes which have escaped from the tilted quantum wells via external selective contacts to the n- and p-type doped quantum wells. Moreover, the built-in electric field, F, can be modified by applying a

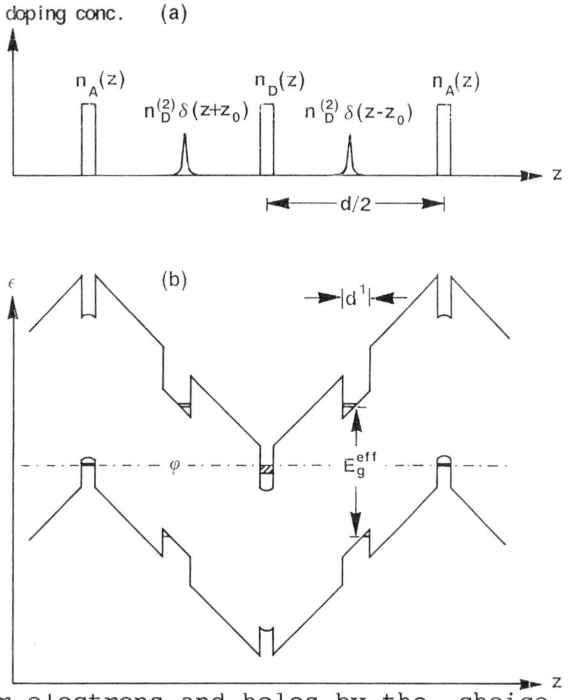

Fig. 6 Hetero n-i-p-i for the optical investigation of impurity and Hubbard bands in 2-diminsional systems. Part (a): Doping profile. $n_A(z)$ and $n_D(z)$ cause the built-in electric fields, F. The doping $n_D^{(2)}$ causes the impurity band which is subject to the optical investigations. Part (b): Band profile in the ground state (all donors in the tilted quantum well ionized, no holes in the valence subbands). The escape probability for electrons and holes generated by optical absorption in the tilted quantum wells determines the steady-state electron/hole concentration ratio $n^{(2)}/p^{(2)}$. It can be tailored independently for electrons and holes by the choice of the band gap in the neighbouring barrier regions. The escape probability can be made small by chosing large values of the superstructure period d. Moreover, it can be influenced strongly by variation of a bias U_{np} applied between selective n- and p- type contacts to the strongly doped n- and p-type quantum wells.

bias between these contacts. This allows to optimize the radiative lifetime within the tilted quantum wells by varying the overlap between the donor impurity band and the heavy hole subband envelope wave functions. Optimization means long enough to avoid broadening of the luminescence spectra by "hot" carriers, recombining before complete relaxation into the donor impurity bands. It also means short enough to observe strong enough luminescence intensity.

The design of the structure shown in Fig. 6 is also chosen such, that the luminescence spectra are not broadened by any close-by ionized impurities. With a thickness of the barriers of the order of 100nm the influence of the donors and acceptors in the n- and p-type quantum wells are negligible. This is true, in particular, as we assume strong enough doping levels in the n- and p-type quantum wells to guarantee screening by free carriers (moderate to high doping levels are required also in order to prevent carrier freeze out for a controlled application of external bias). Finally, we note that the luminescence can be restricted to the tilted quantum wells, if the excitation photon energy is chosen low enough and the width of the n- and p-doped quantum wells is small enough, such that their effective band gap becomes too large for absorption processes in these layers.

Similar to the case of the low doped ideal n-i-p-i structures discussed in section 3.1, the luminescence spectrum is expected to image the density of states distribution of the occupied states of the donor impurity band. Only, in the present case also the Hubbard- or D^--band is included. Thus, these luminescence experiments complete the information obtained from conductivity measurements as discussed in the previous section. In particular, it is expected, that the collapse of the Hubbard gap should be observable if, at a suitable donor density $n_D^{(2)}$ the Hubbard-Mott transition occurs due to a decreasing magnetic field perpendicular to the layers (or in a set of samples of varying $n_D^{(2)}$).

5. Conclusions

We have discussed problems which result from the random distribution of dopant atoms in n-i-p-i doping superlattices as well as those due to the unknown wavefunctions of the acceptor band states which are difficult to treat quantitatively. We have discussed possibilities to avoid those difficulties in weakly doped n-i-p-i's with 2-dimensional doping layers, IV-VI compound- and hetero n-i-p-i's. The first case is found to be an interesting special case of donor acceptor pair luminescence, which should provide valuable information about the impurity band formation (in two dimensions) as a function of impurity concentration. Finally we have proposed new hetero n-i-p-i structures which appear uniquely suited for the observation of both, the Mott-Anderson and the Mott-Hubbard transition. This can be done by investigating the conductivity and optical transitions as a function of the variable carrier density and of the product of 2-dimensional dopant concentration and effective Bohr radius (which can be tuned by a magnetic field perpendicular to the layers).

References

1. G.H. Döhler, Phys. Stat. Sol. 52, 79 (1972)

2. G.H. Döhler, Phys. Stat. Sol. (b) 52, 533 (1972)

3. For a review of the early work see, for instance: G.H. Döhler and K. Ploog, in: Synthetic Modulated Structure Materials, L.L. Chang and B.C. Giessen, Eds. (Academic Press, N.Y. 1985), 163, or, K. Ploog and G.H. Döhler, Advances in Physics 32, 285 (1983), or, G.H. Döhler, in: The Technology and Physics of Molecular Beam Epitaxy, E.H.C. Parker, Ed. (Plenum, N.Y., 1985) 233

4. For a review of the recent work see. for instance: G.H. Döhler, IEEE QE-22, 1682 (1986), or, G.H. Döhler, CRC Review in Solid State and Materials Sciences 13, 97 (1986)

5. P.P. Ruden and G.H. Döhler, Phys. Rev. B27, 3538 (1983)

6. G.H. Döhler and P.P. Ruden, Surface Science 142, 474 (1984)

7. P.P. Ruden and G.H. Döhler, Solid State Commun. 45, 23 (1983)

8. P.P. Ruden and G.H. Döhler, Phys. Rev. B27, 3547 (1983)

9. P.P. Ruden and G.H. Döhler, Proceedings of the 17th Intern. Conf. on the Physics of Semiconductors, J.D. Chadi and W.H. Harrison, Eds. (Springer, N.Y., 1985) 535

10. G.H. Döhler, Superlattices and Microstructures 1, 279 (1985)

11. G.H. Döhler, J. of Non-Crystalline Solids 77 & 78, 1041 (1985)

12. G.H. Döhler, R.A. Street, and P.P. Ruden, Surface Science 174, 240 (1986)

13. G.H. Döhler, NSF Workshop on Optical Nonlinearities, Fast Phenomena and Signal Processing, N. Peyghambarian, Ed. (Tucson, AZ, 1986) 292

14. F. Crowne, T.L. Reinecke, and B.V. Shanabrook, Proceedings of 17th Intern. C onf. on the Physics of Semiconductors, J.D. Chadi and W.H. Harrison, Eds. (Springer, N.Y., 1985) 363

15. G.H. Döhler, unpublished

16. W. Rehm, P. Ruden, G.H. Döhler, and K. Ploog, Phys. Rev. B 28, 5937 (1983)

17. Ref. für Akzeptor Welenflut (Bassani ?, P.R.'s Thesis), 18.Hopfield (DA Pair)

19. U. Heim, O. Röder, H.J. Queisser, and M. Pilkuhn, J. of Luminescence 1,2, 542 (1970)

20. A. Raymond, this workshop

21. J.L. Robert, A. Raymond. L. Konczewicz, C. Bousquet, W Zawadzki, F. Alexandre, I.M. Masson, J.P. Andre, and P.: Frijlink, Phys. Rev. B33, 5935 (1986)

22. The mixing of light- and heavy-hole subbands has bee: treated, for instance, by U. Ekenberg and M. Altarelli Phys. Rev. B30, 3569 (1984); E. Bangert and G. Landwehr Superlattices Microstruct. 1, 363 (1985); D.A. Broido an L.J. Sham, Phys. Rev. B31, 888 (1985); and Y.C. Chang an J.N. Schulman, Phys. Rev. B31, 2069 (1985)

23. R.A. Street, G.H. Döhler, J.N. Miller, and P.P. Ruden, Phys Rev. B33, 7043 (1986)

24. N.F. Mott and E.A. Davis, Electronic Process in Non Crystalline Materials, 2nd ed. (Oxford, 1979)

25. G.P. Srivastava, this workshop

26. G.D. Döhler, G. Hasnain, and J.N. Miller, Appl. Phys. Lett 49, 704 (1986)

27. C.J. Chang-Hasnain, G. Hasnain, N.M. Johnson, G.H. Döhler J.N. Miller, J.R. Whinnery, and A. Dienes, Appl. Phys Letter 50, 915 (1987)

28. E. Abrahams, P.W. Anderson, D.C. Licciardello, and T.V Ramakrishnan, Phys. Rev. Lett. 42, 673 (1979)

29. P.W. Anderson, Phys. Rev. 109, 1492 (1958)

30. See for instance: "Anderson Localization, Y. Nagaoka, Ed (Springer, New York, 1982)

31. R.A. Davies, C.C. Dean, and M. Pepper, Surface Science 142, 25 (1984)

DEEP IMPURITY LEVELS IN SEMICONDUCTORS, SEMICONDUCTOR ALLOYS, AND SUPERLATTICES

John D. Dow, Shang Yuan Ren, and Jun Shen

Department of Physics
University of Notre Dame
Notre Dame, Indiana 46556 U.S.A.

INTRODUCTION

Impurity levels determine the electrical properties of semiconductors and often strongly influence the optical properties as well. Until rather recently it was widely believed that "shallow impurities," namely those impurities that produce energy levels within ≈0.1 eV of a band edge, were well understood in terms of hydrogenic effective-mass theory [1]. However "deep impurities" were regarded as more mysterious, having levels more than 0.1 eV deep in the gap; and several theoretical attempts were made to understand why their binding energies were so large. While specific deep levels were explained rather well by the early theories, most notably the pioneering work of Lannoo and Lenglart on the deep level in the gap associated with the vacancy in Si [2], numerous attempts to explain why the binding energy of a particular level might be large (making the level deep), rather than small, continued until recently, when it was realized that this basic picture of impurity levels was incomplete [3,4].

DISTINCTION BETWEEN DEEP LEVELS AND DEEP IMPURITIES

Now it is recognized that every impurity whose valence differs from that of the host atom it replaces produces both deep and shallow levels, but that the "deep" levels often are not deep energetically. The definitions of deep and shallow levels have been changed [3]: now a deep level is one that originates from the central-cell potential of the defect, and a shallow level originates from the long-ranged Coulomb potential due to the impurity-host valence difference. Impurities that are s-p bonded normally produce four deep levels in the vicinity of the fundamental band gap: one s-like and three p-like. More often than not, the deep levels do not lie in the band gap but are resonant with the host bands. Thus, in terms of the old picture, these resonant levels have negative "binding energies." The four deep levels are normally anti-bonding and host-like in character. As a result, several different deep impurities produce levels with essentially the same wavefunctions [5,6].

In contrast to the deep levels, whose wavefunctions are relatively localized and have multi-band character, shallow levels have wavefunctions that are extended in real space and of single-band character (localized near a band extremum in $\vec{k}$-space). The shallow levels are hydrogenic, originate from the long-ranged Coulomb potential, are associated with the impurity-host valence difference, and have binding energies of reduced Rydbergs, i.e., scaled from 13.6 eV by factors of effective mass and the inverse square of the dielectric constant (resulting in binding energies of typically tens of meV). Shallow levels can be obtained by solving an effective-mass Schrödinger equation for the envelope wavefunction $\phi(r)$, which is (assuming isotropic effective mass m* and dielectric constant ϵ)

$$[(-\hbar^2/2m^*)\nabla^2 - (e^2/\epsilon r)]\phi = (E-E_0)\phi, \tag{1}$$

where E_0 is the band extremum to which the level is "attached."

The attachment to a near-by band edge is the experimental signature of a shallow level, and can be verified by measuring the energy of the level and the band edge, either versus alloy composition x in an alloy such as $A\ell_xGa_{1-x}As$ or $GaAs_{1-x}P_x$, or versus pressure. In contrast to shallow levels, deep levels are not attached to band edges, and their energies (with respect to the valence band maximum) tend to vary more-or-less linearly with alloy composition.

DEPENDENCE ON ALLOY COMPOSITION

Fig. 1 summarizes data [7] for the alloy-composition dependence of levels associated with anion-site impurities in $GaAs_{1-x}P_x$ alloys. The levels associated with S, shallow donor levels, follow the conduction band edge as the composition varies, and are "attached" to it. Se levels exhibit the same behavior as S. The N and O levels vary approximately linearly with alloy composition and are unattached to any band edge. These data, when first obtained, were rather perplexing, because O, S, and Se were expected to behave the same, since all came from Column-VI of the Periodic Table, and all were presumed to occupy a Column-V anion site. By the old definition, the oxygen state was clearly a deep level, lying more than 0.1 eV from the conduction band edge. The N level, however, had the same qualitative dependence on alloy composition as the O state (indicating it should be classified as deep), yet it was close enough to the band edge in GaP to be classified as shallow by the old definition of a shallow level. Moreover it was energetically deep in the gap for $GaAs_{0.5}P_{0.5}$, but invisible (possibly in the conduction band) for GaAs.

The N level data in $GaAs_{1-x}P_x$ provided an important clue for understanding the physics of deep impurities, because N is isoelectronic to As and P, and so the substitutional impurity has no long-ranged Coulomb potential: its level could only be explained in terms of the central-cell defect potential (and, possibly also the modest strain field around the impurity) [4]. The natural explanation for the data of Fig. 1 is that the central-cell potential produces the oxygen deep level and that the N level is similar to the oxygen level, and should also be termed "deep." The shallow S (and Se) levels are caused by the Coulomb potential associated with the valence difference between the Column-VI impurities and the Column-V host. S and Se must also have deep levels similar to the oxygen deep level, but their deep levels must be resonant with the conduction band of $GaAs_{1-x}P_x$ for all alloy compositions x, and hence are invisible (as with the case for N when $x<0.2$). Thus the picture has emerged that the central-cell potential produces deep levels, and the Coulomb potential yields shallow levels (which ideally are infinite in number, because a Coulomb potential has an infinite number of bound states). For s- and p- bonded substitutional impurities, we expect one s-like and three p-like deep levels associated with every band, based on Rayleigh's interlacing theorems [8]; this means that in the vicinity of the

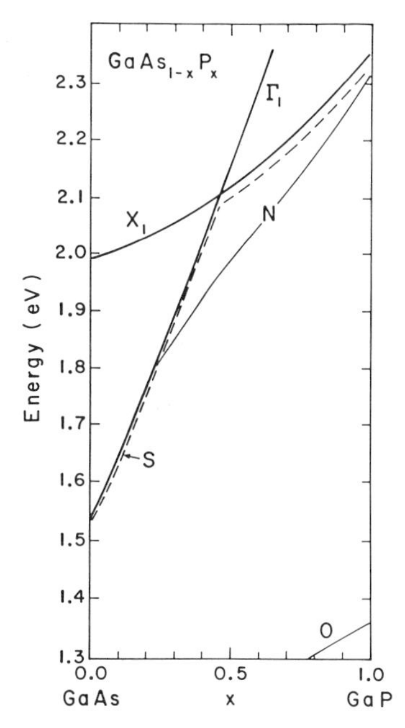

Fig. 1. Schematic illustration of the dependences of the shallow S (or Se) level (dashed) and the deep N and O levels (solid) on alloy composition x in $GaAs_{1-x}P_x$, after Ref. [7]. The energies of the band edges Γ_1 and X_1 are also shown. The zero of energy is the valence band maximum of the alloy.

band gap of a zinc-blende semiconductor we should find one s-like (A_1-symmetric) deep level and one triply degenerate p-like (T_2-symmetric) level for each impurity. The oxygen and nitrogen levels of Fig. 1 are the A_1-symmetric deep levels. In a three-dimensional material, there is no guarantee that any of the deep levels due to the central-cell potential will lie in the fundamental band gap, and so, as in the case of S and Se in $GaAs_{1-x}P_x$, the deep levels may all lie outside of the band gap or else, as in the case of oxygen, one (or more) may lie within the gap. (Fig. 2)

The character of an impurity as "shallow" or "deep" is determined by whether one or more deep levels lies within the fundamental band gap. For example, oxygen in GaP is a deep impurity because its s-like A_1 deep level lies in the band gap, and is occupied by one electron when the oxygen is neutral and in its ground state. This one-electron level can hold two electrons of opposite spin, but when oxygen is neutral, it holds only one, (the extra electron due to the valence difference between O and P). In the case of S or Se, the central-cell potential is weaker than that of O, and the corresponding deep level is degenerate with the conduction band rather than in the gap. The resonant level cannot bind the extra electron, which spills out and decays to the conduction band edge (via the electron-phonon interaction); once at the conduction band minimum, the electron is trapped by the long-ranged Coulomb potential in a shallow impurity level -- and, because the ground state of the S or Se impurity in this host has an electron in the shallow level, the S or Se impurity is termed a shallow impurity.

DEEP-SHALLOW TRANSITIONS

It is possible to change the character of an impurity from deep to shallow by perturbing the host until a deep level passes from the band gap into the host bands. Straightforward ways to achieve such a change are to apply pressure or to vary the alloy composition of the host. N changes its character in $GaAs_{1-x}P_x$ as a function of alloy composition x: it yields a deep level in the gap for $x>0.2$ and so is a deep impurity for such alloy compositions, but for $x<0.2$, neutral N is neither a shallow impurity (because it has no long-ranged Coulomb potential) nor a deep impurity (because its deep levels are in the conduction band), and so we term it an "inert impurity." Thus, in $GaAs_{1-x}P_x$ for $x\simeq0.22$, N undergoes a "deep-inert transition." See Fig. 1. A classic example of a similar "deep-shallow transition" is Si on a cation site in $A\ell_xGa_{1-x}As$, which has a deep-level behavior similar to that of N in $GaAs_{1-x}P_x$. (See Fig. 3.) As a result, neutral Si is a deep impurity, with one electron occupying its A_1-symmetric deep level for $x\geq0.3$, but is a shallow impurity for $x<0.2$ because the deep level is resonant with the conduction band and the electron is autoionized, falls to the conduction band edge, and is trapped by the Coulomb potential of the defect. This Si defect is thought to be the DX center [9-11], or at least a component of it [11,12]. It is noteworthy that when cation-site Si is a shallow impurity (i.e., its deep levels are resonant with the conduction band), it dopes the material n-type. But when it is a deep impurity (with its A_1 level in the gap) the neutral impurity can trap either an electron or a hole in this level, and so tends to make the material semi-insulating.

DEEP-SHALLOW TRANSITIONS IN SUPERLATTICES

Deep-shallow transitions occur in random alloys such as $A\ell_xGa_{1-x}As$ because the deep levels are much less sensitive to changes of alloy composition x than the band edges. Only slightly overstating the point, we may say that the deep levels are almost constant in absolute energy, and that varying x can cause the conduction or valence band edge to move through the deep level. Since deep levels are normally only observed when they lie within the fundamental band gap, varying alloy composition changes the <u>window of</u>

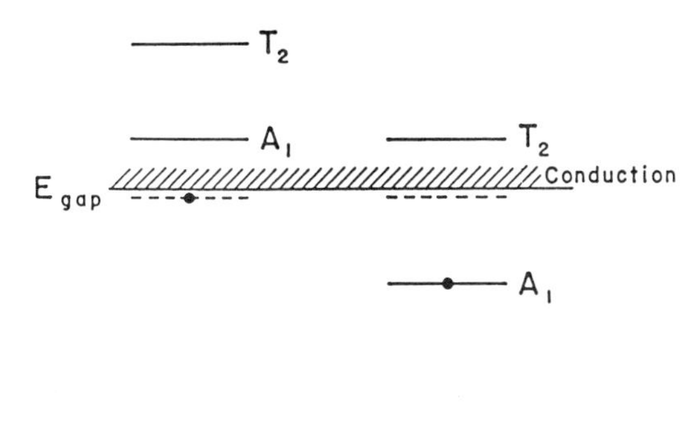

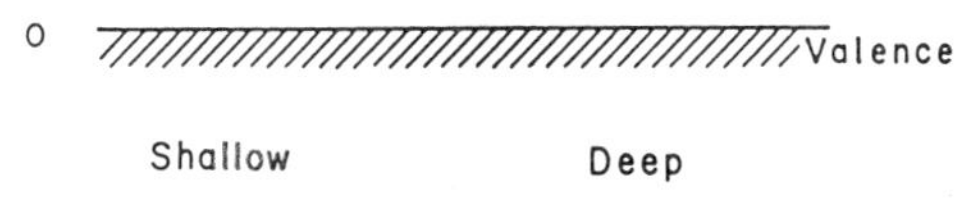

Fig. 2. Illustration of the difference between a shallow impurity such as S on a P site in GaP and a deep impurity such as O. Shallow levels are dashed and deep levels are solid. If the deep levels lie outside of the band gap, the extra electron occupies the lowest shallow level at zero temperature and the impurity is a shallow donor.

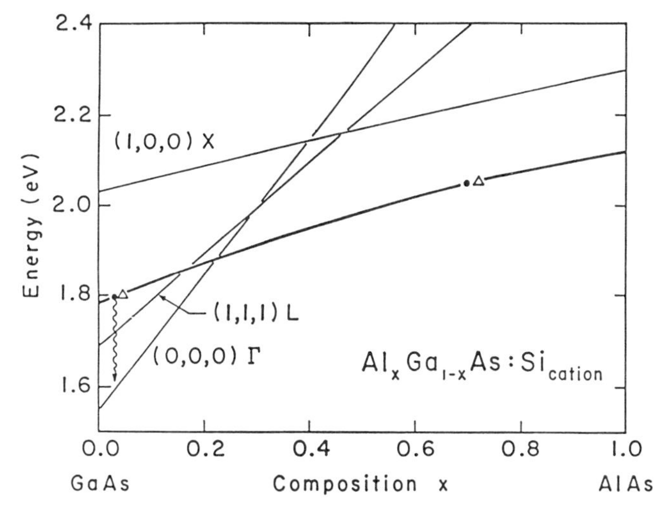

Fig. 3. Chemical trends with alloy composition x in the energies (in eV) of principal band gaps at Γ, L, and X, with respect to the valence band maximum of the alloy, in the alloy $A\ell_xGa_{1-x}As$, as deduced from the Vogl model [15]. Also shown is the predicted energy of the A_1-symmetric cation-site deep level of Si (heavy line), similar to the predictions of Hjalmarson [11]. The Vogl model is known to obtain very little band bending. Moreover the L minimum for $x \simeq 0.45$ is known to be at a bit too low an energy in this model. When the deep level of neutral Si lies below the conduction band minimum, it is occupied by one electron (solid circle) and one hole (open triangle). When this level is resonant with the conduction band, the electron spills out and falls (wavy line) to the conduction band minimum, where it is trapped (at zero temperature) in a shallow donor level (not shown).

observability of deep levels, namely the band gap. Thus selecting the alloy composition so that a given impurity produces a shallow level rather than a deep one is a form of "band-gap engineering."

Similar band-gap engineering can be achieved by altering the atomic ordering in the semiconductor by, for example, making the host a superlattice rather than a random alloy. For example, Si on a Ga site in the center of a GaAs quantum well in a $N_1 \times N_2$ GaAs/$A\ell_xGa_{1-x}As$ superlattice can be made to undergo a shallow-deep transition by reducing the thickness of the quantum well until the conduction band maximum of the superlattice passes above the Si deep level (See Fig. 4.). The dependence on alloy composition of a Ga-site deep level in a 2×2 GaAs/$A\ell_xGa_{1-x}As$ superlattice is given in Fig. 5, and shows that even near the center of a GaAs layer Si can become a deep trap if the layer is sufficiently thin or two-dimensional.

Fig. 6 shows the predictions for the cation-site deep level in bulk GaAs, in a 3×10 GaAs/$A\ell_{0.7}Ga_{0.3}As$ superlattice, and in bulk $A\ell_{0.7}Ga_{0.3}As$ for Si at various sites β in the superlattice. In this superlattice the deep level lies in the gap and Si is therefore a deep impurity. In $N_1 \times N_2$ superlattices such that the GaAs layers are not very thin, $N_1 > 6$, the superlattice conduction band edge approaches the bulk GaAs conduction band edge, and therefore the conduction band of the superlattice covers up the Si deep level, making Si on a Ga site a shallow donor in the superlattice.

T_2 LEVELS

The examples we have given thus far all consider only A_1-symmetric bulk deep levels. The T_2 bulk levels are split in a (001) superlattice into a_1, b_1, and b_2 levels, as illustrated in Fig. 7 for the As vacancy. The valence band maximum is also split into a $(p_x \pm p_y)$-like (or b_1- or b_2-like) maximum and a p_z-like (or a_1-like) edge slightly below it. For sites near the β=0 interface, the b_1 deep levels have orbitals and energies similar to the T_2 levels of bulk GaAs and the b_2 levels are $A\ell_xGa_{1-x}As$-like. The splittings are small even for impurities at the interface, of order 0.1 eV, and decrease rapidly as the impurity moves from the interface. They are non-zero even near the center of the layers, however, because the valence band maximum is split, and so the spectral densities employed in solving Eq. (2) below are also split.

Three factors influence the positions of deep levels in the superlattice layers: (i) the band offset, which tends to move the levels down in $A\ell_xGa_{1-x}As$ relative to in GaAs, (ii) quantum confinement, which tends to cause levels in GaAs and the p_z-like a_1 levels in particular to be farther from the center of the gap, and (iii) the relative electropositivity of $A\ell$ with respect to Ga, which tends to move energy levels up when the associated wavefunctions overlap an $A\ell$-rich layer. To determine the balance among these competing effects, a calculation is required.

FORMALISM

Our basic approach to the problem of deep levels in superlattices is based on the theory of Hjalmarson et al. [3]. This theory is relatively easy to implement and learn, and has been summarized in accessible lecture notes [13] for deep levels in the bulk. Nelson et al. [14] have provided an alternative approach to deep levels in superlattices which is especially well-suited to small-period superlattices.

The deep level energies E in a superlattice can be computed similarly by solving the one-electron Schrödinger equation, using the Green's function method to take advantage of the localized nature of the central-cell defect

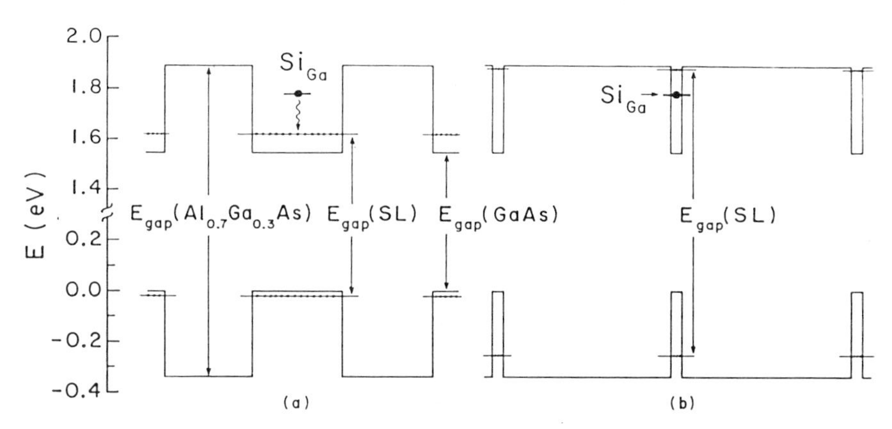

Fig. 4. Schematic illustration of the quantum-well confinement effect on the band gap $E_{gap}(SL)$ of a $N_1 \times N_2$ $GaAs/A\ell_{0.7}Ga_{0.3}As$ superlattice, after Ref. [12]: (a) $N_1=N_2=18$; and (b) $N_1=2$, $N_2=34$. The band edges of the superlattice are denoted by chained lines. For this alloy composition the superlattice gap is indirect for case (b), with the conduction band edge at $\vec{k}=(2\pi/a_L)(1/2,1/2,0)$. Note the broken energy scale. The zero of energy is the valence band maximum of GaAs.

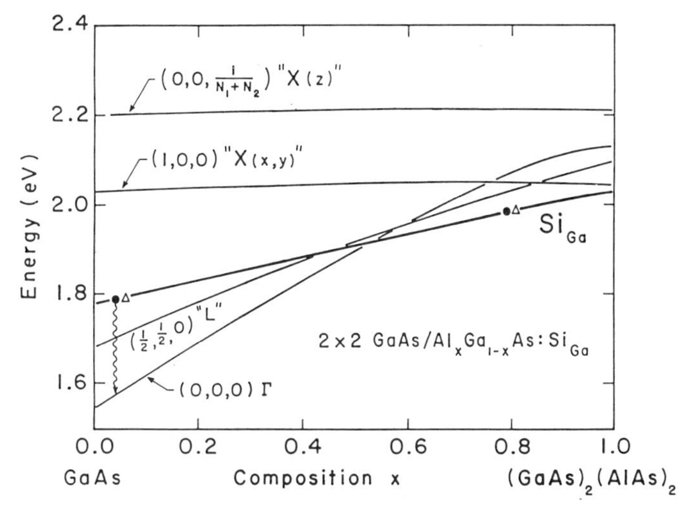

Fig. 5. Predicted chemical trends with alloy composition x in the energies (in eV) of principal band gaps and the Si A_1-symmetric deep level in a $(GaAs)_2(A\ell_xGa_{1-x}As)_2$ superlattice. Compare with Fig. 3 for the alloy. The superlattice wavevectors of the gaps are $\vec{k}=\vec{0}$, $\vec{k}=(2\pi/a_L)(1/2,1/2,0)$, which has states derived from the L point of the bulk Brillouin zone, and the points derived from the bulk X-point: $(2\pi/a_L)(0,0,[N_1+N_2]^{-1})$, (where $N_1=N_2=2$), and $(2\pi/a_L)(1,0,0)$.

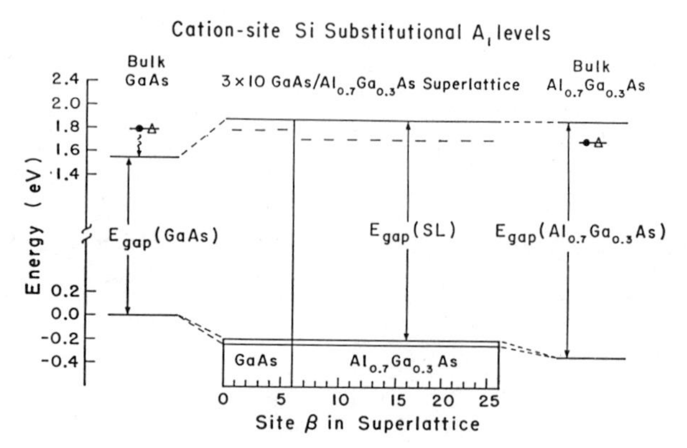

Fig. 6. Predicted A_1-derived deep levels of Si in GaAs, in a 3x10 GaAs/$A\ell_{0.7}Ga_{0.3}As$ superlattice (as a function of β, the position of the Si in the superlattice), and in bulk $A\ell_{0.7}Ga_{0.3}As$. Interfaces (which are As sites) correspond to β = 0, 6, and 26. Note that in bulk GaAs Si is a shallow donor, but that in this superlattice and in bulk $A\ell_{0.7}Ga_{0.3}As$ it is predicted to be a deep impurity.

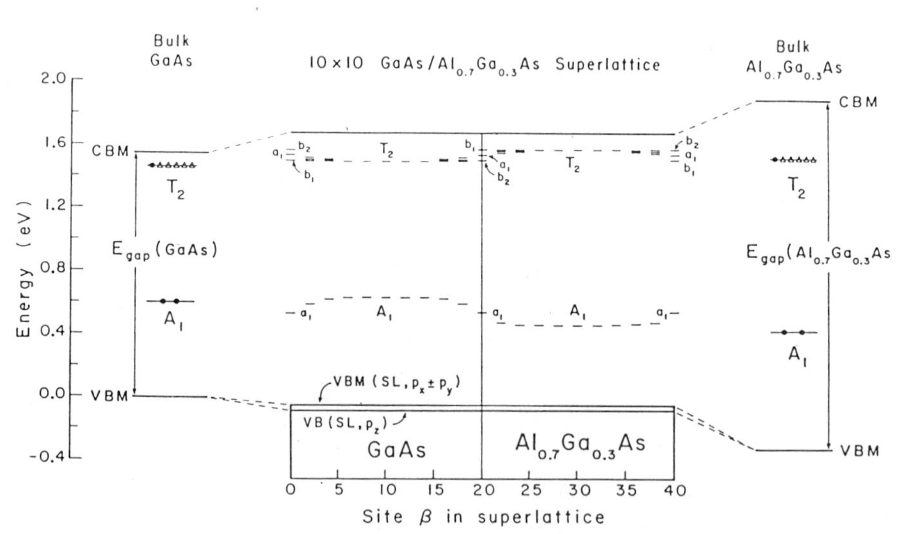

Fig. 7. Predicted energy levels of an As-vacancy in bulk GaAs, in a $(GaAs)_{10}(A\ell_{0.7}Ga_{0.3}As)_{10}$ superlattice (as a function of β, the position of the vacancy: even values of β correspond to As-sites), and in bulk $A\ell_{0.7}Ga_{0.3}As$, after Ref. [12]. Note the splitting of the T_2 levels at and near the interfaces which correspond to β = 0, 20, and 40. The electron (hole) occupancies of the deep levels in bulk GaAs and bulk $A\ell_{0.7}Ga_{0.3}As$ are denoted by solid circles (open triangles).

potential operator V:

$$\det\{1-G(E)V\} = 0 = \det\{1 - P\int[\delta(E'-H)\, V/(E-E')]\, dE'\}. \tag{2}$$

Here we have $G(E) = (E-H)^{-1}$, where H is the host Hamiltonian operator, P denotes the principal value integral over all energies, and $\delta(x)$ is Dirac's delta function. For simplicity of presentation, we consider a periodic $N_1 \times N_2$ $GaAs/A\ell_xGa_{1-x}As$ superlattice grown in the (001) direction. This superlattice has N_1 two-atom thick layers of GaAs and N_2 two-atom thick layers of $A\ell_xGa_{1-x}As$ stacked alternately in a periodic structure. We assume that the layers are perfectly lattice-matched. We describe the superlattice Hamiltonian in terms of an s^*sp^3 basis similar to that of Vogl et al. [15]. Our Hamiltonian is a nearest-neighbor tight-binding model, and, in the limit of $x = 0$, is identical to the Vogl model for GaAs. We treat the superlattice using a superhelix or supercell method: (for the case $x = 1$) a superhelix or supercell is a helical string with axis aligned along the (001)- or z-direction consisting of $2N_1 + 2N_2$ adjacently bonded atoms As, Ga, As, Ga, As, ... Ga, As, $A\ell$, As, $A\ell$, As, $A\ell$, ... As, $A\ell$. (For $x \neq 1$, replace $A\ell$ by the virtual cation $A\ell_xGa_{1-x}$.) The center of this helix is at $\vec{L}$ and each of the atoms of the helix is at position $\vec{L} + \vec{v}_\beta$ (for β=0, 1, 2, ..., $2N_1+2N_2-1$). The superlattice is a stack of superslabs. A superslab of $GaAs/A\ell_xGa_{1-x}As$ consists of all such helices with the same value of L_z and all possible different values of L_x and L_y. The origin of coordinates is taken to be at an As atom and a neighboring cation is at $(1/4,1/4,1/4)a_L$, where a_L is the lattice constant. At each site there are five s^*sp^3 basis orbitals $|n,\vec{L},\vec{v}_\beta)$, where $n = s^*$, s, p_x, p_y, or p_z and the site is specified by $\beta = 0,1,2,\ldots,2N_1+2N_2-1$. From these basis orbitals we form the sp^3 hybrid orbitals at each site $\vec{R} = (\vec{L},\vec{v}_\beta)$. The hybrid orbitals are

$$|h_1,\vec{R}) = [|s,\vec{R}) + \lambda|p_x,\vec{R}) + \lambda|p_y,\vec{R}) + \lambda|p_z,\vec{R})]/2$$

$$|h_2,\vec{R}) = [|s,\vec{R}) + \lambda|p_x,\vec{R}) - \lambda|p_y,\vec{R}) - \lambda|p_z,\vec{R})]/2$$

$$|h_3,\vec{R}) = [|s,\vec{R}) - \lambda|p_x,\vec{R}) + \lambda|p_y,\vec{R}) - \lambda|p_z,\vec{R})]/2$$

and

$$|h_4,\vec{R}) = [|s,\vec{R}) - \lambda|p_x,\vec{R}) - \lambda|p_y,\vec{R}) + \lambda|p_z,\vec{R})]/2 \tag{3}$$

where we have $\lambda = +1$ (-1) for atoms at anion (cation) sites. Introducing the label $\nu = s^*$, h_1, h_2, h_3, or h_4, our hybrid basis orbitals are $|\nu,\vec{R})$, and the related tight-binding orbitals [16] are

$$|\nu,\beta,\vec{k}) = N_s^{-1/2} \Sigma_{\vec{L}} \exp(i\vec{k}\cdot\vec{L}+i\vec{k}\cdot\vec{v}_\beta)\; |\nu,\vec{L},\vec{v}_\beta) \tag{4}$$

where $\vec{k}$ is (in a reduced zone scheme) any wavevector of the mini-zone or (in an extended zone scheme) any wavevector of the zinc-blende Brillouin zone. Here N_s is the number of supercells.

The mini-zone wavevector $\vec{k}$ is a good quantum number. Evaluation of the matrix elements $(\nu,\beta,\vec{k}|H|\nu',\beta',\vec{k})$ leads to a tight-binding Hamiltonian of the block tridiagonal form. For example, the diagonal (in β) 5x5 matrix, $H(\beta,\beta)$ at site β, is

$$H(\beta,\beta)=(\nu,\beta,\vec{k}|H|\nu',\beta,\vec{k}) = \begin{pmatrix} \epsilon_{s*} & 0 & 0 & 0 & 0 \\ 0 & \epsilon_h & T & T & T \\ 0 & T & \epsilon_h & T & T \\ 0 & T & T & \epsilon_h & T \\ 0 & T & T & T & \epsilon_h \end{pmatrix} . \tag{5}$$

where we have the hybrid energy

$$\epsilon_h = (\epsilon_s + 3\epsilon_p)/4 \tag{6}$$

and the hybrid-hybrid interaction

$$T = (\epsilon_s - \epsilon_p)/4 . \tag{7}$$

The energies ϵ_{s*}, ϵ_h, and T in $H(\beta,\beta)$ refer to the atom at the β-th site, and are obtained from the energies w tabulated by Vogl et al. [15]. To account for the observed [17] valence band edge discontinuity of 32% of the direct band gap [18], a constant is added to ϵ_{s*} and ϵ_h for $A\ell_xGa_{1-x}As$; this constant is adjusted to give the valence band maximum of $A\ell_xGa_{1-x}As$ below the valence band maximum of GaAs by 32% of the direct band gap difference in the limit $N_1=N_2 \rightarrow \infty$.

Expressions for the remaining, off-diagonal matrix elements $(\nu,\beta,\vec{k}|H|\nu',\beta',\vec{k})$ or $H(\beta,\beta')$ are given in Ref. [12]. The Hamiltonian matrix has dimension $5(2N_1+2N_2)$ for each $\vec{k}$. We diagonalize this Hamiltonian numerically for each (special-point) $\vec{k}$, finding its eigenvalues $E_{\gamma,\vec{k}}$ and the projections of the eigenvectors $|\gamma,\vec{k}\rangle$ on the $|\nu,\beta,\vec{k})$ hybrid basis: $(\nu,\beta,\vec{k}|\gamma,\vec{k}\rangle$. Here γ is the band index (and ranges from 1 to 200 for $N_1=N_2=10$) and $\vec{k}$ lies within the mini-Brillouin zone in a reduced zone scheme. With these quantitites, we can evaluate Eq. (2) for the Hjalmarson model of the defect potential matrix V [3]: the matrix is zero except at the defect site and diagonal on that site, $(0, V_s, V_p, V_p, V_p)$, in the Vogl s^*sp^3 local basis centered on each atom. The point-group for a general substitutional defect in a $GaAs/A\ell_xGa_{1-x}As$ (001) superlattice is C_{2v}, provided the $A\ell_xGa_{1-x}As$ is treated in a virtual crystal approximation. In the $GaAs/A\ell_xGa_{1-x}As$ superlattice the A_1 and T_2 deep levels of the bulk GaAs or $A\ell_xGa_{1-x}As$ produce two a_1 levels (one s-like, derived from the A_1 level and one T_2-derived p_z-like), one b_1 level [(p_x+p_y)-like], and one b_2 level [(p_x-p_y)-like]. Of course, for impurities far from a $GaAs/A\ell_xGa_{1-x}As$ interface, the s-like a_1 level will have an energy very close to the energy of a bulk A_1 level, and p_z-like a_1 level and the b_1 and b_2 levels will lie close to the bulk T_2 level also.

The secular equation, Eq. (2), is reduced by symmetry to the following three equations:

$$G(b_1;E) = V_p^{-1} \quad \text{for } b_1 \text{ levels,} \tag{8}$$

$$G(b_2;E) = V_p^{-1} \quad \text{for } b_2 \text{ levels,} \tag{9}$$

and

$$\det \begin{pmatrix} G(s,s;E)\,V_s - 1 & G(s,z;E)\,V_p \\ G(z,s;E)\,V_s & G(z,z;E)\,V_p - 1 \end{pmatrix} = 0, \tag{10}$$

for a_1 levels, where we have

$$G(b_1;E)=\Sigma_{\gamma,\vec{k}} \; |(h_1,\beta,\vec{k}|\gamma,\vec{k}\rangle-(h_4,\beta,\vec{k}|\gamma,\vec{k}\rangle|^2/2(E-E_{\gamma,\vec{k}}), \tag{11}$$

$$G(b_2;E)=\Sigma_{\gamma,\vec{k}} \; |(h_2,\beta,\vec{k}|\gamma,\vec{k}\rangle-(h_3,\beta,\vec{k}|\gamma,\vec{k}\rangle|^2/2(E-E_{\gamma,\vec{k}}), \tag{12}$$

$$G(s,s;E) = \Sigma_{\gamma,\vec{k}} \; |(h_1,\beta,\vec{k}|\gamma,\vec{k}\rangle+(h_2,\beta,\vec{k}|\gamma,\vec{k}\rangle+(h_3,\beta,\vec{k}|\gamma,\vec{k}\rangle+(h_4,\beta,\vec{k}|\gamma,\vec{k}\rangle|^2/4(E-E_{\gamma,\vec{k}}), \tag{13}$$

$$G(z,z;E) = \Sigma_{\gamma,\vec{k}} \; |(h_1,\beta,\vec{k}|\gamma,\vec{k}\rangle-(h_2,\beta,\vec{k}|\gamma,\vec{k}\rangle-(h_3,\beta,\vec{k}|\gamma,\vec{k}\rangle+(h_4,\beta,\vec{k}|\gamma,\vec{k}\rangle|^2/4(E-E_{\gamma,\vec{k}}), \tag{14}$$

and

$$G(s,z;E)=\Sigma_{\gamma,\vec{k}} \; [(h_1,\beta,\vec{k}|\gamma,\vec{k}\rangle+(h_2,\beta,\vec{k}|\gamma,\vec{k}\rangle+(h_3,\beta,\vec{k}|\gamma,\vec{k}\rangle+(h_4,\beta,\vec{k}|\gamma,\vec{k}\rangle] \times [(h_1,\beta,\vec{k}|\gamma,\vec{k}\rangle-(h_2,\beta,\vec{k}|\gamma,\vec{k}\rangle-(h_3,\beta,\vec{k}|\gamma,\vec{k}\rangle+(h_4,\beta,\vec{k}|\gamma,\vec{k}\rangle]^*/[4(E-E_{\gamma,\vec{k}})] \tag{15}$$

Here $G(z,s;E)$ is the Hermitian conjugate of $G(s,z;E)$ and β is the site of the defect in the superlattice.

For each site β the relevant host Green's functions, Eqs. (11) to (15), are evaluated using the special points method [19], and the secular equations (8) to (10) are solved, yielding $E(b_1;V_p)$, $E(b_2;V_p)$, and two values of $E(a_1;V_s,V_p)$. The defect potential matrix elements V_s and V_p are obtained using a slight modification of Hjalmarson's approach [20]. For N_1=N_2=10, there are 40 possible sites β, each with four relevant deep levels: two a_1, one b_1, and one b_2; thus there are 160 levels.

SUMMARY

With the basic approach outlined here, one can compute the deep levels associated with substitutional impurities in lattice-matched superlattices and predict the conditions under which deep-shallow and deep-inert transitions are to be expected.

Acknowledgment -- We are grateful to the U.S. Office of Naval Research (Contract No. N00014-84-K-0352) and the U.S. Air Force Office of Scientific Research (Contract No. AF-AFOSR-85-0331) for their generous support.

REFERENCES

[1] W. Kohn, in Solid State Physics (edited by F. Seitz and D. Turnbull, Academic Press, New York, 1957) Vol. 5, p. 258-321; J. M. Luttinger and W. Kohn, Phys. Rev. 97, 969 (1955).
[2] M. Lannoo and P. Lenglart, J. Phys. Chem. Solids 30, 2409 (1969).
[3] H. P. Hjalmarson, P. Vogl, D. J. Wolford, and J. D. Dow, Phys. Rev. Letters 44, 810 (1980); see also Ref. [4] for the concepts that form the foundation of this work.

[4] W. Y. Hsu, J. D. Dow, D. J. Wolford, and B. G. Streetman, Phys. Rev. B 16, 1597 (1977).
[5] S. Y. Ren, W. M. Hu, O. F. Sankey, J. D. Dow, Phys. Rev. B 26, 951 (1982).
[6] S. Y. Ren, Scientia Sinica 27, 443 (1984).
[7] D. J. Wolford, W. Y. Hsu, J. D. Dow, and B. G. Streetman, J. Lumin. 18/19, 863 (1979).
[8] A. A. Maradudin, E. W. Montroll, and G. H. Weiss, Solid State Phys. Suppl. 3, 132 (1963).
[9] H. P. Hjalmarson and T. J. Drummond, Appl. Phys. Letters 48, 657 (1986); see also Ref. [10].
[10] J. C. M. Henning and J. P. M. Ansems, Mat. Sci. Forum 10-12, 429 (1986); Semicond. Sci. Technol. 2, 1 (1987); A. K. Saxena, Solid State Electron. 25, 127 (1982).
[11] See also, J. W. Farmer, H. P. Hjalmarson, and G. A. Samara, Proc. Mater. Res. Soc., 1987, to be published.
[12] S. Y. Ren, J. D. Dow, and J. Shen, "Deep impurity levels in semiconductor superlattices," to be published. See also Ref. [21].
[13] J. D. Dow, in Highlights of Condensed Matter Theory (Proc. Intl. School of Phys. "Enrico Fermi" Course 89, Varenna, 1983) ed. by F. Bassani, F. Fumi, and M. P. Tosi (Societa Italiana di Fisica, Bologna, Italy, and North Holland, Amsterdam, 1985), pp. 465 et seq.
[14] J. S. Nelson, C. Y. Fong, I. P. Batra, W. E. Pickett, and B. M. Klein, unpublished.
[15] P. Vogl, H. P. Hjalmarson, and J. D. Dow, J. Phys. Chem. Solids 44, 365 (1983).
[16] The tight-binding formalism we use here is hybrid-based. For bulk semiconductors the hybrid-based tight-binding formalism is equivalent to the widely used atomic-orbital based tight-binding formalism. But for superlattices these two formalisms are different: because the atomic-orbital based tight-binding parameters are obtained by fitting to the band structure of bulk GaAs and bulk AℓAs, the tight-binding parameters for the As atom at $\beta=0$ (which is considered to be an As atom in GaAs) are usually different from the corresponding tight-binding parameters for the atom at $\beta=2N_1$ (which is considered to be an As atom in AℓAs). But both of these As atoms are interfacial As atoms and physically are completely equivalent to each other. The way we correct for this problem is that we consider the hybrids h_1 and h_4 of the As atom at $\beta=0$ to be hybrids of GaAs, while h_2 and h_3 are taken to be AℓAs hybrids. Similarly, h_2 and h_3 of the As atom at $\beta=2N_1$ are considered to be hybrids in GaAs, and h_1 and h_4 are AℓAs hybrids. We believe that this properly accounts for the correct physics and the nature of interfacial bonds.
[17] D. J. Wolford, T. F. Kuech, J. A. Bradley, M. A. Grell, D. Ninno, and M. Jaros, J. Vac. Sci. Technol. B4, 1043 (1986).
[18] In our model, which uses low-temperature band gaps, we take the valence band offset for $A\ell_{0.7}Ga_{0.3}As$ to be 0.334 eV below the valence band edge of GaAs.
[19] S. Y. Ren and J. D. Dow, "Special points for superlattices and strained bulk semiconductors," Phys. Rev. B, in press.
[20] The defect potential matrix elements are related to the difference in atomic energies of the impurity and the host atom it replaces:

$$V_\ell = \beta_\ell \, (w_{impurity}(\ell) - w_{host}(\ell)\,) + C_\ell,$$

where we have ℓ = s or p, $\beta_s = 0.8$, and $\beta_p = 0.6$, and the atomic orbital energies w can be found in Table 3 of Ref. [15]. The constant C_ℓ is zero in Ref. [3], but here we take C_s = 1.434 eV in order to have the Si deep level in GaAs appear 0.234 eV above the conduction band minimum, where

it is observed [12], instead of 0.165 eV below it, thereby compensating for the small theoretical uncertainty.
21] R.-D. Hong, D. W. Jenkins, S. Y. Ren, and J. D. Dow, Proc. Materials Research Soc. 77, 545-550 (1987), Interfaces, Superlattices, and Thin Films, ed. J. D. Dow and I. K. Schuller.

"PINNING" OF TRANSITION-METAL IMPURITY LEVELS

J. Tersoff

IBM Thomas J. Watson Research Center
Yorktown Heights, N.Y. 10598

This summary abstract reviews the conclusions of recent work by Tersoff and Harrison,[1] using the defect-molecule approach of Picoli, Chomette and Lannoo.[2] For a more detailed presentation, the reader is referred to Ref. 1, and to references therein.

In III-V and II-VI semiconductors, transition-metal (TM) impurities are cation-substitutional. We can imagine starting with a cation vacancy, and filling it with a TM atom. If we think in a simple picture of local sp^3 hybrid orbitals,[3] and if moreover the vacancy level were completely localized on the anion dangling bonds around the vacancy, then the TM atom would interact almost solely with the vacancy level. (The delocalization of the vacancy level will be included roughly later.)

It turns out[1,2] that the interesting physics here involves only the levels of T_2 symmetry. Thus the problem reduces approximately to a two-level system, with the T_2 vacancy level interacting with the T_2-symmetry TM *d* orbitals. The feature which was neglected in an earlier analysis[4] is the charge transfer between these two states, and its effect on the TM impurity level.

A detailed analysis[1,2] reveals that any shift in the relative positions of the TM *d* and vacancy levels changes the charge transfer between them by polarizing the bond. However, because of the large electrostatic interaction within the TM *d* shell ($U \simeq 10$ eV), any net excess charge in the *d* shell raises the energy of the *d* shell enough to polarize the bonds and "spill" most of the excess charge back into the vacancy level (which has a much smaller U). A charge deficit has the opposite effect. The only stable energy for the TM level is then that which yields approximate local neutrality within the *d* shell. Thus the energy of the *d* levels (and therefore also of the impurity level, which is an antibonding combination of the *d* and vacancy levels) is "pinned" with respect to the vacancy level.

If one imagines, for a given TM atom, changing the host semiconductor from one III-V to another, the main change[1] is simply to shift the vacancy level up or down in energy. According to the argument above, the impurity levels will then shift so as to follow the change in the vacancy level. Thus, up to some additive constant, the observed TM level gives a measure of the position of the ideal vacancy level. However, changing from a III-V to a II-VI semiconductor involves much more than just shifting the energetic position of the vacancy level,[1] so a comparison of TM levels between III-V's and II-VI's has no such simple interpretation.

The conclusions above are in marked contrast to an earlier theoretical analysis,[4] which concluded that the impurity level was pinned to the vacuum level, rather than to the vacancy level. That earlier conclusion, however, was based on a model which neglected the effect of electrostatic self-consistency, an unjustified approximation in the case of TM atoms.

While the arguments so far only relate the TM impurity level to the ideal vacancy level, one can argue further[1] that, because of the delocalization of the vacancy state, the vacancy level is approximately the same as the average of anion and cation dangling bond levels. Since Harrison and Tersoff showed[5] that the self-consistent band lineup at a heterojunction could be obtained by aligning the average dangling bond energy in the respective semiconductors, it should also then be true that aligning the TM impurity levels in the respective semiconductors gives the correct band lineup. Experimentally, this result had in fact been pointed out earlier by Langer and Heinrich[6] and by Zunger,[7] and provided the motivation for the present work.

According to Harrison and Tersoff,[5] the average dangling bond energy should also correlate with the Fermi-level pinning position at a Schottky barrier, and so the observed TM impurity levels should correlate not only with observed band lineups,[6] but also with observed Schottky barriers. This expectation is in fact confirmed by experimental data.[8]

In conclusion, TM impurity levels in zincblende-type semiconductors are "pinned" relative to the cation vacancy level, or average dangling bond energy, due to the self-consistent charge transfer between the host semiconductor and the TM *d* shell. This pinning is quite analogous to that which has been suggested as determining heterojunction band lineups[9] and Schottky barriers.[10]

References

1. J. Tersoff and W. A. Harrison, Phys. Rev. Lett. *58*, 2367 (1987).
2. G. Picoli, A. Chomette and M. Lannoo, Phys. Rev. B *30*, 7138 (1984).
3. W. A. Harrison, "Electronic Structure and the Properties of Solids", Freeman (New York, 1980).
4. M. J. Caldas, A. Fazzio and A. Zunger, J. Appl. Phys. *45*, 671 (1984).
5. W. A. Harrison and J. Tersoff, J. Vac. Sci. Technol. B *4*, 1068 (1986).
6. J. M. Langer and H. Heinrich, Phys. Rev. Lett. *55*, 1414 (1985).
7. A. Zunger, Phys. Rev. Lett. *54*, 849 (1985); and Ann. Rev. Mater. Sci. *15*, 411 (1985).

8. J. Tersoff, Phys. Rev. Lett. *56*, 675 (1986).
9. J. Tersoff, Phys. Rev. B *30*, 4874 (1984); and Phys. Rev. Lett. *56*, 2755 (1986).
10. J. Tersoff, Phys. Rev. Lett. *52*, 465 (1984); and references therein.

III. QUANTUM WELL STATES

THEORY OF IMPURITY STATES IN SUPERLATTICE SEMICONDUCTORS

G.P. Srivastava

Physics Department
University of Ulster
Coleraine
Northern Ireland BT52 1SA, U.K.

KEYWORDS/ABSTRACT: Superlattices/superlattice semiconductors/semiconductors/impurity states/impurities in semiconductors/theory of impurity states/effective mass theory/nipi crystals/impurity states in superlattices.

This paper presents a review of the theory of impurity states in superlattice semiconductors. First of all, a brief description is given of the effective-mass theory for shallow impurities in bulk semiconductors. A modification of the method is described to deal with hydrogenic donor states in a quantum well. A further modification of the method is described which allows for donor level calculations in a superlattice or a multiple quantum well. For calculating acceptor levels in quantum wells a multiband effective-mass theory is described. Finally, a self-consistent method, within the local-density and effective-mass approximations, is described for calculating electronic subband structure in the n-type layers of a n-i-p-i crystal.

INTRODUCTION

It is well known that shallow impurities play an important role in bulk semiconductors. Development of growth techniques such as molecular-beam epitaxy and metal-organic chemical vapour deposition has lead to fabrication of superlattice semiconductors. The role of shallow impurities in the electronic and optical properties of these two-dimensional semiconductors has become a subject of great interest in recent years.

In this paper we present a brief review of the theory of shallow impurity states in superlattice semiconductors[1].

The one-electron energy E and wave-function ϕ of the impurity problem are obtained by solving the Schrödinger's equation

$$(H^o + U)\Phi = E\Phi, \tag{1}$$

where H^o is the one-electron Hamiltonian of the pure and perfect crystal. The impurity potential is usually quite weak outside the immediate vicinity of the impurity and is expressed as

$$U(r) = -\frac{e^2}{\varepsilon_o r}, \tag{2}$$

where ε_o is the static dielectric constant of the host. The impurity wave function is mainly contributed by the host Bloch function $\psi(\underline{k}_o, \underline{r})$ at some reference point $\underline{k}_o$ in the Brillouin zone. Thus we express

$$\Phi = \frac{1}{\sqrt{V}} \sum_{n\underline{k}} F_n(\underline{k}) \quad \psi_n(\underline{k}_o, \underline{r}) \; e^{i(\underline{k}-\underline{k}_o)\underline{r}}$$

$$= \sum_n F_n(\underline{r}) \quad \psi_n(\underline{k}_o, \underline{r}) \tag{3}$$

i.e. the impurity wave function is an oscillatory band wave function modulated by an envelope function, $F_n(\underline{r})$, which is the Fourier transform of the coefficients $F_n(\underline{k})$.

For slowly varying potential U and the wave function of the type (3), eq. (1) can be transformed to [2]

$$[E_n(-i\nabla) + U(r)]\, F_n(\underline{r}) = E\, F_n(\underline{r}), \tag{4}$$

where E is now measured from the host band energy at $\underline{k}_o$. Eq. (4) is known as the effective mass equation (EM eq) and the approximation implied is called the effective mass approximation (EMA). It is much easier to solve eq. (4) than the original eq. (1). We will consider three different cases for the solution of eq (4):

(i) Donor states: the simple case of non-degenerate conduction band minimum at $\underline{k}_o = 0$

This is a particularly simple case. The non-degenerate conduction band minimum is at $\underline{k}_o = 0$ (Γ point) and the impurity wave function in (3) is expressed in terms of this band. Taking the zero of the energy at the conduction band minimum at Γ, eq (4) becomes the simple EM eq.

$$\left[- \frac{\hbar^2}{2m^*} \nabla^2 - \frac{e^2}{\varepsilon_o r} \right] F(r) = E\, F(r) \tag{5}$$

This is a scaled hydrogenic problem which can be solved exactly. It is also possible to obtain the true ground-state wavefunction and energy of this problem by using a trial function of the type

$$F\ (r) = \frac{1}{\sqrt{\pi r_o^{\ 3}}}\ e^{-r/\lambda} \tag{6}$$

where λ is a variational parameter.

The hydrogenic ground state is

$$E = \min \langle F/H/F \rangle \tag{7}$$

where $H=H^o+U$, and the minimization is done with respect to λ. The resulting expressions are

$$\lambda = a_o^* = \frac{\varepsilon_o \hbar^2}{m^* e^2} = \text{effective Bohr radius} \tag{8}$$

and

$$E = R_o^* = \frac{m^* e^4}{2\varepsilon_o^{\ 2} \hbar^2} = \text{effective Rydberg constant} \tag{9}$$

(ii) Donor states: intervalley mixing

If the minimum of the conduction band is at $\underline{k}_o \neq o$, then there are a number of equivalent minima at all points of the star $R\underline{k}_o$, R being a symmetry operator. For N equivalent minima eq (3) has to be expressed as

$$\Phi(\underline{r}) = \sum_{j=1}^{N} \alpha_j F_j(\underline{r}) \Psi(\underline{k}_o^j, \underline{r}) \tag{10}$$

where again we are using the one-band approximation. The coefficients $\{\alpha_j\}$ obey the point group symmetry of the host. The intervalley mixing splits the degeneracy of the ground state of the donor into states which belong to the irreducible representations of the symmetry group. To obtain the various splitted states one can use the perturbation theory over and above the EMA[2].

(iii) <u>Acceptor states</u>: the case of degenerate bands

Acceptor states are derived from the top valence bands which are usually degenerate at Γ. The problem of valence band mixing at cannot be dealt with the ordinary perturbation theory. Kane[3] showed that this problem can be solved by using the $\underline{k}.\underline{p}$ perturbation method. The acceptor states are obtained by solving a set of coupled equations which generalise the effective mass theory

$$\sum_{n'} \left[\sum_{\alpha\beta} D^{\alpha\beta}_{nn'} (-i\nabla_\alpha)(-i\nabla_\beta) - \lambda_{so} \Theta_n \delta_{nn'} \right] F_{n'}(\underline{r}) = E F_{n'}(\underline{r}), \tag{11}$$

$\alpha, \beta = x,y,z;\ n,n'=1, \ldots N.$ $\Theta_n=0$, $n=1, \ldots M$, for states which do not split due to spin-orbit interaction parameter λ_{so}. $\Theta_n=1$, $n=M+1, \ldots N$, for the spin-split-off states. $D_{nn'}$ are parameters related to the matrix elements of the momentum operator.

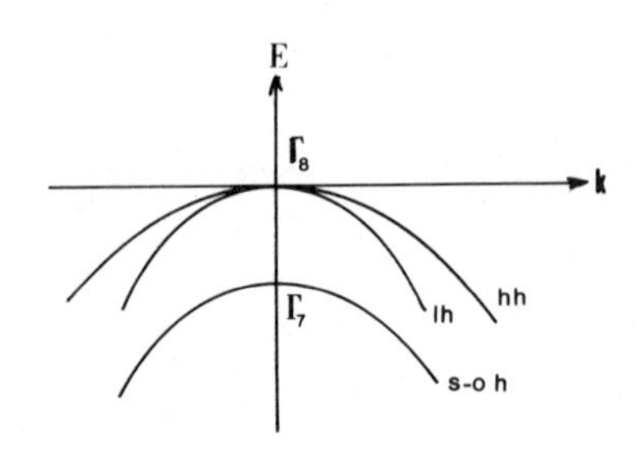

Fig. 1. Schematic representation of the top of the valence band structure in typical cubic semiconductors. The six-fold degenerate state Γ_{25}', (including spin degeneracy) splits into a four-fold state Γ_8 and a two-fold state Γ_7. hh and lh represent heavy hole and light hole, respectively.

In typical cubic semiconductors the top of the valence band is six-fold degenerate Γ_{25}' (including spin-degeneracy), so that N=6. In the limit of infinite spin-orbit interaction ($\lambda_{so} \rightarrow \infty$) eq (11) splits into a 4x4 eq. (M=4) which corresponds to Γ_8 for $\underline{k}=0$, and a 2x2 eq. which corresponds to Γ_7 for $\underline{k}=0$ (see Fig 1). The form of eq (11) depends on the choice of the perturbed valence band functions at Γ. A particularly convenient form of acceptor Hamiltonian of Γ_8 symmetry was suggested by Luttinger[4]

$$H_h = \frac{p^2}{2m^*}(\gamma_1 + \frac{5}{2}\gamma_2) - \frac{\gamma_2}{m^*}\sum_\alpha p_\alpha^2 J_\alpha^2$$

$$- \frac{2}{m^*}\gamma_3\left[\{p_x p_y\}\{J_x J_y\} + CP\right] - \frac{e^2}{\varepsilon_o r}, \tag{12}$$

where α = x,y,z, CP= cyclic permutation, $\{ab\}=(ab+ba)/2$, J_α is the 4 by 4 angular momentum operator matrix for J=3/2 and γ_i are dimensionless parameters to describe the valence band of the host.

The impurity wave function is expressed as

$$\phi(\underline{r}) = \sum_{n=1}^{M} F_n(\underline{r})\, \psi_n(o,\underline{r})$$

$$= \sum_{n=1}^{4} {}^n\Gamma_8 \sum_\ell f_{n\ell}(\hat{r})\, \psi(o,\underline{r}), \tag{13}$$

where M=4 = degeneracy of the top valence band, $\psi_n(o,\underline{r})$ is the Bloch function for band n at $\underline{k}$=o. $F_n(\underline{r})$ is an envelope function for band n and can be expressed in terms of spherical harmomics $f_{n\ell}(\hat{r})$ and a spin functions ${}^n\Gamma_8$. Since the EM eqs. (11) have inversion symmetry, only even angular momentum components (ℓ =0,2,...) contribute to the ground-state wave function $F_n(\underline{r})$. The four spin-3/2 spinors $\{{}^n\Gamma_8\}$ are deduced from the valence band symmetry Γ_8.

HYDROGENIC DONOR IMPURITY IN A QUANTUM WELL (QW)

Having briefly described the theory of impurity states

in bulk semiconductors, we now describe some of the theoretical models used in the literature to calculate impurity states in low dimensional systems such as quantum wells (QWs) and superlattices or multiple quantum wells (MQWs).

(i) Infinite potential barrier

Following Bastard [5], consider a QW of width L formed between two semi-infinitely thick barriers. Assume that image forces are negligible. The impurity problem can be described by the following modification of the hydrogenic EM Hamiltonian (one-band, spherically symmetric model) in (5)

$$H = \frac{-\hbar^2\nabla^2}{2m^*} - \frac{e^2}{\varepsilon_o(\rho^2 + z^2)^{\frac{1}{2}}} + V_B(z) \tag{14}$$

Here the impurity is assumed to be at the centre of the QW, $\rho^2=x^2+y^2$, z is the QW axis, and $V_B(z)$ is the potential energy barrier which confines the carrier in the 1D square well. We consider the origin at the centre of the well and assume the barrier of infinite height (Fig. 2). Thus

$$V_B\ (z) = \begin{cases} \infty & |z| > L/_2 \\ o & |z| \leq L/_2 \end{cases} \tag{15}$$

In the absence of the impurity potential the one-electron eigenvalues in the QW can be expressed as [6]

$$E_{nk} = \frac{\hbar^2 k_\perp^2}{2m^*} + \frac{\hbar^2 k_n^2}{2m^*}, \tag{16}$$

where $\underline{k}_\perp = (k_x, k_y)$, and $k_n = \frac{n\pi}{L}$, $n \geq 1$

Eq. (14) does not admit solutions in closed form as the variables ρ and z are non-separable. Therefore we consider the impurity wave function as a modified form of the variational trial function in eq. (6).

$$F(r) = \begin{cases} N \cos k_1 z \exp(-\frac{1}{\lambda}[\rho^2 + z^2]^{\frac{1}{2}}) & \text{if } |z| \leq L/_2 \\ o & \text{if } |\ | > L/_2 \end{cases} \tag{17}$$

N is a normalisation constant which depends on λ and L

$$N^2(\lambda,L) = \frac{2}{\pi\lambda^3}\left[1 - e^{-L/\lambda} + (1 + k_1^2\lambda^2)^{-2}\right.$$

$$\left. + \left\{(1 + k_1^2\lambda^2)^{-2} - \frac{L}{2\lambda} k_1^2\lambda^2(1 + k_1^2\lambda^2)^{-1}\right\} e^{-L/\lambda}\right]^{-1} \quad (18)$$

If E(L) denotes the eigen-energies of the Hamiltonion in (14), then the hydrogenic donor binding energy is given by

$$E_b(L) = \frac{\hbar^2 k_1^2}{2m^*} - E(L) = \frac{\hbar^2}{2m^*}\frac{\pi^2}{L^2} - E(L) \quad (19)$$

With (17) the result for E(L) is

$$E(L) = \frac{\hbar^2 k_1^2}{2m^*} + \frac{\hbar^2}{2m^*\lambda^2} - \frac{e^2\pi N^2 \lambda^2}{2\,\varepsilon_o}\left[1 + \frac{1}{1+k_1^2\lambda^2} - \frac{k_1^2\lambda^2 e^{-L/\lambda}}{1 + k_1^2\lambda^2}\right] \quad (20)$$

which can be extremised with respect to λ.

In the limit $L \to \infty$, the trial function in (17) reduces to (6) and $E_b(\infty)$ reduces to the result (9), thus reducing to the 3D case.

For L=o, eqs (17) and (19) reduce to

$$F(r,L=o) = \sqrt{\frac{2}{\pi\lambda^2}}\; e^{-\rho/\lambda} \quad (21a)$$

with

$$\lambda = a_o^*/2 \quad (21b)$$

and

$$E_b\ (L\to o) = 4R_o^*. \quad (21c)$$

This is the ground state situation of the 2D hydrogen atom.

The variation of the reduced binding energy of the hydrogenic donor level as a function of the well thickness is shown in Fig. 2. Clearly for $L \to \infty$, $E_b \to R_o^*$ the 3D result, and for $L \to o$, $E_b \to 4R_o^*$ the 2D result.

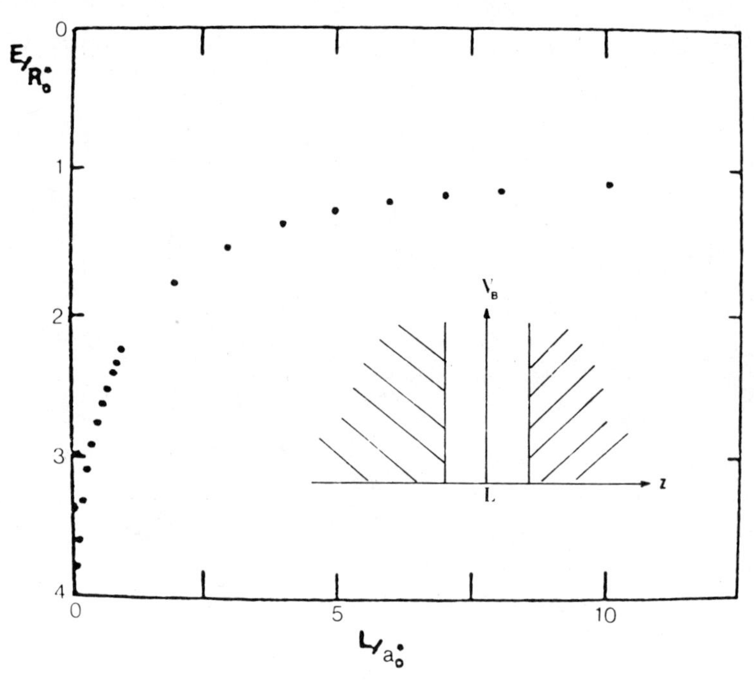

Fig. 2. Reduced binding energy of a hydrogenic impurity as a function of reduced well thickness. In the inset is shown a model QW with infinite barrier height.

(ii) Finite potential barrier

Greene and Bajaj[7] and Mailhiot et al[8] considered a GaAs QW sandwiched between two semi-infinite slabs of $Ga_{1-x}Al_xAs$. For a given size (L) of the QW they calculated the ground-state binding energy of a hydrogenic impurity, as a function of a finite potential barrier:

$$V_B(z) = \begin{cases} V_o & |z| > L/_2 \\ o & |z| \leq L/_2 \end{cases} \tag{22}$$

They used the variational method with a more general function. The main findings of Greene and Bajaj, shown in Fig 3 are: (a) a finite potential well shows a turning point in the binding energy for small well size (L), and (b) the binding energy E_b of a donor increases linearly as the square root of the well height, expect for very small potentials:

$$E_b \propto (V_o)^{\frac{1}{2}}. \tag{23}$$

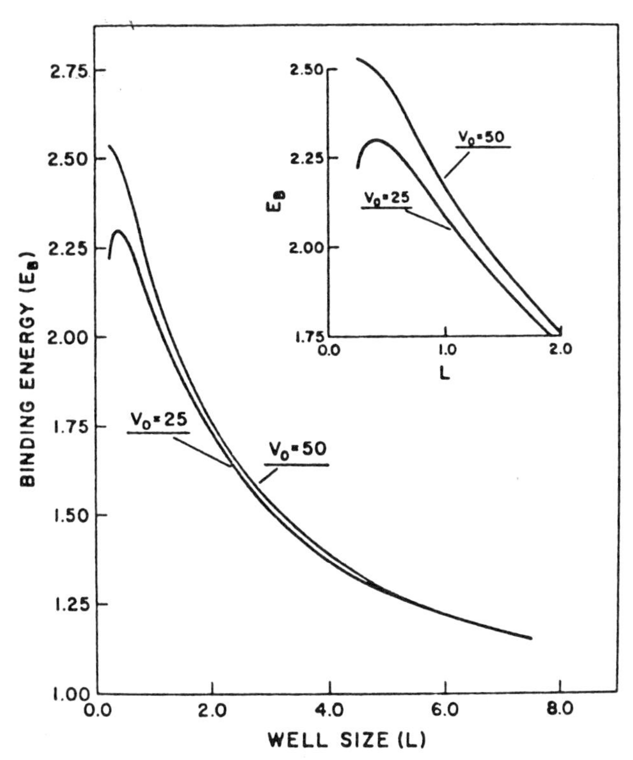

Fig. 3. Variation of the binding energy of the ground state of a hydrogenic donor as a function of the GaAs well size L for two different potential barriers. E_b in effective Rydberg (R_o*) and L in units of effective Bohr radius (a_o*).

The occurence of maximum in E_b for a finite V_o and very small L is due to the spread of the donor wave function into the surrounding $Ga_{1-x} Al_x As$ regions. This therefore may give a value of E_b which is more characteristic of a bulk donor in $Ga_{1-x} Al_x As$.

HYDROGENIC DONOR IMPURITY IN A MULTIPLE QUANTUM WELL (MQW)

The effective-mass -variational method for impurity level calculation described in the previous section can be extended to the case of a multiple quantum well (MQW). The method described here is due to Chaudhuri.[9]

Rather than considering a GaAs QW of size L surrounded by $Ga_{1-x} Al_x As$ of semi-infinite thickness, we consider a system as shown in Fig. 4: GaAs QW of size L surrounded by $Ga_{1-x} Al_x As$ barrier of thickness L_B, and height V_o. To avoid complicated mathematics, we further assume semi-infinite barriers beyond the adjacent wells.

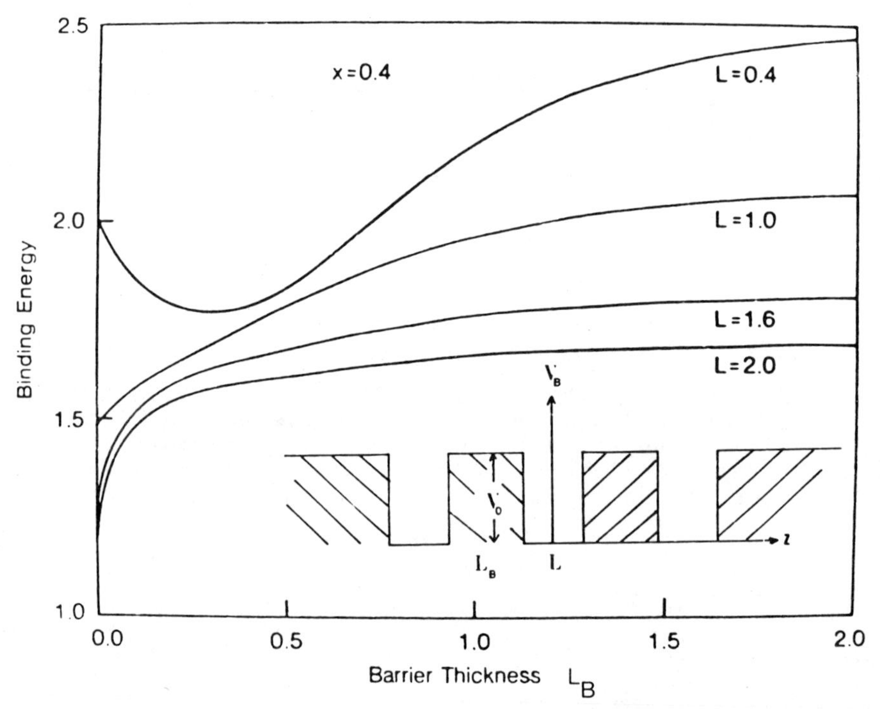

Fig. 4. Binding energy for a hydrogenic impurity at the centre of the GaAs well in a $GaAs/Ga_{1-x}Al_x As$ MQW. Energy in units of R_o* and thickness in units of a_o*. The calculations are done for a model MQW which is shown in the inset: the well and barrier widths are L and L_B respectively, and barriers beyond the adjacent wells are semi-infinite.

For the donor impurity atom at the centre of the central GaAs QW, the potential barrier is expressed as

$$
V_B(z) = \begin{cases} V_o & L/2 < |z| < L/2 + L_B \\ & \text{or } 3/2 L + L_B < |z| < \infty \\ o & o \leqq z \leqq L/2 \\ & \text{or } L/2 + L_B \leqq z \leqq 3/2 L + L_B \end{cases} \qquad (24)
$$

For this system the impurity wave function is chosen as a modification of eq. (17):

$$F(r) = N\, G(r)\, e^{-r/\lambda}$$

with

$$
G(r) = \begin{cases} \cos \alpha z & o \leqq z \leqq L/2 \\ A\, e^{\beta z} + B e^{-\beta z} & L/2 < z < L/2 + L_B \\ C \cos \alpha z + D \sin \alpha z & L/2 + L_B \leqq z \leqq 3/2 L + L_B \\ F\, e^{-\beta z} & 3/2 L + L_B < z < \infty \end{cases}
$$

(25b)

G(r) is a trial function appropriate to the Hamiltonian H without the Coulomb potential U(r), and as we have seen earlier $e^{-r/\lambda}$ is a hydrogenic trial wave function in the absence of the potential barrier $V_B(z)$. The quantities α and β are given by

$$\alpha = \sqrt{\frac{2m^*}{\hbar^2} E_o}\,, \qquad \beta = \sqrt{\frac{2m^*}{\hbar^2} (E_o - V_o)}\,,$$

(25c)

where E_o is the eigenenergy of an electron in the potential given by eq. (24).

Following the variational procedure outlined earlier,

the ground-state binding energy is given by

$$E_b\ (L, L_B, \lambda) = E_o - \min < F/H/F>, \tag{26}$$

where the minimisation is done with respect to λ. We will not present the detailed expression for E_b, but will only discuss the main features of the results obtained from this procedure.

Chaudhuri considered that the band gap difference, ΔE_g, between GaAs and $Ga_{1-x}\ Al_x$ As as a function of Al concentration x is given by

$$\Delta E_g = 1.555\ x + 0.37\ x^2 \quad \text{eV}. \tag{27}$$

He further assumed that the conduction band discontinuity is about 0.85 ΔE_g, so that the barrier height, for a given x, is

$$V_o = 0.85\ (1.555\ x + 0.37\ x^2) \quad \text{eV}. \tag{28}$$

Fig. 4 shows the binding energy for a hydrogenic impurity at the centre of a GaAs well for the barrier height corresponding to Al concentration x=0.4. The behaviour of the binding energy E_b as a function of the barrier thickness L_B depends on the values of the well thickness L and barrier height V_o. We have seen in Fig. 2 that for $L_B \to \infty$ a decrease in L increases the binding energy. In general, an increase in L_B tends to strongly localise the wave function around the impurity, which increases the binding energy. On the other hand, for finite barrier height V_o, increase in the barrier thickness L_B from zero allows for the wave function penetration in the barrier which reduces the binding energy.

It should be noted that this model is not expected to work for very thin SLs ($L, L_B < 0.5\ a_o^*$), because in that situation the approximations made in eqs. (24) and (25) of incorporating only the adjacent wells will break down.

The one-band, spherically symmetric hydrogenic model described earlier can be used to make a simple calculation of an acceptor ground state in a QW of infinite barrier. However, as we have mentioned earlier, a proper calculation of an acceptor state in semiconductors is more involved than the calculation of a donor state. In bulk cubic semiconductors the ground acceptor state is of Γ_8 symmetry defined by the Hamiltomian and wave function in eqs (12) and (13), respectively.

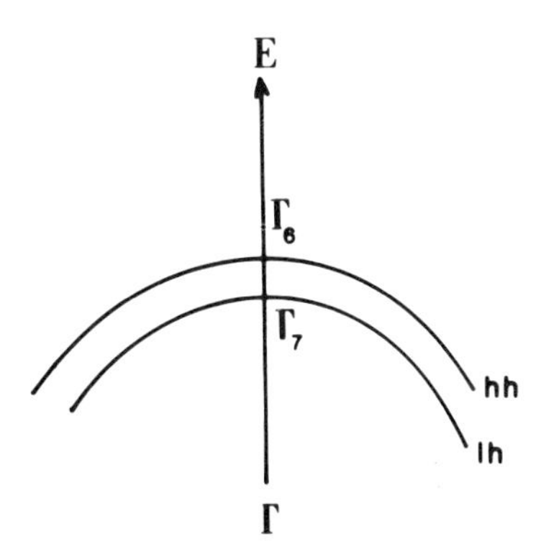

Fig. 5. Schematic representation of light and heavy hole states at the top of the valence band of a GaAs QW.

In bulk GaAs the point group symmetry is T_d. For a GaAs QW between semi-infinite $Ga_{1-x}Al_xAs$ barrier the point group symmetry is reduced to D_{2d}. The top of the bulk valence band Γ_8 is split into two two-fold degenerate states Γ_6 and Γ_7, shown schematically in Fig. 5, which correspond to the spin-3/2 spinors. The Γ_6 state is predominantly heavy hole (hh) like and the Γ_7 state is predominantly light hole (lh) like. There can be acceptor states corresponding to both symmetries.

We mentioned earlier that the acceptor envelope wave function will be a product of either an ℓ =o or ℓ =2 polynomial and a spin - 3/2 spinor. Masselink etal[10] used a 36-parameter variational wave function to describe an acceptor state of Γ_7 symmetry in a GaAs QW:

$$F_{\Gamma_7}(r) = \sum_{n=0}^{4} {}^{n}\Gamma_7 \sum_{i=1}^{7} c_n(i) e^{-\alpha_i (x^2 + y^2 + \mu z^2)} \quad (29)$$

where the 35 $c_n(i)$'s are variational parameters, the seven α_i are exponents to describe the extent of the wave function, μ (the 36th parameter) is an anisotropy factor which allows for the compression of the wave function in z-direction, and ${}^{n}\Gamma_7$ are the five spinor-polynomial products, obtained from group theory, as follows

$${}^{o}\Gamma_7 = \begin{bmatrix} s \\ o \\ o \\ o \end{bmatrix}, \qquad {}^{1}\Gamma_7 = \begin{bmatrix} z^2 - \frac{1}{2}(x^2+y^2) \\ o \\ o \\ o \end{bmatrix}$$

$${}^{2}\Gamma_7 = \begin{bmatrix} o \\ o \\ \sqrt{3}/2\ (x^2-y^2) \\ o \end{bmatrix}, \qquad {}^{3}\Gamma_7 = \begin{bmatrix} o \\ o \\ ixy \\ o \end{bmatrix}$$

$${}^{4}\Gamma_7 = \begin{bmatrix} o \\ \frac{1}{\sqrt{2}}\ (xz+iyz) \\ o \\ o \end{bmatrix} \quad (30)$$

The Γ_6 basis is obtained by exchanging the 1st and 2nd entries and the third and fourth entries in the Γ_7 basis.

The acceptor Hamiltonian in the QW is obtained by adding the acceptor barrier term V^v_B to the acceptor bulk Hamiltonian (12): Thus

$$H_h^{QW} = \frac{p^2}{2m^*} \left(\gamma_1 + \frac{5}{2}\gamma_2\right) - \frac{\gamma_2}{m^*} \sum_\alpha p_\alpha^2 J_\alpha^2$$

$$- \frac{2\gamma_3}{m^*} \left[\{p_x p_y\}\{J_x J_y\} + CP \right] - \frac{e^2}{\varepsilon_o r} + V_B^v (z). \quad (31)$$

For a finite barrier height

$$V_B^v (z) = \begin{cases} V_o^v & |z| \geqq L/2 \\ o & |z| < L/2 , \end{cases} \quad (32)$$

with $V^v{}_o$ as the valence-band discontinutiy between the well material of (thickness L) and the barrier material (assumed of infinite thickness).

Masselink et al took

$$V_o^v = 0.35 \quad \Delta E_g(x) \quad (33a)$$

where

$$\Delta E_g \;(x) = 1.247\; x \quad \text{eV for } x < 0.45 \quad (33b)$$

for the GaAs/Ga_{1-x} Al_x As system. [Notice that this is a more recent result for the discontinuity split than is suggested in eg. (28)]. The Luttinger parameters were taken to be:

for GaAs ε_o = 12.35, γ_1 = 7.65, γ_2 = 2.41, γ_3 = 3.28
for AlAs ε_o = 9.80, γ_1 = 4.04, γ_2 = 0.78, γ_3 = 1.57

For the alloys, weighted averages of the values for GaAs and AlAs were used.

The eigensolutions of the Hamiltonian in eq. (31) are obtained by solving a 35x35 matrix equation $H_h{}^{QW}F = E\,S\,F$, where S is the overlap matrix. The parameter μ is varied until E is minimised. The resulting lowest energy eigenvalue is the Γ_6 or Γ_7 ground-state energy relative to the bulk valence-band edge. The Γ_6 binding energy is measured from the top of the heavy-hole subband and the Γ_7 binding energy is measured from the top of the light-hole subband.

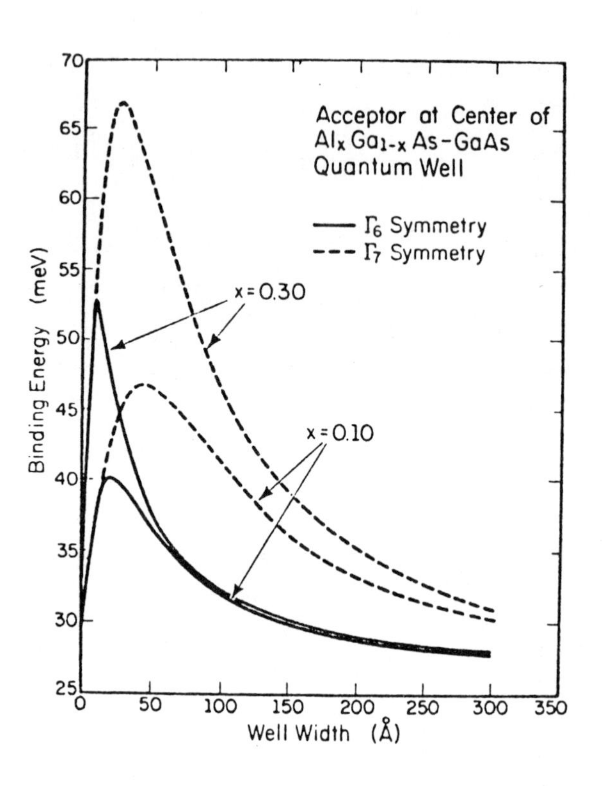

Fig. 6. Calculated binding energies of ideal Γ_6 (hh) and Γ_7 (lh) acceptor ground states in a GaAs QW between semi-infinite $Ga_{1-x}Al_x$ As barriers of finite height.

Fig. 6 shows the binding energies of acceptor ground states of both Γ_7 and Γ_6 symmetries in a GaAs QW for two barrier heights. For very thick wells the Γ_7 and Γ_6 acceptor ground states merge into the bulk result (27.1 meV) for the Γ_8 acceptor ground state. As the QW effects become important, the binding energy of the Γ_7 (lh) ground state becomes greater than that of the Γ_6 (hh) ground state. In analogy with the hydrogenic donor case, for narrower wells the effect of the hole confinement becomes pronounced and the binding energies increase. Also, for a finite barrier height there is a maximum acceptor ground state binding energy for a narrow well width.

To account for binding energy shift for different

acceptors, Masselink et al found it necessary to add a short-ranged core potential of the form $H_c = U \exp[-(r/r_o)^2]$, with $r_o = 1\text{Å}$ and adjustable U. For the Γ_6 centre acceptor in GaAs - $Ga_{0.7}Al_{0.3}As$, U was adjusted to be 8.0 eV for C and -5.55 eV for Be. As shown in Fig 7, in the limit of very thick QWs, the C acceptor and Be acceptor binding energies, 26.0 meV and 28.0 meV, respectively, are the bulk results. This figure also shows experimentally measured data for the binding energies of C and Be in the QWs. Clearly, the agreement between theory and experiment is very good.

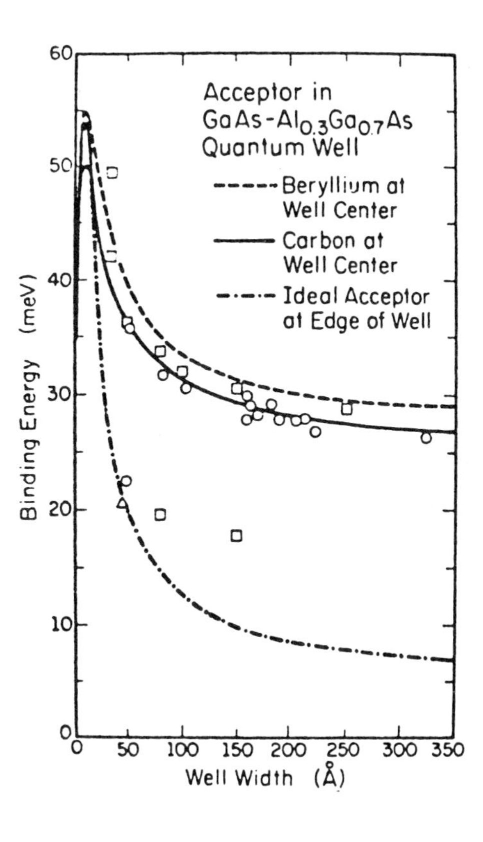

Fig. 7. Calculated binding energies of the heavy-hole centre doped Be and C acceptors in GaAs - $Ga_{0.7}Al_{0.3}As$ QWs. The experimental data are indicated by squares and triangles for Be and by circles for C.

A comment on the use of the splitting of the total band gap discontinuity ΔE_g at GaAs - $Ga_{1-x}Al_xAs$ between the conduction and valence band discontinuities, V_o and V^v_o, is in order. The acceptor binding energies in wells wider than about 100 Å do not seem to be affected by the choice of the valence-band discontinuity V^v_o. For narrower wells the theoretical results are in better agreement with experiment for the valence-band discontinuity of 35% (eq. 33) rather than 15% as suggested from the use of eq. (28).

ELECTRONIC SUBBANDS IN n-i-p-i CRYSTALS

The variational - EM theory we have discussed so far uses a linearly screened Coulomb impurity potential and tacitly assumes very low impurity concentration. Here we briefly review the work of Döhler and co-workers [11-13] on self-consistent calculations of the electronic states in doping superlattices. More details of this work can be found Professor Döhler's review in this workshop.

We consider a particular type of doping SLs, a n-i-p-i crystal, consisting of periodic n- and p- type doped semiconductor layers which are separated by undoped or intrinsic (i-) layers as shown in Fig. 8.

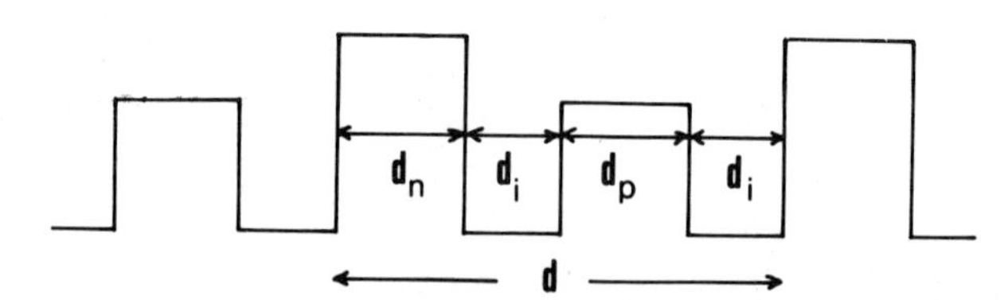

Fig. 8. A n-i-p-i crystal of periodicity d in z-direction.

Consider periodic rectangular homogeneous doping profiles with concentrations n_A and n_D in the p- and n-layers, respectively,

$$n_D\,(z+d) = n_D(z)$$
$$n_A\,(z+d) = n_A(z) \qquad (34)$$

where $d = d_p + 2d_i + d_n$ is the periodicity in the z, or SL, direction. Consider the origin in the middle of an n-type doped layer. In the ground state all the impurities are ionised. This gives rise to a space-change ionic potential which is determined by the solution of Poisson's equation

$$\frac{\partial^2 v_i(z)}{\partial z^2} = \frac{4\pi e^2}{\varepsilon_o} [n_D(z) - n_A(z)] \tag{35a}$$

subject to the boundary conditions

$$\left.\frac{\partial v_i(z)}{\partial z}\right|_{z=o} = o, \qquad v_i(o) = o \tag{35b}$$

If the crystal is not macroscopically compensated, then the condition of macroscopic neutrality requires a periodic electron or hole space-charge distribution - en(z) or ep(z), respectively. If we are only interested in the electronic subband structure in the n-type layers, then the Coulomb repulsion can be expressed by the Hartree contribution of the electrons which is given by the solutions of Poisson's equation

$$\frac{\partial^2 v_H(z)}{\partial z^2} = -\frac{4\pi e^2 n(z)}{\varepsilon_o} \tag{36a}$$

subject to the boundary conditions

$$\left.\frac{\partial v_H(z)}{\partial z}\right|_{z=o} = o, \qquad v_H(o) = o \tag{36b}$$

The quantum mechanical exchange and correlation potentials can be combined together and expressed, within the local density approximation [14,15] as

$$V_{xc}(z) = \varepsilon_{xc}(n(z)) + n(z)\frac{\delta\varepsilon_{xc}}{\delta n}, \tag{37}$$

where $\varepsilon_{xc}(n)$ is the exchange and correlation energy per electron of a homogeneous electron gas of the (local) density n. As the electronic densities of interest in n-i-p-i crystals are usually very high, one can approximately take

$$\varepsilon_{xc} \cong \varepsilon_{x} = \frac{-0.916}{r_s} \cdot \frac{e^2}{2\varepsilon_o a_o^*} \tag{38a}$$

with

$$r_s = \left(\frac{4\pi}{3n}\right)^{\frac{1}{3}} \frac{1}{a_o^*} \tag{38b}$$

and a_o^* = effective-mass Bohr radius.

We will neglect donor impurity band formation. Within the EMA, and assuming MQW effects, the electronic charge density n(z) is calculated from

$$n(z) = 2 \sum_{\mu,\underline{k}} \left| F_{n,\mu} (k_z,z) \right|^2 n_{\mu}^{(2)} \tag{39a}$$

where

$$n^{(2)} = \sum_{\mu} n_{\mu}^{(2)} = n_D d_n - n_A d_p \tag{39b}$$

is the 2D electron concentration, and $F_{n,\mu}$ is the envelope function of the subband μ corresponding to n-th band of the undoped SL at a reference $\underline{k}_o$. The envelope function is obtained from the solution of the equation

$$\left[\frac{-\hbar^2}{2m^*} \frac{d^2}{dz^2} + v_i(z) + v_H(z) + v_{xc}(z)\right] F_{n,\mu}(z) = \varepsilon_{n,\mu} F_{n,\mu}(z) \tag{40}$$

This has to be solved self-consistently. The results $\varepsilon_{n,\mu}$ define sub-band energies of electrons in the n-i-p-i crystal. We shall discuss the results for two specific cases.

(i) Compensated doping SL with $d_n = d_p$, $d_i = o$, and $n_D = n_A$

As shown in Fig 9, in this case the system has an effective energy gap

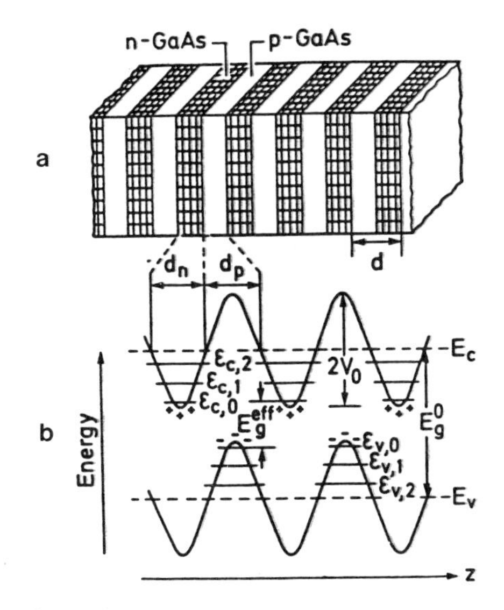

Fig. 9.
(a) A GaAs doping SL with layer thicknesses d_n and d_p and doping concentrations n_D and n_A.

(b) Schematic real-space energy band structure of a macroscopically compensated GaAs doping SL (wth $d_n = d_p$, $d_i = o$, $n_D = n_A$). Plus signs indicate ionized donor levels in the n layers near the conduction band edge and minus signs the negatively charged acceptor levels in the p layers above the valence band edge.

$$E_g^{eff} = E_g - 2V_o + \varepsilon_{c,o} + |\varepsilon_{vh,o}| \tag{41}$$

where

$$V_o = \frac{4\pi e^2}{\varepsilon_o} \cdot \frac{n_D \, d_n^2}{8} \tag{42}$$

is the amplitude of $v_i(z)$, E_g is the bulk band gap,

$$\varepsilon_{c,\mu} = \hbar \left(\frac{4\pi e^2 \, n_D}{\varepsilon_o \, m^*} \right)^{\frac{1}{2}} (\mu + \tfrac{1}{2}) \tag{43}$$

for conduction subbands, and a similar expression for valence hole subbands $\varepsilon_{vh,\mu}$. The spacing and width of the subbands can be tailored by appropriate design parameters of the structure.

The compensated GaAs doping SL becomes a material with an <u>indirect band gap in real space</u>. With a proper choice of the design parameters, the electron-hole recombination lifetimes may become extremely large.

(ii) <u>Non-compensated doping SL</u>

For a doping SL with $n_D \, d_n > n_A \, d_p$, there will be a finite 2D electron concentrations $n^{(2)}$ in the n layers. In the ground state of this n-type SL one defines an effective band gap

$$\tilde{E}_g^{eff} = \phi_n - \phi_p \tag{44}$$

where ϕ_n and ϕ_p are electron and hole quasi-Fermi levels:

$$\phi_n = E_c + v_{sc} \, (o) + \varepsilon_{c,o} + E_{F,o} \tag{45}$$

with $v_{sc} = v_i + v_H + v_{xc}$, and $E_{F,o}$ is the finite Fermi energy in the subband o, and

$$\phi_p \cong E_v + v_{sc} \, (d/2) + E_A \tag{46}$$

with E_A as the threshold of occupied acceptor impurity band states.

An important consequence is that one can modulate the effective gap of a given doping SL by varying the electron and hole concentrations in n- and p- layers, respectively.

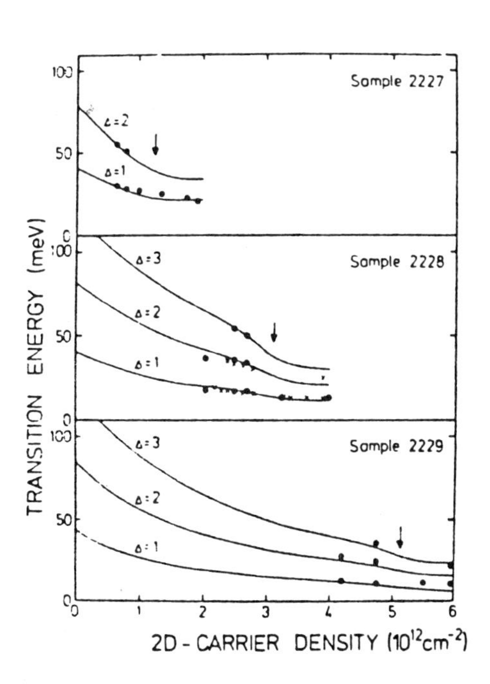

Fig. 10. Experimental subband transition energies in GaAs doping SLs compared with the results of self-consistent calculations. In all cases $n_D = n_A = 10^{18}$ cm^{-3}, and $d_i = o$. Sample 2227: $d_n = d_p = 20$ nm; Sample 2228: $d_n = d_p = 40$ nm; Sample 2229: $d_n = d_p = 60$ nm.

The theoretical predictions of a tunable effective band gap, quantisation of carriers and spacing in subbands have been validated by the results of photoluminescence and Raman measurements (Döhler etal[11], Zeller etal[16]). Fig. 10 shows such a comparison.

SUMMARY

In this article we have reviewed some of the existing theories of impurity states in superlattice semiconductors. The EMA- variational method predicts results which are in good agreement with experimentally measured impurity states in QWs. The self-consistent theory, within the EMA, of 2D electronic subband structure and tunable effective energy band gap of doping SLs predicts results which are in very good agreement with experiment.

REFERENCES

1. The reference list is not intended to be exhaustive.
2. see, e.g. F. Bassani, G. Pastori Parravicini, and R.A. Ballinger, "Electronic States and Optical Transitions in Solids", Pergamon Press, Oxford (1975).
3. E.O. Kane, J. Phys. Chem. Solids 1: 83 (1956).
4. J.M. Luttinger, Phys. Rev. 102:1030 (1956).
5. G. Bastard, Phys. Rev. B24:4714 (1981).
6. see, e.g. L. Schiff, "Quantum Mechanics", McGraw-Hill Book Company, Inc., New York (1955).
7. R.L. Greene and K.K. Bajaj, Solid State Commn., 45:825 (1983).
8. C. Mailhiot, Y.-C. Chang and T.C. McGill, Phys. Rev. B26: 4449 (1982).
9. S. Chaudhuri, Phys. Rev. B28:4480 (1983).
10. W.T. Masselink, Yia-Chung Chang, and H. Morkoc, Phys. Rev. B32:5190 (1985).
11. G.H. Döhler, H. Künzel, D. Olego, K. Ploog, P. Ruden, H.J. Stolz, and G. Abstreiter, Phys. Rev. Lett. 47:864 (1981).
12. P. Ruden and G.H. Döhler, Phys. Rev. B27:3538 (1983).
13. K. Ploog and G.H. Döhler, Adv. in Phys. 32:285 (1983).
14. P. Hohenberg and W. Kohn. Phys. Rev. 136:B864 (1964).
15. W. Kohn and L.J. Sham, Phys. Rev. 140: A1133 (1965).
16. Ch. Zeller, B. Vinter, G. Abstreiter, and K. Ploog, Phys. Rev. B26:2124 (1982).

EFFECTIVE-MASS THEORY OF ELECTRON STATES IN HETEROSTRUCTURES AND QUANTUM WELLS*

U. Rössler, F. Malcher, and A. Ziegler

Institut für Theoretische Physik
Universität Regensburg
D-8400 Regensburg, F.R.G.

ABSTRACT

For describing near band edge states of intrinsic (subbands) and extrinsic (impurity states) character in semiconductor structures effective-mass theory is as important as in bulk semiconductors. We 1) reconsider the concept of effective-mass theory in this context, 2) demonstrate, how in a self-consistent calculation of subband states in modulation doped heterostructures the implications of the bulk band structure and doping profiles are considered, and 3) suggest a new concept for treating the impurity problem.

INTRODUCTION

Over the last 30 years experimental investigations of near band cedge states in bulk semiconductors and their quantitative interpretation in terms of effective-mass theory have contributed much to our understanding of electron states in semiconductors. The essential idea of the effective-mass theory is, that band edge states in semiconductors can be considered as of free particles with kinematic properties defined by the band edge dispersion, i.e. the effect of the periodic crystal potential is contained in effective-mass parameters.[1-4] The Hamiltonian, whose eigenvalues give the band edge dispersion, can be taken as kinetic energy operator in order to describe electron states in an extrinsic potential, which varies slowly on the scale of the crystal lattice constant.[4]

Details of the band edge structure, like many-valley aspect (in Si, Ge) and valence band degeneracy, were recognized early[1-4] and lead to a series of very detailed studies of donors and acceptors,[5] Landau levels,[6] excitons and exciton polaritons.[7] As far as these investigations were connected with quantitative interpretations, they resulted for some semiconductors (like Ge and GaAs), in highly consistent sets of valence and conduction band parameters.[8] Nonparabolicity, i.e. the deviation of the energy dispersion relation from a simple quadratic dependence on the particle momentum, was known to be important for small-gap semiconductors,[3] but has become of interest recently also in wide gap materials like GaAs.[9] While in narrow-gap semiconductors like InSb and HgCdTe nonparabolicity is dominated by the $\mathbf{k}\cdot\mathbf{p}$ interaction across

*Work supported by Deutsche Forschungsgemeinschaft

the small gap between lowest conduction band Γ_{6c} and topmost valence bands $\Gamma_{8v} + \Gamma_{7v}$, contributions from the coupling to the higher conduction bands, $\Gamma_{7c} + \Gamma_{8c}$, are important as well in GaAs (and other wide-gap materials).

The effective-mass theory, which has been so successful in bulk semiconductors, has proved to be powerful also in semiconductor structures like heterojunctions, quantum wells and superlattices. A derivation of the effective-mass Hamiltonians for this case will be given in section 2. In section 3 we report on self-consistent subband calculations in heterostructures of a wide-gap (AlGaAs/GaAs) and a narrow-gap system (HgCdTe). In section 4, we critically discuss frequently used effective-mass Hamiltonians for impurities in quantum wells, and conclude with a brief summary.

EFFECTIVE-MASS HAMILTONIANS FOR SEMICONDUCTOR STRUCTURES

In order to derive the effective-mass Hamiltonian for semiconductor structures we proceed in complete analogy to the bulk case[4] and start with the single-particle Hamiltonian

$$H_0 = -\frac{\hbar^2}{2m}\Delta + U(\mathbf{r}) \tag{1}$$

where $U(\mathbf{r})$ is the crystal potential, which in the present case does not have the translational symmetry with respect to lattice vectors in all directions. Instead

$$U(\mathbf{r}) = \begin{cases} U^A(\mathbf{r}) & |z| < 0, \quad |z| < \frac{L_z}{2}, \quad |z - nd| < \frac{L_z}{2} \\ U^B(\mathbf{r}) & |z| > 0, \quad |z| > \frac{L_z}{2}, \quad |z - nd| > \frac{L_z}{2} \end{cases} \tag{2}$$

for heterostructures, quantum wells, and superlattices, respectively, where $U^{A,B}(\mathbf{r})$ is the lattice potential of semiconductors A, B, of which the structure is built up. Certainly, assuming the potential to change abruptly at the interface, is an idealization, but it is not too far from the reality of well-fabricated semiconductor structures.

In the vein of effective-mass theory[4] we expand the solution of the eigenvalue problem

$$H_0\psi = E\psi \tag{3}$$

by using the set

$$\chi_{n\mathbf{k}}(\mathbf{r}) = e^{i\mathbf{k}\mathbf{r}} \begin{cases} u_{n0}^A(\mathbf{r}) & z \in A \\ u_{n0}^B(\mathbf{r}) & z \in B \end{cases} \tag{4}$$

where $u_{n0}^{A,B}(\mathbf{r})$ are band edge eigenfunctions (at $\mathbf{k} = 0$) of bulk semiconductors A or B. Because of lacking translational symmetry in the growth direction (to be taken as z-direction) the solutions of eq. 3 are wave packets with respect to k_z

$$\psi_{\nu\mathbf{k}_\|}(\mathbf{r}) = \sum_{nk_z} C_{n\nu}(k_z)\chi_{n\mathbf{k}}(\mathbf{r}) \tag{5}$$

having ν and $\mathbf{k}_\| = (k_x, k_y, 0)$ as a complete set of quantum numbers. It should be mentioned, that in contrast to ref. 4, the $\chi_{n\mathbf{k}}$ of eq. 4 are only piecewise continuous and eventually do not form a complete set. The envelope function $f_\nu(z)$, which is the Fourier transform of $C_{n\nu}(k_z)$, is expected to be slowly varying in z, i.e. $C_{n\nu}(k_z)$ will be different from zero only for $|k_z| \ll \frac{2\pi}{a}$, where a is the lattice constant. Therefore, we can perform averaging over unit cells[10] after multiplication with $(u_{n0}^{A,B})^*$ from the left and obtain

$$\sum_{nk_z} e^{ik_z z}\left[(E_n^{A,B}(0)+\frac{\hbar^2 k^2}{2m}-E)\delta_{n'n}+\frac{\hbar}{m}\mathbf{k}\cdot\mathbf{p}_{n'n}^{A,B}\right]C_{n\nu}(k_z)=0, \tag{6}$$

where

$$\mathbf{p}_{n'n}^{A,B}=\int d^3r\left(u_{n'o}^{A,B}(\mathbf{r})\right)^*\mathbf{p}\,u_{n0}^{A,B}(\mathbf{r}) \tag{7}$$

is the z-dependent ($z\in A,B$) momentum matrix element. Eq. 6 is arrived at by making use of the fact, that $u_{n0}^{A,B}(\mathbf{r})$ are solutions of H_0 for the band edge energies $E_n^{A,B}(0)$ in semiconductors A and B and that the integral in eq. 7 is taken over a crystal unit cell which is assumed to be entirely in A or B.

The next step in deriving the effective-mass equation is to eliminate (e.g. by a unitary transformation[4]) the off-diagonal momentum matrix elements in lowest order to obtain for a given band n:

$$\sum_{k_z} e^{ik_z z}\left[\frac{\hbar^2 k^2}{2m}+\frac{\hbar^2}{m^2}\sum_{n''\neq n}\frac{\mathbf{k}\cdot\mathbf{p}_{nn''}^{A,B}\,\mathbf{p}_{n''n}^{A,B}\cdot\mathbf{k}}{E_n^{A,B}(0)-E_{n''}^{A,B}(0)}+E_n^{A,B}(0)-E\right]C_{n\nu}(k_z)=0. \tag{8}$$

If we consider as band n the lowest (spin-degenerate) conduction band Γ_{6c} of a standard semiconductor, the Fourier transform of this equation is the well known effective-mass equation for the semiconductor structure

$$\left\{-\frac{\hbar^2}{2}\partial_z\frac{1}{m(z)}\partial_z+V(z)-\left(E_\nu+\frac{\hbar^2 k_\parallel^2}{2m(z)}\right)\right\}f_\nu(z)=0. \tag{9}$$

The expectation value of the last term with respect to $f_\nu(z)$ gives the dispersion of the νth subband.

The z-dependent band edge mass has been introduced in accordance with

$$\frac{m}{m(z)}=\begin{cases}1+\frac{2}{m}\sum_{n''\neq n}\frac{|\mathbf{p}_{nn''}^A|^2}{E_n^A(0)-E_{n''}^A(0)} & z\in A\\ 1+\frac{2}{m}\sum_{n''\neq n}\frac{|\mathbf{p}_{nn''}^B|^2}{E_n^B(0)-E_{n''}^B(0)} & z\in B\end{cases} \tag{10}$$

and the z-dependent band edge energies have been formulated as potential

$$V(z)=\begin{cases}E_n^A(0) & z\in A\\ E_n^B(0) & z\in B.\end{cases} \tag{11}$$

It is important to note, that this potential appears in the effective-mass eq. 9, although it does not vary slowly in space. This is not in contradiction to effective-mass theory, but a consequence of the inhomogenity of the semiconductor structure. In addition, for modulation doped heterostructures and quantum wells the electrostatic potential of the built-in charges, which varies slowly in space and fulfills the requirements of the effective-mass concept, has to be added and determined self-consistently.

If in eq. 8 n denotes the topmost fourfold degenerate valence band Γ_{8v} with angular momentum states $M=\pm\frac{1}{2},\pm\frac{3}{2}$, then

$$H_{\Gamma_8} = \left(\frac{\hbar^2 k^2}{2m} + E_n^{A,B}(0)\right)\delta_{MM'} + \frac{\hbar^2}{m^2}\sum_{n''\neq n} \frac{\mathbf{k}\cdot\mathbf{p}_{nM,n''}^{A,B}\,\mathbf{p}_{n'',nM'}^{A,B}\cdot\mathbf{k}}{E_n^{A,B}(0) - E_{n''}^{A,B}(0)} \tag{12}$$

corresponds to the Luttinger Hamiltonian,[2] and by taking the Fourier transform becomes a set of four coupled differential equations.[11] To solve these equations is already a considerable task, which has been achieved so far only approximately.[11] As a consequence of the coupling between heavy ($M = \pm\frac{3}{2}$) and light holes ($M = \pm\frac{1}{2}$) at finite $k_{\|}$ the hole subbands show a complex dispersion, with consequences on Landau levels,[12] exciton binding energies,[13] and polarization of luminescence.[14]

The approximation that lead from eq. 6 to eqs. 9 and 12 consisted in eliminating the $\mathbf{k}\cdot\mathbf{p}$ coupling between different energy bands to lowest order. The resulting effective-mass Hamiltonians, consequently, contain only terms bilinear in $\mathbf{k}$. Nonparabolicity, which is important in narrow-gap materials[3] but shows up even in GaAs,[9] can be considered either by eliminating the interband coupling to higher order — this results in higher order (than bilinear) in $\mathbf{k}$ terms in the Hamiltonian for a single band[15] — or by considering valence and conduction bands as quasi-degenerate — which gives larger matrix Hamiltonians with linear in $\mathbf{k}$ coupling between different bands as in eq. 6[16,17] (see next section). Because in the bulk band structure the nonparabolicity increases with the kinetic energy of the quasiparticles, its effect on subband states in a quantum well or heterostructure increases with the confinement (or subband) energy and results also in an increase of the subband masses over the band edge mass.[15] Moreover, due to the nonspherical symmetry of the zincblende lattice, the effective-mass Hamiltonian depends on the crystallographic orientation of the semiconductor structure even for the Γ_{6c} conduction band.

SELF-CONSISTENT CALCULATION OF SUBBAND STATES IN MODULATION DOPED HETEROSTRUCTURES

In order to demonstrate the combined effect of bulk band structure and doping profile on subband energies, we report in this section on our calculations for n-inversion channels in AlGaAs/GaAs[15,18] heterostructures and on p-doped HgCdTe.[17]

AlGaAs/GaAs heterostructures

The potential profile of a modulation-doped AlGaAs/GaAs heterostructure with slightly p-doped GaAs ($N_A \simeq 10^{15}\mathrm{cm}^{-3}$) and n-doped AlGaAs ($N_{S_i} \simeq 10^{18}\mathrm{cm}^{-3}$) is shown in Fig. 1.

The thermodynamic equilibrium (constant Fermi energy over the interface) requires transfer of electrons from the AlGaAs side (taken from the Si donors) to GaAs, where they occupy acceptor states and electron states in the inversion channel, which forms due to the band bending by the electrostatic potential connected with these charges. Usually an undoped spacer (thickness d) is introduced on the AlGaAs side in order to separate the ionized impurities from the electrons in the inversion channel, which makes possible very high mobilities of these carriers for motion parallel to the interface. The charge distribution for the heterostructure is

$$\rho(z) = \begin{cases} eN_{S_i} - e\sum_\nu N_\nu f_\nu^2(z) & -d' < z < -d \\ -e\sum_\nu N_\nu f_\nu^2(z) & -d < z < 0 \\ -eN_A - e\sum_\nu N_\nu f_\nu^2(z) & 0 < z < z_d \end{cases} \tag{13}$$

$-e\sum_\nu N_\nu f_\nu^2(z)$ is the charge distribution of electrons in the n-channel, described by the subband eigenfunctions $f_\nu(z)$ and partial density N_ν, which have to be determined by self-consistent solution of the Schroedinger and Poisson equations. The latter is

$$\frac{d^2}{dz^z} V_H(z) = \frac{e}{\varepsilon_0 \varepsilon(z)} \rho(z) \tag{14}$$

where the boundary conditions $\frac{d}{dz} V_H(z) = 0$ for $z = z_d$ and $z \to -d'$ and $V_H(z = 0) = 0$ for the Hartree potential guarantee the charge neutrality

$$N_A z_d + \sum_\nu N_\nu = N_{S_i}(d' - d) \tag{15}$$

and fix the zero point of the Hartree potential, respectively. $N_S = \sum_\nu N_\nu$ is the total areal electron density in the n-channel. In actual calculations the Schroedinger eq. 9 is solved for a potential

$$V(z) = V_0 \theta(-z) + V_H(z) + V_{xc}(z) \tag{16}$$

which contains also an exchange-correlation part in the local density approximation.[19] $V_0 = E^A(0) - E^B(0)$, A: $Al_xGa_{1-x}As$ and B: GaAs, is the conduction band-offset, which we have chosen to be 65% of the bandgap difference.

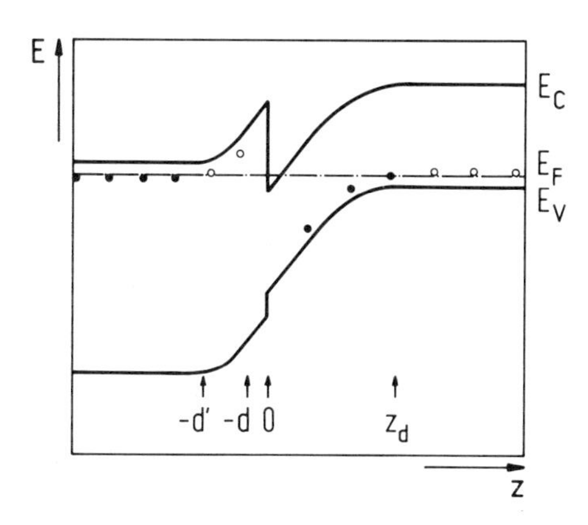

Fig. 1. Potential profile (z-dependence of band edges) of a modulation-doped AlGaAs/GaAs heterostructure. z_d is the depletion length, d the spacer thickness. The potential is flat for $z > z_d$ and $z < -d'$.

In our self-consistent calculation we keep $m^{A,B}$, V_0, $\varepsilon^{A,B}$ fixed and vary the input parameter $N_D = N_A z_d$ and N_S. The dependence of the energy difference E_{10} between the lowest subbands and of the Fermi energy E_{F0} with respect to the bottom of the lowest subband as function of N_S is shown for two different N_D in Fig. 2. With increasing N_S or N_D the channel gets narrower and the confinement effect, i.e. the subband separation E_{10}, increases. The Fermi energy increases (almost) linearly with N_S, due to the (almost) constant density of states, because nonparabolicity does not show up in this drawing. The dependence of E_{10} on N_D, in particular the point, at which the second subband touches the Fermi energy, can be used to determine the acceptor concentration, which is usually not accurately known for MBE or MOCVD grown GaAs.

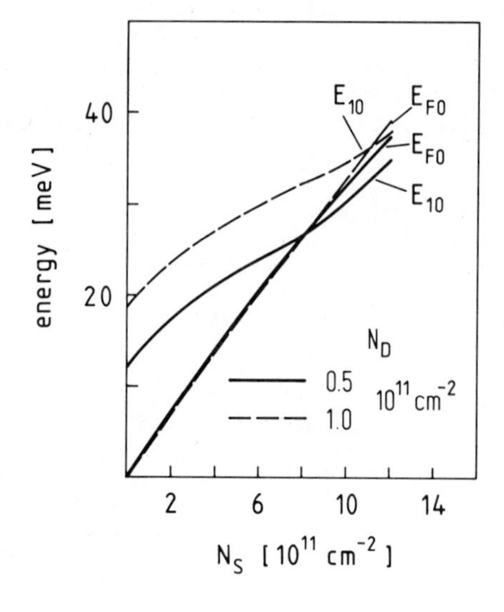

Fig. 2. Dependence of subband separation E_{10} and Fermi energy E_{F0} on N_S for $Al_{0.3}Ga_{0.7}As/GaAs$ heterostructures with $N_D = 0.5$ and $1.0 \times 10^{11} \text{cm}^{-2}$.

Nonparabolicity corrections due to terms of third and fourth order in the electron momentum have been considered on the basis of this self-consistent calculation by perturbation theory[15] and give only small changes in the subband separation, which can hardly be detected by intersubband spectroscopy. However, some aspects of cyclotron and spin-resonance experiments as oscillations of the cyclotron mass with magnetic field[20] and reduced g-factors (as compared with the bulk value)[18] can clearly be ascribed to nonparabolicity.

Inversion layers on narrow-gap semiconductors

Expected nonparabolicity effects have been a strong motivation for a long time to study n-inversion layers on narrow-gap semiconductors like InSb and HgCdTe.[21–25] These calculations used a 6x6 Kane model which includes the $\mathbf{k}\cdot\mathbf{p}$ coupling between Γ_{6c} and Γ_{8v} in all orders by treating these states as quasi-degenerate. We want to present here results of a more complete calculation for the 8x8 Kane model, which considers also the split-off band.[17]

Kane's 8x8 Hamiltonian, whose eigenvalues describe the dispersion of the lowest conduction band and the topmost valence bands in the vicinity of the Γ point ($\mathbf{k} = 0$), is given by

$$H_{8\times 8} = \begin{pmatrix} H_{cc} & H_{cv} & H_{cs} \\ H_{cv}^{+} & H_{vv} & H_{vs} \\ H_{cs}^{+} & H_{vs}^{+} & H_{ss} \end{pmatrix} \tag{17}$$

where

$$H_{cc} = \left(\frac{E_g}{2} + \frac{\hbar^2 k^2}{2m}\right) 1_{2\times 2}, \qquad H_{vv} = \left(-\frac{E_g}{2} + \frac{\hbar^2 k^2}{2m}\right) 1_{4\times 4},$$

$$H_{ss} = \left(-\frac{E_g}{2} - \Delta + \frac{\hbar^2 k^2}{2m}\right) 1_{2\times 2}$$

are diagonal blocks for conduction band (Γ_6, c), valence band (Γ_8, v) and split-off band (Γ_7, s). The off-diagonal blocks of (17) are

$$H_{cv} = \begin{pmatrix} -\frac{1}{\sqrt{2}}Pk_x & \sqrt{\frac{2}{3}}Pk_z & \frac{1}{\sqrt{6}}Pk_- & 0 \\ 0 & -\frac{1}{\sqrt{6}}Pk_+ & \sqrt{\frac{2}{3}}Pk_z & \frac{1}{\sqrt{2}}Pk_- \end{pmatrix}, \quad H_{cs} = \begin{pmatrix} -\frac{1}{\sqrt{3}}Pk_z & -\frac{1}{\sqrt{3}}Pk_+ \\ -\frac{1}{\sqrt{3}}Pk_- & \frac{1}{\sqrt{3}}Pk_z \end{pmatrix}.$$

The off-diagonal block H_{vs} is zero. E_g is the energy gap, Δ the spin-orbit coupling, $k_\pm = k_x \pm ik_y$ and $P = \frac{\hbar}{m} < S|p_z|Z >$ is Kane's momentum matrix element coupling between valence and conduction band. Note that this Hamiltonian does not contain any contributions from states outside of this set. For the subband problem we replace k_z by $\frac{1}{i}\partial_z$ and add the interface potential $V(z)$ (see eq. 16) in the diagonal

$$H_{\text{subband}} = H_{8\times8}(k_z \to \frac{1}{i}\partial_z) + V(z)1_{8\times8}. \tag{18}$$

Eq. 18 corresponds to the Fourier transform of the Hamiltonian appearing in eq. 6. As the band edge discontinuity is quite large for MIS systems as which the inversion layers on InSb and HgCdTe are usually prepared, we assume an infinite potential barrier which requires a zero subband function at $z = 0$. In the self-consistent procedure it is important to consider the energy dependence of the subband density of states caused by the nonparabolicity.[17]

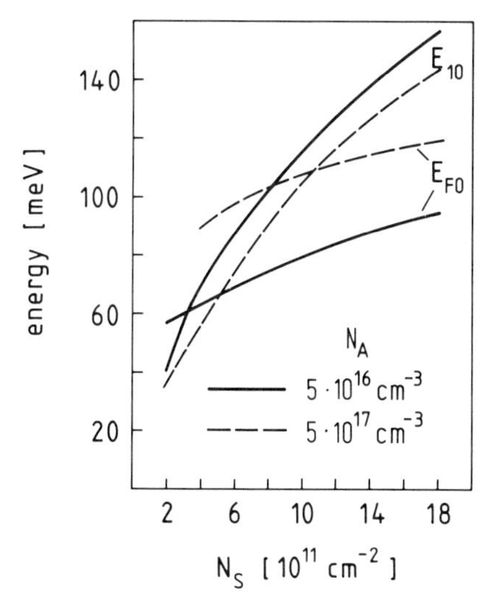

Fig. 3. Dependence of subband separation E_{10} and Fermi energy E_{F0} on N_S for n-inversion layers on $p-Hg_{0.8}Cd_{0.2}Te$ with $N_A = 5 \times 10^{16} cm^{-3}$ and $5 \times 10^{17} cm^{-3}$.

The integration of eq. 18 is performed by using the knowledge of the solutions for $V(z) = \text{const}$, which for an energy above (below) the conduction band edge are oscillating (exponential) functions. Because the interface potential can be considered as piecewise constant, we combine these solutions at each potential step by the requirement of flux conservation. This procedure converges very fast with decreasing step width. The integration is started at $z = 0$ for a trial value of the energy E and the given potential with the boundary condition $f(z = 0) = 0$. For large enough $z > z_d$ we check the behaviour of $f(z)$: if it approaches zero, than E is an eigenvalue, if not, the integration is repeated with different E. For energies smaller than the bulk valence band edge the channel is degenerate with the continuum of bulk valence band states. The subband states become resonant (Zener tunnelling). As a rigorous treatment of this situation in a self-consistent subband calculation is subject of future work, we neglect tunnelling and require an exponentially decaying solution as characteristic for an eigenstate.

Results of our self-consistent calculation show in Fig. 3 the variation of subband separation E_{10} and Fermi energy E_{F0} with N_S and N_A. The qualitative features, increase of subband separation with N_S and N_A, are the same as for AlGaAs/GaAs heterostructures, but the confinement effect is much more pronounced. The strong nonparabolicity can be seen in the nonlinear increase of E_{10} with N_S, which also depends on N_A. Because of the short depletion length of about 400 Å (compared to about 10 000 Å in AlGaAs/GaAs) the distribution of charged acceptors has to be taken into account explicitly. It is interesting to note that almost idential results as in Fig. 3 can be obtained from a 2x2 model including higher order terms in $k_z = \frac{1}{i}\partial_z$.[17]

EFFECTIVE-MASS HAMILTONIANS FOR IMPURITIES IN QUANTUM WELLS

While in the last section we have demonstrated the use of effective-mass theory for calculation of subband states in inversion-layers, in which impurities contribute essentially to the band bending potential, we consider now the effective-mass concept for individual impurities in quantum wells. For this problem the potential of an impurity at $\mathbf{R}_i$ has to be added to the Hamiltonian H_0 introduced in the second section. The Schroedinger equation then reads

$$H\phi = (H_0 + V(\mathbf{r}-\mathbf{R}_i))\phi = E\phi. \tag{19}$$

An expansion of the solution ϕ is possible using the set of eq. 4

$$\phi(\mathbf{r}) = \sum_{n\mathbf{k}} B_n(\mathbf{k})\chi_{n\mathbf{k}}(\mathbf{r}) \tag{20}$$

where the wavepacket is now formed by summing over all components of $\mathbf{k}$, as the impurity destroys the translational symmetry in all directions of space. Averaging over unit cells and assuming, that $V(\mathbf{r}-\mathbf{R}_i)$ is slowly varying on the scale of a lattice constant we find after eliminating the interband $\mathbf{k}\cdot\mathbf{p}$ coupling to lowest order as before

$$\begin{aligned}\sum_{\mathbf{k}}\Bigg[\frac{\hbar^2 k^2}{2m} + \frac{\hbar^2}{m^2}\sum_{n'\neq n}\frac{\mathbf{k}\cdot\mathbf{p}^{A,B}_{nn'}\,\mathbf{p}^{A,B}_{n'n}\cdot\mathbf{k}}{E^{A,B}_n(0)-E^{A,B}_{n'}(0)} + E^{A,B}_n(0)\\ + \int d^3r\, e^{i(\mathbf{k}'-\mathbf{k})\cdot\mathbf{r}}V(\mathbf{r}-\mathbf{R}_i) - E\Bigg]B_n(\mathbf{k}) = 0\end{aligned} \tag{21}$$

which is just the Fourier transform of the effective-mass equation for the impurity in a quantum well. For the case of donors connected with the Γ_{6c} conduction band it is

$$\left\{-\frac{\hbar^2}{2}\nabla\frac{1}{m(z)}\nabla + V(z) + V(\mathbf{r}-\mathbf{R}_i) - E\right\}F(\mathbf{r}) = 0 \tag{22}$$

where $V(z)$ and $m(z)$ are the quantum well potential of eq. 11 and the z-dependent band edge mass of eq. 10, respectively. $F(\mathbf{r})$ is the Fourier transform of $B_n(\mathbf{k})$. Usually[26,27,28] this equation (and the corresponding one for the acceptor problem[29]) is solved by a variational calculation to obtain the lowest impurity state of a given symmetry with respect to the conduction band edge of the well material. A second variational calculation is required without the impurity potential to obtain the corresponding lowest subband energy, relative to which the impurity binding energy can be determined in an experiment.

We would like instead to suggest a somewhat different procedure, which is based on an expansion of the impurity function of eq. 19 using the solutions $\psi_{\nu\mathbf{k}_\parallel}$ of eq. 5 for the quantum well

$$\phi(\mathbf{r}) = \sum_{\nu \mathbf{k}_{\|}} A_\nu(\mathbf{k}_{\|}) \psi_{\nu \mathbf{k}_{\|}}(\mathbf{r}) = \sum_{\nu \mathbf{k}_{\|}} A_\nu(\mathbf{k}_{\|}) \sum_{n k_z} C_{n\nu}(k_z) \chi_{n\mathbf{k}}(\mathbf{r}). \tag{23}$$

The advantages of this ansatz are: i) $\phi(\mathbf{r})$ fulfills by construction the correct matching conditions at the interfaces, ii) impurity states connected with different subbands of same symmetry can be calculated and iii) it includes explicitly the mixing of several subbands, which becomes important for binding energies comparable or even larger than the subband separation. The last point is relevant for acceptors and for large quantum well widths.

Using eq. 23 in eq. 19 and averaging over unit cells with the assumption of a slowly varying envelope function leads to

$$\begin{aligned} \sum_{\nu \mathbf{k}_{\|}} \sum_{\substack{k_z' \\ k_z}} C^*_{n\nu'}(k_z') \Bigg[\Big(\frac{\hbar^2 k^2}{2m} + \frac{\hbar^2}{m^2} \sum_{n' \neq n} \frac{\mathbf{k} \cdot \mathbf{p}^{A,B}_{nn'} \, \mathbf{p}^{A,B}_{n'n} \cdot \mathbf{k}}{E^{A,B}_n(0) - E^{A,B}_{n'}(0)} + E^{A,B}_n(0) \Big) \delta_{k_z k_z'} \\ + \int d^3r \, e^{-i(\mathbf{k}' - \mathbf{k})\mathbf{r}} V(\mathbf{r} - \mathbf{R}_i) - E \Bigg] C_{n\nu}(k_z) A_\nu(\mathbf{k}_{\|}) = 0. \end{aligned} \tag{24}$$

The first part of eq. 24 is just the subband problem for the quantum well (eq. 8), i.e. we can replace the first three terms by the subband energy $E_\nu(\mathbf{k}_{\|})$, which after performing the Fourier transform $(\mathbf{k}^2_{\|} \to -(\partial^2_x + \partial^2_y))$ is the kinetic energy operator of the two-dimensional effective-mass equation for the impurity in a quantum well. The potential energy term of this equation is

$$\sum_{k_z k_z'} C^*_{n\nu'}(k_z') C_{n\nu}(k_z) \int d^3r \, e^{-i(\mathbf{k}' - \mathbf{k})\mathbf{r}} V(\mathbf{r} - \mathbf{R}_i) = -\frac{e^2}{2\varepsilon_0 \varepsilon \kappa} F_{\nu'\nu}(\kappa, Z_i) \tag{25}$$

where $\kappa = |\mathbf{k}'_{\|} - \mathbf{k}_{\|}|$ and

$$F_{\nu'\nu,}(\kappa, Z_i) = \int dz \, f^*_{n\nu'}(z) f_{n\nu}(z) e^{\kappa |z - Z_i|}. \tag{26}$$

We thus obtain

$$\sum_{\nu' \mathbf{k}'_{\|}} \left\{ E_{n\nu}(\mathbf{k}_{\|}) \delta_{\nu\nu'} \delta_{\mathbf{k}_{\|} \mathbf{k}'_{\|}} - \frac{e^2}{2\varepsilon_0 \varepsilon \kappa} F_{\nu'\nu}(\kappa, Z_i) - E \right\} A_{\nu'}(\mathbf{k}'_{\|}) = 0 \tag{27}$$

which corresponds to eq. 14 of ref. 13 for quantum well excitons. Eq. 27 demonstrates the possibility of inter subband coupling by the impurity potential, an aspect of impurities in quantum wells, which to our knowledge has not been studied so far.

SUMMARY

The concepts of effective-mass theory, which for near-band edge states in bulk-semiconductors has been so successful in the past, have been reviewed in connection with semiconductor structures. We derive the effective-mass equations for calculating subbands in these structures and demonstrate its application to heterostructures with emphasis on the implications of bulk-band structure and doping profiles. Finally we discuss the effective-mass concept in connection with impurity states in quantum wells and suggest a treatment, which allows to study inter subband coupling due to the impurity potential and impurity states connected with higher subbands.

REFERENCES

[1] G. Dresselhaus, A.F. Kip, C. Kittel Phys. Rev. **98** 368 (1955)
[2] J.M. Luttinger Phys. Rev. **102** 1030 (1956)
[3] E.O. Kane J. Phys. Chem. Solids **1** 249 (1957)
[4] J.M. Luttinger, W. Kohn Phys. Rev. **97** 869 (1955)
[5] for reviews see S. Pantelides Rev. Mod. Phys. **50** 797 (1978); A.K. Ramdas, S. Rodriguez Rep. Prog. Phys. **44** 1297 (1981)
[6] K. Suzuki, J.C. Hensel Phys. Rev. **B 9** 4184 (1974); H.R. Trebin, U. Rössler, R. Ranvaud Phys. Rev. 20 686 (1979)
[7] A. Baldereschi, N.O. Lipari Phys. Rev. **B 8** 2697 (1973); U. Rössler Festkörperprobleme XIX / Adv. in Solid State Physics (ed. J. Treusch, Vieweg, Braunschweig 1979) p. 77
[8] Landolt-Börnstein, New Series, Semiconductors Vol. 22a (ed. O. Madelung, M. Schulz, Springer, Heidelberg 1987)
[9] U. Rössler Solid State Commun. **49** 943 (1984)
[10] W. Pötz, D.K. Ferry Superlattices & Microstructures **3** 57 (1987)
[11] D.A. Broido, L.J. Sham Phys. Rev. **B 31** 888 (1985)
[12] U. Ekenberg, M. Altarelli Phys. Rev. **B 32** 3712 (1985); E. Bangert, G. Landwehr Superlattices & Microstructures **1** 363 (1985)
[13] D.A. Broido, L.J. Sham Phys. Rev. **B 34** 3917 (1986)
[14] A. Twardowski, C. Hermann Phys. Rev. **B 35** 8144 (1987)
[15] F. Malcher, G. Lommer, U. Rössler Superlattices & Microstructures **2** 267 (1986)
[16] R. Lassnig Phys. Rev. **B 31** 8076 (1985)
[17] F. Malcher, I. Nachev, A. Ziegler, U. Rössler Z. Phys. **B 68** 437 (1987)
[18] G. Lommer, F. Malcher, U. Rössler Phys. Rev. **B 32** 6965 (1985) and Superlattices & Microstructures **2** 273 (1986)
[19] F. Stern, S. DasSarma Phys. Rev. **B 30** 840 (1984)
[20] F. Thiele, U. Merkt, J.P. Kotthaus, G. Lommer, F. Malcher, U. Rössler, G. Weimann Solid State Commun. **62** 841 (1987)
[21] F.J. Okhawa, Y. Uemura J. Phys. Soc. Japan **37** 1325 (1974)
[22] Y. Takada, K. Arai, N. Uchimura, Y. Uemura J. Phys. Soc. Japan **49** 1851 (1980)
[23] G.E. Marquez, L.J. Sham Surf. Sci. **113** 131 (1982)
[24] T. Ando J. Phys. Soc. Japan **54** 2676 (1985)
[25] W. Brenig, H. Kasai Z. Phys. B - Condensed Matter **54** 191 (1984)
[26] G. Bastard Phys. Rev. **B 24** 4714 (1981)
[27] C. Mailhiot, Yia-Chung Chang, T.C. McGill Phys. Rev. **B 26** 4449 (1982)
[28] R.L. Greene, K.K. Bajaj Solid State Commun. **53** 1103 (1985)
[29] W.T. Masselink, Yia-Chung Chang, H. Morkoc Phys. Rev. **B 28** 7373 (1983) and **B 32** 5190 (1985)

HOT ELECTRON CAPTURE IN GaAs MQW:

NDR AND PHOTO-EMISSION

N. Balkan and B.K. Ridley

University of Essex
Department of Physics
Colchester, UK

ABSTRACT

We present experimental results on the hot electron distribution and energy relaxation process in doped GaAs/AlGaAs quantum wells. The experiments make use of steady state hot electron photoluminescence spectroscopy and fast pulse transport measurement techniques. Our results reveal a negative differential resistance (NDR) with threshold fields of a few hundred volts cm^{-1} and a surface emission of photons when the electric field is applied along the quantum well layers. These observations are shown to be associated with the quality of the quantum wells.

INTRODUCTION

The interest in hot carriers in 2D semiconductors has received a great deal of attention in recent years (1,2). The research in the field has been motivated not only by the possibility of obtaining direct information about the effects of two dimensionality on the fundamental interactions of semiconductor physics but also by the invention and further predictions of ultra fast, ultra small novel devices which operate in the hot carrier regime (3,4,5).

The starting point of our understanding of hot carrier phenomena is the determination of the energy distribution of hot carriers. Optical spectroscopy is the most widely used technique which gives the distribution function in the presence of an applied field directly (6,7). This technique, together with the usual transport measurements, provides information about the microscopic scattering mechanisms.

Observations of hot carrier spectra in 2D semiconductors are wide-spread (7,8). A brief review of the literature, however, reveals that there is a large discrepancy in the energy loss rates obtained by different research groups (9), and also most measurements are carried out on highly modulation doped good quality material. Although such samples provide an ideal system for experimental studies and for device operations, there is a lack of information about how perturbations in the ideal system such as poor interface quality, traps and non-uniform doping might effect the hot carrier relaxation. In this work our aim was to investigate hot carrier effects in some 'poor quality' GaAs-GaAlAs quantum wells.

Our results are compared with those reported and expected in high quality material. A phenomenological model is proposed to account for the differences.

EXPERIMENTAL

The GaAs-$Ga_{1-x}Al_xAs$ quantum well samples used for the experiments were grown in different laboratories using MOCVD and MBE techniques. Unless otherwise stated the results presented here were taken on sample CPM279, but they are representative of other samples of similar quality. To enhance the absorption of the exciting light for the photo-luminescence measurements, the samples contained more than 50 similar layers (10 in MB669).

The specimens were fabricated into Hall bridges and simple bar shapes and ohmic contacts were formed by alloying Au-Ge or In.

Table 1. Sample description

Sample	Growth	x %	Well width/ Barrier width	Doping cm^{-3}	Carrier Concentration	μ $cm^2(Vs)^{-1}$ (10K)	Capture/ Photo-emission
CMP279	MOCVD	30	60Å/100Å	n,10^{17}	5.5 10^{15}	1300	Yes/Yes
NJ188	"	"	80 /300	p,?	6 10^{17}	3000	Yes/?
OC82	"	"	75 /100	n,background	1.2 10^{16}	3700	No/Yes
MB669	MBE	"	50 /125	" "	4.4 10^{16}	2.2$10^{4}$	No/No

An orthodox Hall set up comprising of a 7 Tesla magnet and a cryostat with optical windows was employed for the electrical characterization at temperatures between 1.3K and 300K. Hot carrier studies were performed usign the set-up shown in Figure 1. The 676 nm line of a krypton laser, plus a selection of neutral density filters, was used for the excitation. A high resolution 1/3m monochromator and a cooled GaAs photomultiplier were used for the dispersion and collection of the luminescence.

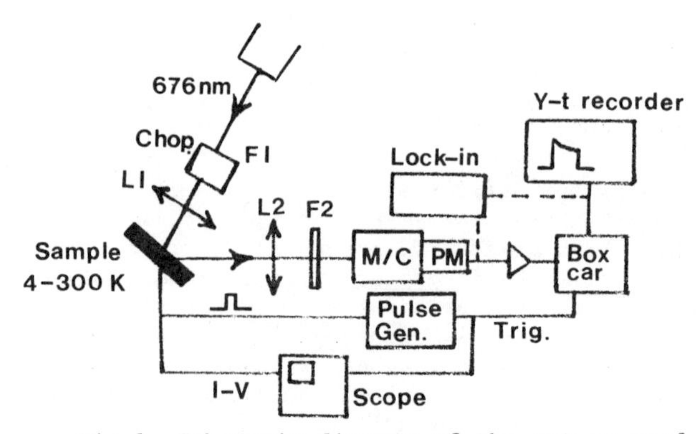

Fig.1. Schematic diagram of the set-up used in the hot electron experiments.

Cold electron luminescence was measured using lock-in techniques. For the hot electron studies, short electric field pulses were applied parallel to the quantum well layers with a maximum duty cycle of less than 0.5% to minimize Joule heating. Both the hot electron photoluminescence and the field-induced photoemission were obtained by synchronizing the detection with the electric field pulse. I-V characteristics were simultaneously measured using a fast scope.

EXPERIMENTAL RESULTS

i) Optical and Electrical Assessment

Figure 2.A shows the photoluminescence (PL) spectra of sample CPM279, when the PL is excited and collected from two different spots on the sample. CPM279 was nominally uniformly modulation doped with silicon (Si ~ $10^{17}cm^{-3}$) and had a similarly doped GaAs capping layer.

The broad band emission (FWHM > 15meV) at the high energy side of the spectra is due to radiative recombination in the quantum well. Before attempting to assess the nature of this recombination, however, we will first look closely at the two lower energy emissions. These are separated by $\Delta E \sim 22$ meV in energy and the broad bands peak at $h\nu_1 \sim 1.515$ eV and $h\nu_2 \sim 1.493$ eV respectively. These emissions are due to intrinsic and extrinsic luminescence in the n-doped GaAs capping layer, and probably associated with the Si^o h and e-h ($h\nu_1$) and e-A^o and Si^o-A^o($h\nu_2$) transitions, where the acceptors are the inadvertently introduced carbon impurities. Carbon is known to be present in large concentrations in similarly grown material (10) and indeed we have measured carbon densities as high as $P > 5 \times 10^{16}$ cm^{-3} in some unintentionally doped MOCVD grown GaAs quantum wells. The ionization energy of carbon in GaAs is ΔE_c = 25 meV (11). This is in the correct range as far as the observed separation of the emission peaks ($\Delta E \sim 22$ meV) is concerned. When the temperature is increased above T ~ 35 K (due to the partial ionization of the impurity centers) the extrinsic emission tends to disappear (Figure 2.C). The interesting feature of the GaAs emission is that when the PL is collected from a different spot on the same sample the intensity changes considerably (Fig.2.A). This strongly suggests that both the intentional and the background doping are not uniform along the plane of the layers in the sample. Similar behaviour is also present in the quantum well emission: the relative intensity of the low energy shoulder ($h\nu_3 \sim 1.567$ eV) follows the behaviour of the extrinsic emission ($h\nu_2 \sim 1.493$ eV) and increases considerably when the PL is sampled from spot B. The prominent peak for spot A ($h\nu_4 \sim 1.576$ eV) is, however, only a weak shoulder for spot B. This observation is not expected for purely free-excitonic emission in GaAs quantum wells (12,13) but suggests the presence of extrinsic emission. Further evidence for the extrinsic behaviour comes from the temperature dependence of the luminescence spectra (Figure 2.C). When the temperature is increased the peak emission shifts towards higher energies and the width of the luminescence, after an initial increase, decreases dramatically at high temperatures. For a purely intrinsic free exciton emission the PL peak energy is expected to follow the effective bandgap variation to lower energies with increasing temperature. The observed high energy shift clearly demonstrates the bound nature of the excitons (13,14). The term bound implies trapping of free excitons either at randomly distributed C^o centers and/or at potential fluctuations associated with a non-uniform distribution of intentionally introduced and background impurities as exists in our sample as well as those associated with monolayer fluctuations in the quantum well width (14,15). If there is a continuous distribution of potential fluctuations in the sampled volume of the quantum wells, this in turn will result in a continuous distribution of bound states in energy. Obviously within the sampled volume some statistical averaging will take place. If this averaging is in favour of smaller fluctuations, and hence bound states with higher energies, then at low temperatures at low excitation intensities the greatest contribution to bound exciton emission will come from the high-energy bound states. However, if the statistical averaging is in favour of larger fluctuations in another sampled volume of the quantum wells, and hence in favour of lower lying bound states in energy, then the contribution to emission will shift to lower energies as observed in Figure 2.A.

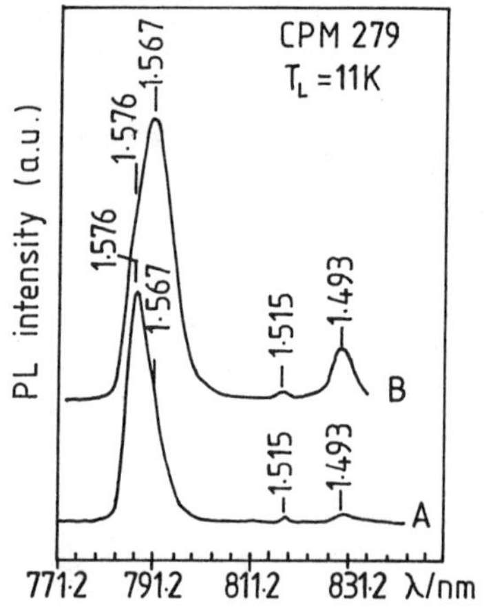

Fig.2A. PL spectra collected from two different spots on the sample.

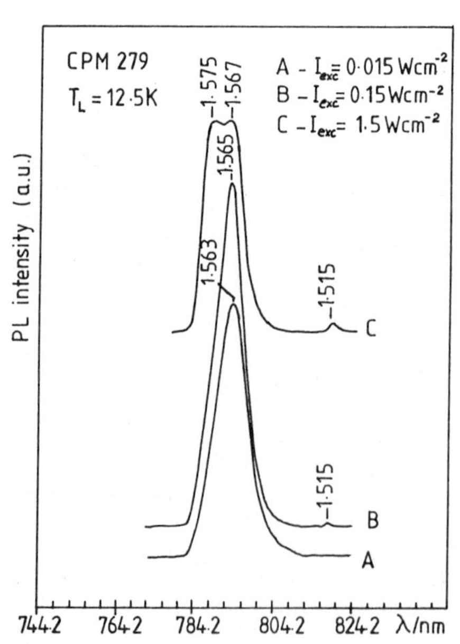

Fig.2.B. Excitation intensity dependence of the PL spectra.

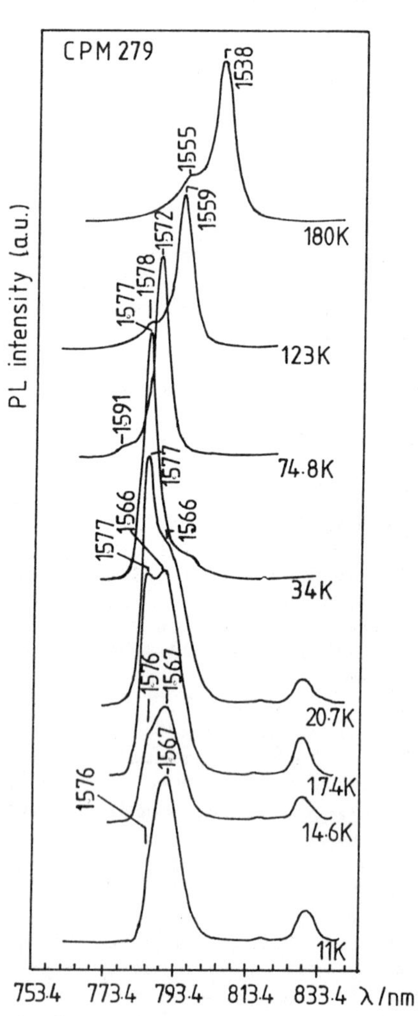

Fig.3.C. Temperature dependence of the PL spectra.

When the temperature is increased excitons trapped at lower energy states will be detrapped and populate the states at higher energies. The maximum contribution to PL will therefore shift towards excitons bound to higher energy states, as clearly observed in Fig. 2.C. A possible explanation for the observed behaviour of the luminescence width is that with increasing temperature the excitons become more mobile and sample a larger fraction of the potential fluctuations, as opposed to low temperatures where more of the lower lying states are occupied. Eventually when the temperature is high enough for the emission to be dominated by free exciton recombination, the luminescence width decreases to that of free exciton emission and the peak energy follows the effective gap shrinkage with increasing temperature.

Also, when the laser intensity is increased, as shown in Fig. 2.B, saturation of the lower energy states and a shift towards higher energy states occurs.

In order for free exciton trapping to occur at centers randomly distributed in energy and the sample volume the rate of trapping must be comparable with that of the free exciton recombination (14,15). This, in turn, requires the average separation of such centers to be comparable with the free exciton diameter, $d \sim 300A^o$. Hence an approximate concentration of $10^{17} cm^{-3}$ trapping centers must be available, at least locally. Due to the random nature of the spatial distribution of these centers intrinsic free exciton recombination might dominate the PL spectra when it is collected from some arbitrary spots on the sample, while the bound exciton recombination might be more favourable for some other spots. Whether bound excitons will recombine before detrapping or not will depend on the relative rates of recombination and the transition associated with the hopping to a higher energy state, which is in turn determined by the spatial separation of the states. It is not unreasonable therefore to assume that for regions with a high density of trapping centers the transition and bound exciton recombination times might be comparable (14,15).

The carrier concentration and the Hall mobility of the same sample, CPM279, are plotted against temperature using a logarithmic scale in Fig. 3. At room temperature the sample has a carrier concentration of $n = 1.8 \times 10^{16}$ cm^{-3} and a mobility of $\mu = 3.2 \times 10^3$ cm^2 $(Vsec)^{-1}$. As the temperature is decreased both the carrier concentration and, after a slight rise, the mobility sharply decrease with decreasing temperature. In a high quality modulation doped quantum well structure such a temperature dependence for either the mobility or the carrier concentration is not expected (16). A possible explanation for the observed behaviour might be that there is a complete lack of depletion of the barriers and the available free carriers at high temperatures are due to unintentionally introduced quantum well donors. This explanation, however, fails merely because the lack of depletion of the GaAlAs barriers is extremely unlikely. These barriers are only $100A^o$ thick and $n \sim 10^{17}$ cm^{-3} doped and therefore a complete depletion is expected (16). On the other hand, if there is a large concentration of deep potential fluctuations as we have observed in our PL studies, these can localize the carriers (electrons in n type regions, and holes in p type regions), the observed drop in the free carrier concentration and mobility can therefore occur as a result of trapping. If such is the case the maximum contribution to the conductivity at low temperatures and low electric fields should arise from hopping of the carriers between traps, similar to those in highly disordered and amorphous semiconductors (17). In Fig. 4 we have plotted the conductivity using a logarithmic scale against the inverse of the temperature and also against $T^{-1/4}$.

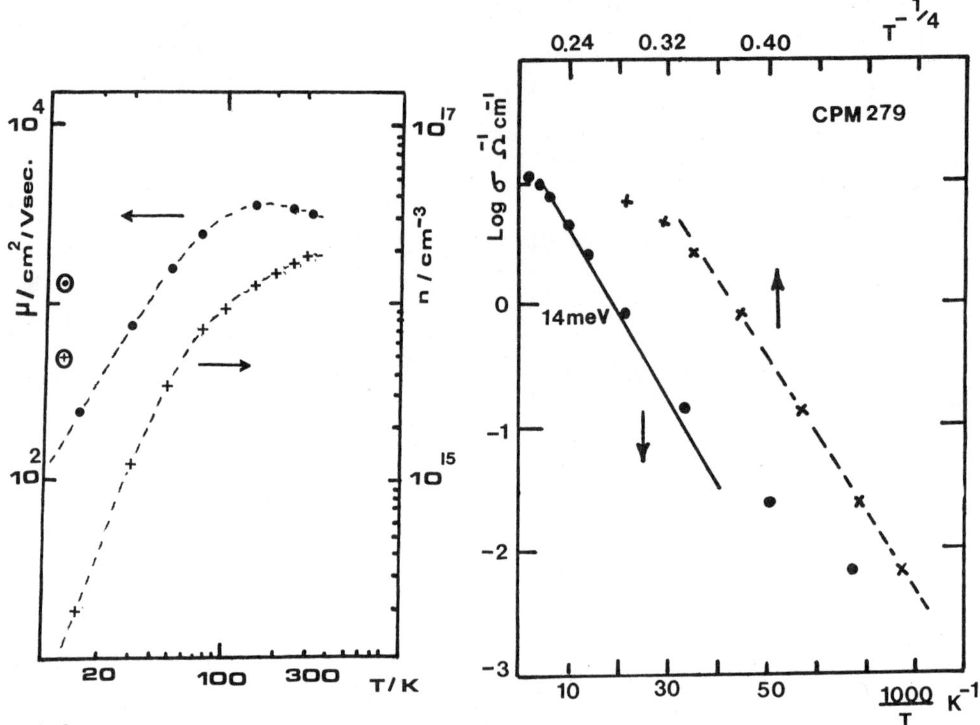

Fig3. Temperature dependence of the carrier concentration and the Hall mobility,circled points represent the"light on"values.

Fig.4. "Dark" conductivity vs inverse temperature and $T^{-1/4}$. 14 meV is the nearest neighbour activation energy.

The observed behaviour is rather striking. At intermediate temperatures the conductivity is thermally activated, $\sigma \sim \sigma_0 \exp(-\Delta/kT)$ where Δ = 14 meV is the average nearest neighbour hopping energy (16,17). At temperatures below T ~ 40K deviations from the nearest neighbour hopping start and the logarithm of the conducitivty fits a $T^{-1/4}$ plot perfectly down to the lowest measured temperature, T = 14K. This clearly demonstrates that variable-range hopping (17) is the dominant conduction mechanism in our sample at low temperatures and that the sample is indeed highly disordered confirming our PL results. The question of the exact nature of the disorder still remains to be answered. In a 3D semiconductor in order to have localization of the carriers, and hence for thermally activated hopping conduction to occur, Anderson's localization criteria must be satisfied (17). This states that $P = V_0/B$ must exceed some critical value,P_0, where V_0 is the range over which the depth of the random potential fluctuations are distributed and B is the bandwidth. There have been many theoretical calculations of P_0 and values as high as 5.5 and as low as 0.5 have been reported (17). Any predictions for Anderson's criteria for transport along the layers of a 2.D system is beyond the scope of this paper. We shall, however, assume that the Anderson localization criteria is satisfied in our sample as our experimental results suggest.

ii) Hot Electron Measurements

Once it had been established that the sample quality was rather poor we studied how such disorder, as has been discussed in the previous section, would affect the hot-carrier phenomena. We first measured the current-field characteristics of the sample as a function of both temperature and light intensity. Electric field pulses were applied parallel to the quantum well layers and care was taken to ensure a long time lapse between successive pulses to establish zero field equilibrium between pulses. At temperatures $T > 100$ K the I-V curves were ohmic but at low temperatures deviations from ohmic behaviour were observed, as we discuss below. The I-V characteristics were symmetrical upon reversing the polarity at all temperatures.

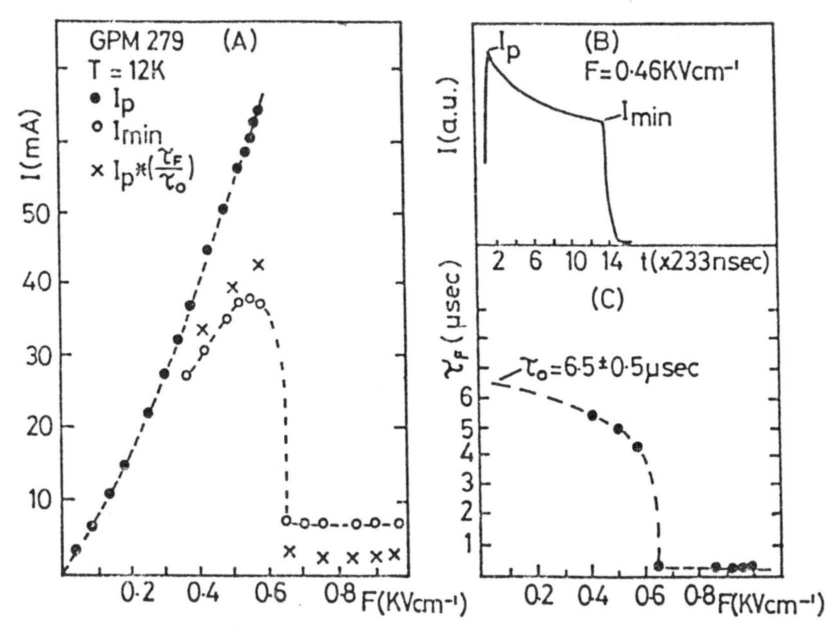

Fig.5. A,Current - Field characteristics of the sample when the sample is illuminated,B, Current-pulse decay and C, field dependence of the decay constant.

Fig. 5.A shows the current-field characteristics of sample CPM279 at T = 12K, when the sample is illuminated. Under illumination the low field carrier concentration and mobility were $n = 5.5 \times 10^{15}$ cm^{-3} and $\mu = 1.3 \times 10^{3}$ $cm^{2}(Vsec)^{-1}$ respectively. In the figure I_p represents the peak current and I_{min} is the current at the end of the 3.5 μsec wide pulse. At electric fields $F > 200$ V cm^{-1} the current pulse decays exponentially, as shown in Fig. 5.B. The time constant associated with this decay decreases as the field strength is increased as shown in Fig. 5C. This exponential decay of the current suggests that the carriers contributing to the high field conductivity are lost, the rate of loss of carriers increasing with increasing field. Indeed, when the field is increased I_{min} drops sharply defining a negative differential resistance (NDR), followed by a plateau indicating the formation of a highly resistive region in the specimen. The zero field capture time $\tau_o = 6.5$ μsec is obtained by extrapolating the high field capture time τ_F, as shown in Fig. 5.C. When the peak current I_p, is multiplied by the ratio τ_F/τ_o we obtain the steady state current $I_s = I_p \times \tau_F/\tau_o$, which is plotted in Fig. 5.A. The agreement between the observed and the calculated curves is excellent. NDR occurs at an electric field $F \sim 0.6$ kVcm^{-1}, with a peak to valley ratio of $I_p/I_{min} = 23$. The observed threshold field $F \sim 0.6$ kV cm^{-1} is the lowest field for any NDR reported to date in n-doped GaAs (18). When the electron temperature at the observed NDR threshold field is obtained from the hot electron PL spectra (see Fig. 7) we find that $T_e = 56$ K, which is too low for any significant proportion of the hot electrons to transfer either into the barrier layers or into the L valley of GaAs (19). Besides, both real space transfer and intervalley space transfer are much faster processes than that which is observed here. The mechanism responsible for the present NDR might be similar to that which occurs in 3D semiconductors as a result of impurity-barrier capture (20). A schematic diagram of this process is shown in Fig. 6. The only requirement in this simple model is the presence of a Coulombic barrier and any negatively charged impurity center would satisfy this requirement. As the electric field strength is increased the electrons gain more and more energy from the field and hence become hot. The rate of their capture by the repulsive center will therefore increase, as observed. If W is the barrier height, we can then express the capture rate as $\nu_F = \nu_o \exp(-W/kT_c)$, where $\nu_F = 1/\tau_F$, $\nu_o = 1/\tau_o$ and T_c is the carrier temperature. In the context of the impurity-barrier capture model, negative differential resistance can occur if the rate of capture of hot carriers is greater than their thermal regeneration rate, namely $\nu_F > \nu_g$. The thermal regeneration rate $\nu_g \alpha \exp(-\emptyset/kT_L)$, where $\emptyset$ is the depth of the center and T_L is the lattice temperature. A detailed study fo these processes has been reported by us previously (21), where we obtained the barrier height W = 21 meV, and the depth of the capturing center $\emptyset = 50$ meV. A simple calculation then reveals that the condition for NDR, $\nu_F/\nu_g >> 1$ is easily satisfied at the observed threshold field.

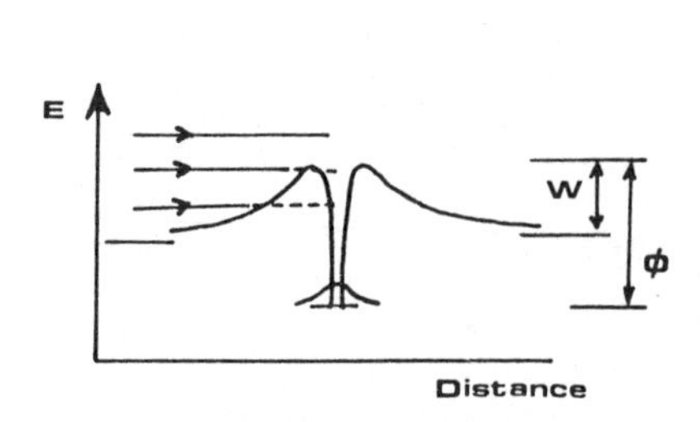

Fig.6. Model for the impurity-barrier capture of hot electrons.

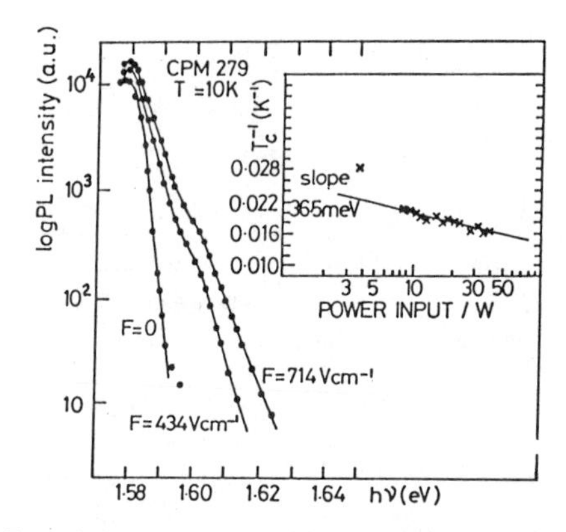

Fig.7. High energy tail of the PL spectrum at various electric fields. Inset: Power input vs the inverse of carrier temperat

The next step in our hot-carrier studies was an investigation of the hot-electron luminescence. Fig. 7 shows the high energy tail of the photoluminescence spectra which was collected by gating the signal at the peak of the current pulse. The intensity of the high energy tail decreases exponentially with increasing photon energy indicating a Maxwellian distribution of carriers (22). The carrier temperature, therefore, determines the decrease as

$$I(h\nu) \sim \exp(-(h\nu - E_g)/kT_c) \quad (1)$$

where $I(h\nu)$ is the luminescence intensity at photon energy $h\nu$, E_g is the energy gap and T_c is the carrier temperature. As shown in Fig. 7, increasing the field populates the high energy states further and decreases the slope of the line as a result of increasing the carrier temperatures. Since the lattice temperature is low (T_L = 10K) we can ignore optical phonon absorption and assume that for the carrier temperatures observed, LO phonon emission by the electrons is the dominant loss mechanism. We then express the energy loss per carrier as

$$dE/dt = \hbar\omega_{Lo}/\tau_{avg} \exp(-\hbar\omega_{Lo}/kT_c) \quad (2)$$

where $\hbar\omega_{Lo}$ is the optical phonon energy (36.5 meV) and τ_{avg} is the average emission rate of an optical phonon (22). In the steady state the energy loss should equal the power input to the carrier system by the electric field, $dE/dt = P/n$ where n is the number of carriers and P is the total power input. We therefore expect the inverse of the carrier temperature to depend on the input power as shown in the inset of Fig. 7. The straight line in the figure has a slope of $\hbar\omega_{Lo}/k$ as expected. We conclude, therefore, that the energy loss of hot electrons in our sample is by optical phonon emission at carrier temperatures above T_c = 45K. For the average phonon emission rate we obtain τ_{avg} = 0.1 psec. This is much more rapid than emission rates reported in high quality samples having high carrier densities (23). In our calculation of τ_{avg} we estimated the carrier concentration $n = 1 \times 10^{10}$ cm^{-2} per well. This value is obtained from the low field conductivity. As is clearly shown in Fig. 5.A the current-voltage characteristics are not linear over the whole range. The carrier concentration at high electric fields might, therefore, be somewhat larger than the zero field value. Also, it was assumed in our calculations that all 50 layers of the quantum wells were contacted. If the contacts diffused into only half of the total quantum well thickiness, this would result in an error of a factor of two in the value of τ_{avg}.

We also investigated the PL pulse shapes at various applied electric fields. This was to justify the model which assumes that the observed NDR is indeed associated with the capture of hot carriers. When the detection wavelength was fixed at the high energy tail of the PL spectra, the PL pulse was found to decay exponentially with time as did the current pulse, as shown in the inset of Fig.8. As the PL intensity is a direct measure of the number of carriers at the photon energy sampled, the relationship between the current decay and the PL decay, as shown in Fig. 8, strongly supports the assumption that the NDR is associated with the loss of the hot carriers. At electric fields above the plateau region instabilities in both the current and the PL pulse are observed, as shown in Fig. 9. The oscillations in both the current and the PL pulses are found to have a period T ~ 0.35 μsec. If these oscillations are associated with domain motion then the estimated domain velocity in the 5mm sample is $V_D = 1.4 \times 10^6$ cm/sec.

The PL pulse at the high energy tail of the spectra decays exponentially with the same time constant as the current pulse. However, when the detection wavelength is fixed at photon energies around or below the luminescence peak a completely different picture is obtained. This is illustrated in Fig.10.

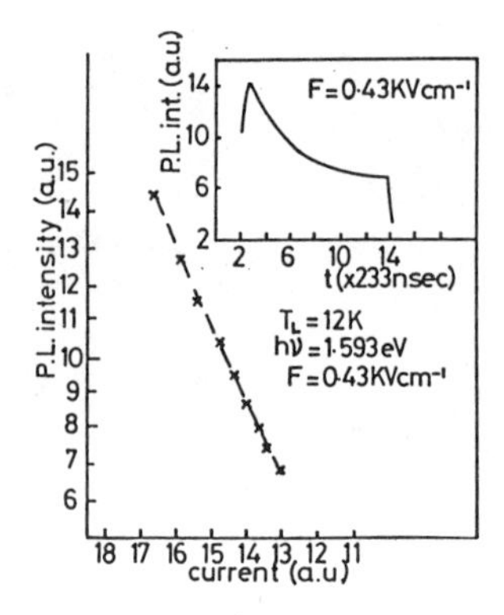

Fig.8. PL Intensity versus current during the pulse. Inset:PL pulse shape.

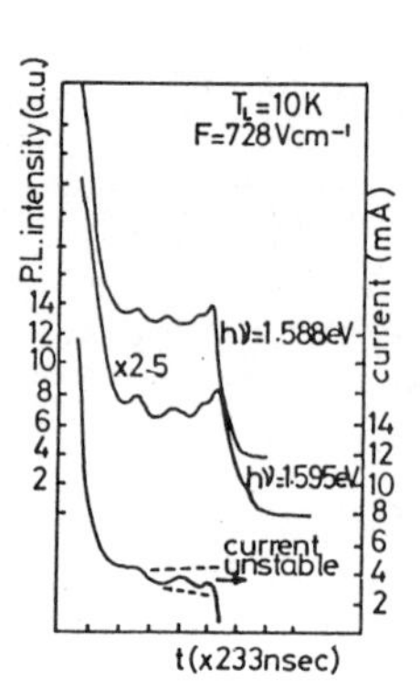

Fig.9. Instabilities in the current and the PL pulses.

At intermediate electric fields the PL pulse, after an initial decay, develops a broad kink and then slowly decays towards the trailing edge of the pulse. When the electric field is increased the kink becomes more prominent and moves towards the leading edge of the pulse. This rather surprising behaviour suggests that the observed PL pulse is in fact the sum of two components, one following the decay of the current pulse and the other initially rising and then falling with time as the first one. As the field is increased the rise time decreases and the amplitude increases. The observed behaviour results from the superimposition of the second component on the first. Since the intensity of the second component is a strong function of the applied electric field one is led to believe that this component is due to some radiative recombination mechanism associated with the field. To see if this was the case we kept the sample in the dark and without altering the focusing of the detection system applied varying electric fields along the layers.

Fig.11 shows the light pulse detected at $h\nu$=1.576eV synchronous with the applied field. The pulse rises exponentially with a time constant τ_r, which decreases with increasing field. The delay between the rise time of the field pulse, t_F and the rise time of the light pulse τ_r-t_F defines the time constant associated with the 'build-up' of the radiative process. The intensity of the emitted light decreases sharply as the detection wavelength is decreased towards the high energy tail but τ_r remains unchanged. These observations are in accord with the 'two-component' interpretation of the PL pulse shapes as shown in Fig.10.

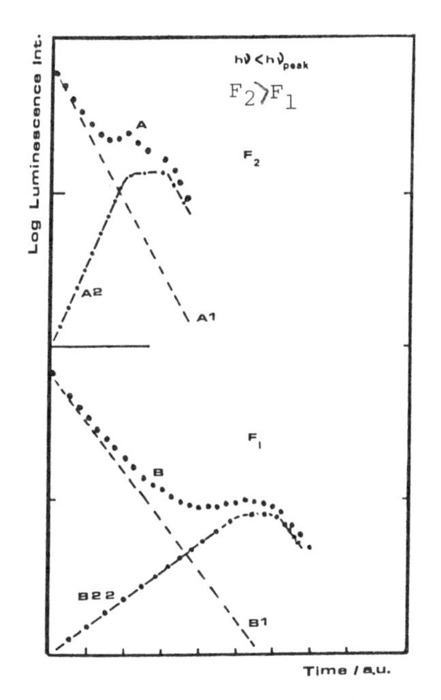

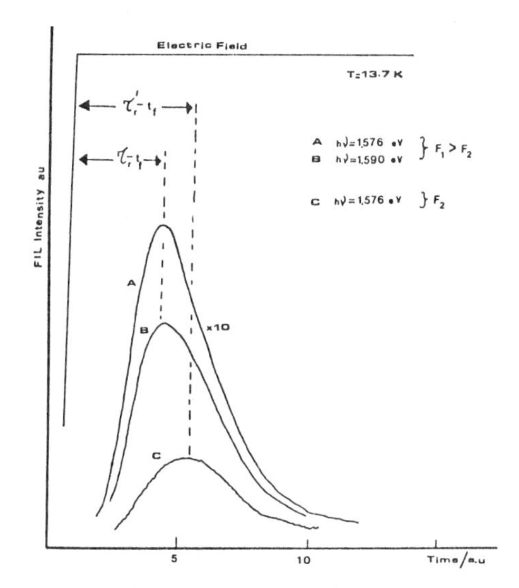

Fig.10 Two component model for the PL pulses at two fields.The components A2 and B2 are due to FIL emission.

Fig.11. FIL pulse shapes at two photon energies.

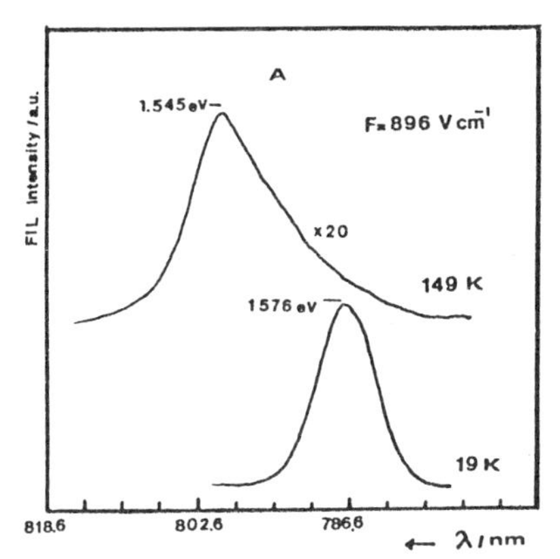

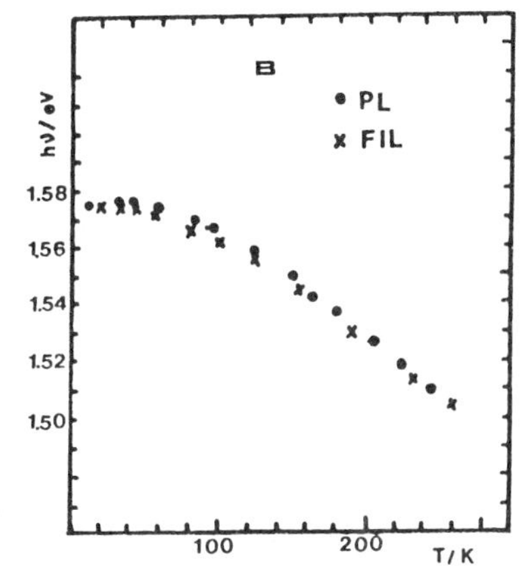

Fig.12. A, The FIL spectrum plotted at two temperatures. B,FIL and PL peak emission energies vs. temperature.

In Fig.12.A we have plotted the field-induced light spectra (FIL) at two temperatures for $F=896Vcm^{-1}$. Fig.12.B shows the temperature dependence of the peak emission energy of the FIL spectra as well as the excitonic PL emission energy for the excitation intensity $I_{exc}=2.9Wcm^{-2}$. It is evident from the figure that the FIL spectra has a rather broad low energy tail, and the temperature dependence of the FIL peak energy is similar to that of the (high excitation level) PL spectra. The peak emission energy and the low energy broadening of the FIL spectra change with field in a similar way to the change of the PL spectra with light intensity. As the field is decreased the emission peak shifts to lower energies and the low energy tail shrinks as shown in Fig.13. The high energy tail of the FIL spectra at high fields also exhibits a similar behaviour to that of the hot electron spectra; namely the intensity varies with the photon energy in the manner expressed by equation 1 (see Fig.14). This suggests that FIL emission also involves hot electrons at high electric fields. We have obtained electron temperatures as high as T_c=62K at the maximum applied field, F>1kVcm^{-1}. It should be noted that in the hot electron PL studies electron temperatures of T_c~62K were obtained at fields of F~0.7kVcm^{-1}. This suggests that the optically injected (free) excess electrons absorb more energy from the field than those involved in the FIL recombination. The reason for this will become clear when we discuss the nature of the FIL emission.

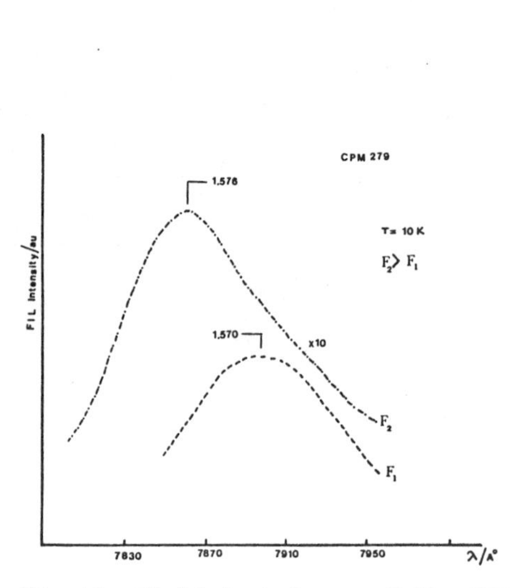

Fig.13. Field dependence of the FIL emission peak energy.

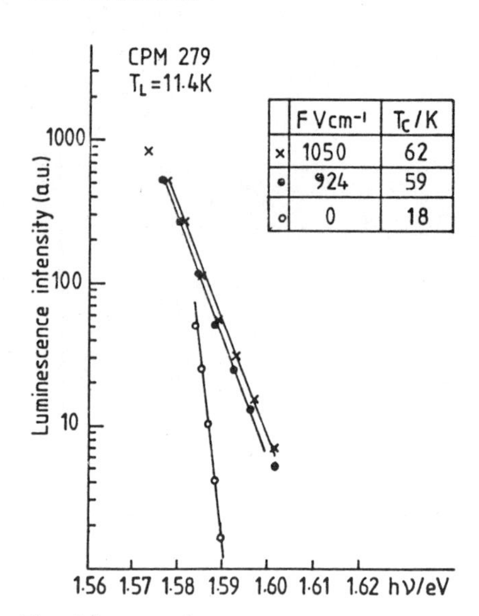

Fig.14.Hot electron FIL spectra.

To see if the observed FIL emission was a surface emission, or was associated with contacts, we studied the FIL emission as a function of the sampled spot size and sampled spot-contact separation. The results were as follows:

a) When the contacts were covered and the optics for the light collection were modified to enlarge the sampled spot, the FIL emission intensity increased with increasing sampled area.

b) When the sampled spot (d~100μm) was systematically moved between the two contacts of the bar shaped sample, the FIL emission intensity changed arbitrarily but the overall pattern was a slight decrease towards the middle point of the sample. The changes, if any, in the FIL pulse risetime could not be detected as they were within the limit of our experimental error (Δt~20nsec).

c) The observations were, within the experimental accuracy, independent of the polarity of the applied field.

These observations clearly indicate that the FIL emission is associated with the whole volume of the quantum well sample. This behaviour is somewhat similar to the tunable luminescence reported in doping superlattices (24,25), where the radiative lifetimes and emission spectra are determined by the space charge induced separation of the injected carriers. Obviously, due to the periodicity and uniformity of the p and n layers in a doping superlattice, a periodic space charge is present. In our case, however, there is neither the periodicity nor the uniformity. As we have observed in our PL and conductivity measurements there are p and n rich regions. The concentration and spatial distribution of these regions is random along the layers. Large potential fluctuations associated with this heterogeneous structure therefore exist. Holes are localized in regions where the local acceptor density is excessively high. Electrons transferred from the barriers are then localized predominantly where there are few holes. The potential profile along the quantum well layers is therefore an extremely distorted version of a doping superlattice structure. Within this picture the predicted behaviour of various conduction and optical processes are as follows:

(1) The conductivity at low temperatures and low electric fields is due to the carriers hopping between localized states. The overall conductivity type is determined by the relative average densities of the contributing carrier types. This is n-type in our case.

(2) The photoluminescence spectrum is extremely sensitive to the sampled volume of the specimen. At low excitation intensities and low temperatures, the PL spectrum is dominated by bound exciton recombination when the concentration of trapping centers is high enough. Increasing the temperature and excitation intensity results in a detrapping of the bound excitons and a saturation of the bound states, and hence results in a shift of the emission peak towards free exciton recombination, as observed. Also, at very low excitation intensities, there should be a contribution to the PL from the radiative recombination of the spatially separated excess carriers. This should result in a broadening of the low energy side of the spectrum. If the electron and hole quasi Fermi level separations of the spatially separated excess carriers are in the same range as the emission energy of the bound excitons, however, the two effects cannot be separated.

(3) Consider the situation when electric field pulses are applied along the layers of the sample in the dark. At low fields the conductivity occurs by hopping, as described above, and the transition rate for the hopping process is (17),

$$\nu_h = \nu_o \exp(-2\alpha R) \exp -\Delta/kT \qquad (3)$$

where α^{-1} is the spatial extent of the localized wavefunction, Δ is the energy separation, R is the spatial separation between the two states concerned and ν_o is a characteristic frequency. As the field is increased not only does the excess carrier concentration increase but the hopping rate increases as well. According to (26):

$$\nu_h' = \nu_o \exp(-2\alpha R) \exp \Delta/kT \qquad (4)$$

where, $\Delta' = \Delta - eFR\cos\Theta$

and F is the field making an angle Θ with R. The conductivity will therefore increase. The radiative recombination, which is proportional to both the majority and excess carrier concentrations, and also the rate of recombination, should increase. The recombination rate should also be determined by the hopping rate, as the hopping motion takes the carriers into regions where they recombine. The photon emission should be distributed in energy around the quasi-Fermi level separation of the holes and electrons, by as much as twice the energy distribution of the contributing localized states, $2\Delta'$.

Increasing the field further shifts the emission peak towards higher energies because of the increasing quasi-Fermi level separation, as observed in fig. (13). At much higher electric fields, when the majority of carriers become hot, the emission should develop a high energy tail due to the recombination of minority holes with hot electrons. This behaviour is shown in fig. (14). Exactly how hot the electrons become depends on their initial mobility. When the sample is illuminated both the carrier concentration and mobility are considerably higher than in the dark, as shown in fig. (3). It is not surprising, therefore, that we obtain lower carrier temperatures from FIL measurements than from hot electron PL measurements. At high temperatures the conductivity occurs by nearest neighbour hopping, where there are enough phonons for the localized carriers to absorb and hop to the nearest neighbouring states (α^{-1}~R) so that the first exponential term in equation 4 becomes a constant, A. Hence, at low fields, the transition rate is simply $\nu_h \sim A\nu_o \exp-\Delta/kT$. In figure 15 the build up time of the FIL emission is plotted as a function of field. By assuming that this is directly related to the hopping rate, and taking $\Delta \sim 14$ meV, we find the characteristic hopping frequency $\nu_o = 1.2\ 10^6$ sec-1.

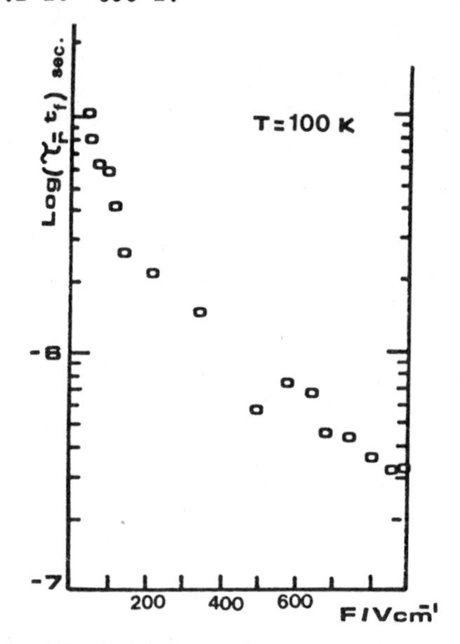

Fig.15.Field dependence of the FIL emission" Build up" time.

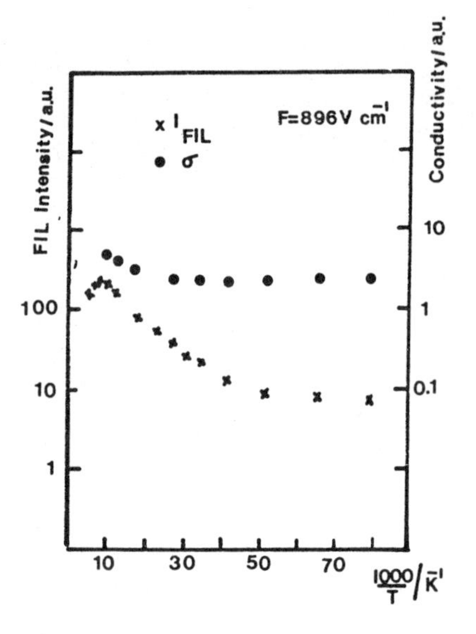

Fig.16. Temperature dependence of the FIL peak emission intensity, and the high field conductivity.

(4) The hot electron photoluminescence is determined by two processes;

(a) Excitonic recombination of optically excited carriers

(b) Recombination of spatially separated excess carriers, injected by both photoexcitation and the application of a field.

At high excitation intensities and low fields process (a) is expected to dominate, as described in section (2). At intermediate fields, when the field induced excess carriers are of comparable concentrations to those optically injected, we might expect the PL spectra to be determined by a mixture of both processes. Free exciton emission, however, having recombination rates much faster than those associated with the spatially separated excess carriers, should dominate the peak emission. Lower energy emission, on the other hand, can be either bound excitons or spatially separated carriers. When the field is high enough to heat the majority electrons the high energy tail of the spectrum will be due to the recombination of hot electons with minority holes, excited either optically or electrically.

(5) According to the model described above a potential build up between any two contacts placed along the layers would be expected, due to the heterogeneous structure of the quantum well layers and the subsequent charge separation. Also this potential difference should increase upon illumination and its temperature dependence should follow that of the FIL emission as shown in fig. (16). Indeed experiments confirming these precictions have been carried out (27) and will be published in the near future.

SUMMARY

We have investigated the influence of sample quality and impurites on hot electron transport in GaAs multiple quantum wells. We have presented the results obtained on an extremely disordered system. The hot electron energy relaxation is not effected much by the sample quality, but we have shown that poor quality quantum well material can reveal rather unexpected low and high field conduction effects, namely hopping conduction along the quantum well layers and bulk emission of photons, intensity of which increases iwith increasing temperature.

ACKNOWLEDGEMENTS

We would like to express our thanks to Dr. P.J. Bishop and Dr. M.E. Daniels for the Hall data and for many useful discussions. We also express our gratitude to the SERC for a research grant. We are indebted to GEC Hirst Research Laboratories and Plessey Caswell Research Laboratories for supplying some of the samples.

REFERENCES

1) C.H. Yand and S.A. Lyon, Physica B and C, 134: 305 (1985).
2) J. Shah, A. Pinczuk, A.C. Gossard and W. Wiegmann, Physica B and C, 134: 174 (1985).
3) S. Luryi and A. Kastalsky, Physica, 134: 453 (1985).
4) S. Luryi, Physica, 134: 466 (1985).
5) F. Capasso, K. Mohammed and A.Y. Cho, Physica B and C, 134: 487 (1985).
6) J. Shah and R.C.C. Leite, Phys. Rev. Letters, 22: 1304 (1969).
7) J. Shah, A. Pinczuk, A.C. Gossard and W. Wiegmann, Phys. Rev. Letters, 54: 2054 (1985).
8) J.F. Ryan, Physica 134B, 403 (1985).
9) S.A. Lyon, Journ. Lumin, 35: 121 (1986).
10) N. Balkan, B.K. Ridley, J. Frost, D.A. Andrews, I. Goodridge, J. Roberts, Superlatt. and Microstructures 2: 357 (1986).
11) S.M. Sze, Physics of Semiconductor Devices , John Wiley and Sons (1981).
12) R.C. Miller, A.C. Gossard, W.T. Tsang and O. Munteanu, Phys. Rev. B 25: 3871 (1982).
13) B. Deveaud, J.Y. Emergy, A. Chomette, B. Lambert, M. Baudet, Superlatt. Microstruct. 3: 205 (1985).
14) M.S. Skolnick, P.R. Tapster, S.J. Bass, A.D. Pitt, N. Apsley, S.P. Aldred, Semicond. Sci. Technol 1: 29 (1986).
15) L. Goldstein, Y. Horikoshi, S. Tarucha, H. Okamoto, Jap. J. Appl. Phys. 22: 1489 (1983).
16) C. Guillemot, M. Baudet, M. Gauneau, A. Regreny, J.C. Portal, Superlatt. and Microstruct. 2: 445 (1986).
17) M.L. Knotek, A.I.P. Proc. No. 20: 297 (1974) and N.F. Mott, E.A. Davis, Electronic Process in Non Crystalline Materials ,

Clarendon Press, Oxford (1979).

18) K. Hess, H. Markoc, B.G. Streetman, Appl. Phys. Lett. 35: 460 (1979).
19) R.A. Hopfel, J. Shah, A.C. Gossard, W. Wiegman, Physica 134B 509 (1985).
20) See e.g. B.K. Ridley, R.G. Pratt, Phys. Lett 4: 300 (1968).
21) N. Balkan, B.K. Ridley and J. Roberts, to be published in the Proceedings of the 5th Int. Conf. on Hot Electrons, Boston (1987).
22) S.A. Lyon, Jour. Lumin. 35: 121 (1986).
23) J. Shah, A. Pinczuk, H.L. Stormer, A.C. Gossard and W. Wiegman, App. Phys. Lett. 42: 55 (1983).
24) K. Ploog and G. Dohler, Adv. in Phys. 32: 285 (1983).
25) K. Kohler, G.H. Dohler, J.N. Miller and K. Ploog,1 Superlatt. and Microstruct. 2:339 (1986).
26) N. Apsley, H.P. Hughes, Phil. Mag. 30: 963 (1974).
N. Apsley, H.P. Hughes, Phil. Mag. 31: 1327 (1975).
27) P.J. Bishop and M.E. Daniels (private communication), to be published.

IN-PLANE ELECTRONIC EXCITATIONS IN GaAs/GaAlAs MODULATION DOPED QUANTUM WELLS

N. Mestres*, G. Fasol** and K. Ploog

Max-Planck-Institut für Festkörperforschung, Heisenbergstrasse 1
D-7000 Stuttgart 80, Fed. Republic of Germany

*Present Address: University of Michigan, Ann Arbor, Michigan, USA

**Permanent Address: Cavendish Laboratory, University of Cambridge
Madingley Road, Cambridge CB3 0HE, UK

ABSTRACT

We study the in-plane motion of electrons in modulation doped GaAs/AlGaAs multiple quantum wells by resonant Raman scattering. Electrons from donors incorporated in the barriers of these structures drop into the GaAs wells, forming a multilayered 2D electron gas with very high mobility. The long range *interlayer Coulomb interaction* couples the in-plane motion of electrons in different wells, even if there is no wave function overlap, as in our present samples. The *interlayer Coulomb interaction* thus leads to a band of N non-degenerate coupled layer plasmon eigenmodes for an electron system with N sheets. Using resonant Raman scattering we measure the dispersion of these discrete layer plasmon eigenmodes and their coupling with intersubband excitations yielding a very detailed instrument for the characterisation of multilayered electron systems. We measure the single particle excitation spectra as a function of $k_{||}$. We fit the single particle excitation spectra with the Lindhard formula for the dielectric response function of the 2D electron gas and we show that the electron density and the scattering time can thus be determined optically.

1. INTRODUCTION

Electrons in modulation doped heterojunctions and quantum wells are separated from the region of the ionised donors by spacer layers. Scattering of the carriers in the wells by the donors (or acceptors) is therefore weak and the carriers are free to move with exceptionally high mobility parallel to the layers[1]. Therefore these structures are used for many fundamental investigations and for many applications. The response of the electrons to applied electromagnetic fields is characterised by single particle excitations, and by collective excitations, the plasmons [2]. Single particle excitations correspond in simple language to acceleration of single electrons. This corresponds to excitation of a single electron from below the Fermi sphere (in two dimensions: the Fermi circle) to above the Fermi sphere leaving a hole behind. The character of single particle excitations is shown schematically in Figure 1. Plasmons on the other hand are collective excitations - they involve motion of all electrons in the multiple quantum well system as shown in Figure 2. Plasmons are charge density oscillations in the quantum well system. The *interlayer Coulomb interaction* couples the motion in different wells, so that the charge density oscillations in different wells are correlated. Plasmons also may have some direct applications: they may be made to emit infrared light by manufacturing a grating onto the surface of the superlattice, and they represent a loss mechanism for hot electrons. The present paper describes work we have done to measure the single particle excitation spectra and the dispersion of the plasmon

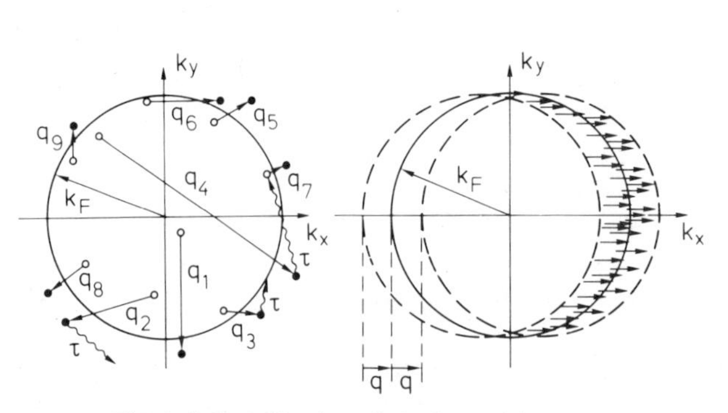

Fig. 1. Schematic view of single particle excitations. The right hand part of the figure shows excitations all with the same q, demonstrating the calculation of the Lindhard function.

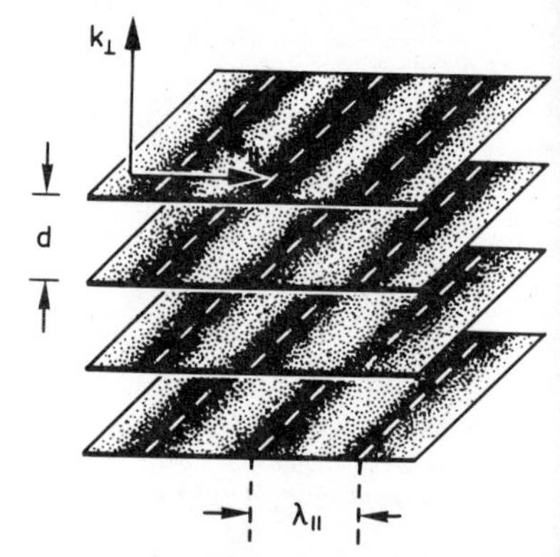

Fig. 2. Example of the charge density oscillation associated with the $k_\perp = 0$ plasmon mode for a system with four electron layers. For the mode with $k_\perp = 0$ the charges on all layers oscillate in phase.

eigenmodes in layered 2D electron systems as a function of energy and in-plane wavevector $k_{||}$ using resonant Raman scattering. Measuring the characteristics of both types of excitation helps to determine the electronic structure of quantum wells. By combining Raman measurements of the single particle excitation spectra and of the plasmon dispersion with calculations we obtain a very detailed contactless characterisation method of modulation doped quantum well structures. From the single particle excitation measurements the carrier density and the carrier scattering time are determined *optically*. The measured plasmon dispersion yields a detailed microscopic characterisation of the multisheeted electron gas.

2. IN-PLANE ELECTRONIC EXCITATIONS OF MULTI QUANTUM WELLS

The electronic properties of multilayered electron systems has been studied by a large number of authors[4,5,6,7,8,9,10,11]. The response of a *non-interacting* electron system to an external perturbing field (i.e. for example the electric field of an impurity, the applied DC or AC field in a FET transistor, or the electric field of a light wave) is expressed by the Lindhard function $\chi_o(\omega, k)$. The response function $\chi_o(\omega, k)$ for a *non-interacting* system has been calculated for the three dimensional (3D) electron gas by Lindhard[3] and for the two-dimensional (2D) electron gas by Stern[4] . The Coulomb interaction between the electrons can be taken into account in lowest order RPA approximation by considering the dielectric function $\varepsilon(\omega,k)$, where

$$\varepsilon(\omega,k) = 1 + V_k\, \chi_o(\omega,k). \tag{1}$$

$V_k = 4\pi e^2/k^2$ is the Fourier transform of the Coulomb interaction between the particles. The response function $\chi(\omega,k)$ for the interacting system is related to the non-interacting Lindhard function by:

$$\chi(\omega, k) = \chi_o(\omega,k) / \varepsilon(\omega,k). \tag{2}$$

The zeros of the inverse dielectric function $\varepsilon^{-1}(\omega,k)$ determine the plasmon resonance frequencies $\omega_p(k)$. The single particle excitation spectra are equal to the imaginary part of the dielectric response function Im $\chi(\omega,k)$.

Figure 3 shows the dispersion of various elementary excitations of a layered electron gas. The shaded area shows the continuum of single particle excitations. At E = 0 they extend from $k_{||} = 0$ to $k_{||} = 2k_F$ (the k-axis in Figure 3 does not extend out to $2k_F$). For a particular $k_{||}$ the spectrum of single particle excitations is sharply peaked near $E = \hbar k_{||} v_F$ as shown for two particular $k_{||}$ values in Figure 3. This behaviour is different than in 3D, where the single particle spectra show a triangular shape.

The plasmon dispersion in a single 2D sheet of electrons is proportional to $k_{||}^{1/2}$. In a multiple quantum well system we have N parallel sheets of electrons. Typically the separation from one quantum well to the next is of the order of 500 Å. Therefore the electron wavefunction overlap between the electrons on different layers is negligible. Thus in the (hypothetical) absence of Coulomb interaction between the electrons on different layers we would simply have a single, though N fold degenerate $k_{||}^{1/2}$ plasmon dispersion curve. In the real world the *Coulomb interaction* between electrons on different layers correlates the motion of electrons in the N layer system and the simple N fold degenerate $k_{||}^{1/2}$ dispersion fans out into a band of layer plasmons for $k_{||}<2\pi/d$ where d is the interlayer separation. In the case of a system with a finite number of N layers, there are N discrete plasmon layer eigenmodes, as shown for the case N = 20 in Figure 3. The discrete layer plasmon modes lie in the region of the dispersion which would be filled by the modes of a system of infinitely many layers (shown as bounded by the line of crosses in Figure 3). At large $k_{||}$ ($k_{||} \gg 1/d$, where d is the separation between the layers), the dispersions of all plasmon eigenmode branches converge towards the $k_{||}^{1/2}$ behaviour of a single 2D sheet of electrons.

Each particular plasmon eigenmode "n" of the whole layer system can be labelled by a particular value of $k_{\perp n}$:

$$k_{\perp n} = 2\pi n/Nd, \text{ where } n = 1,\ldots\ldots, N. \tag{3}$$

N is the number of layers in the system and d is the layer separation. $k_{\perp n}$ describes the phase relationship caused by the *interlayer Coulomb interaction* for the charge oscillations of a particular plasmon eigenmode. For sample structures with a large number of layers, $k_{\perp n}$ plays the rôle of a wavevector. Thus from a system with a large number of layers (N > 20), only those modes are expected to have considerable Raman intensity, where $k_{\perp n}$ is equal to the wavevector transfer of the Raman process perpendicular to the layers. This is indeed seen in experiment[12]. In systems with a few number of layers (up to around N = 20) $k_{\perp n}$ still describes the phase relationship of the electron density oscillations locked together in a particular layer plasmon eigenmode. But $k_{\perp n}$ loses it's meaning as a wavevector. Thus for small N the wavevector component perpendicular to the layers is not a quantity conserved during scattering processes. We exploit this fact for our experiments to map out the dispersion as a function of $k_{||}$ of discrete layer plasmons.

3. RAMAN SCATTERING FROM IN-PLANE EXCITATIONS

With resonant Raman spectroscopy both the plasmon resonances[12,13,14,15,16] and the single particle excitation spectra[15,16,17] can be measured as a function of the in-plane wavevector $k_{||}$ for

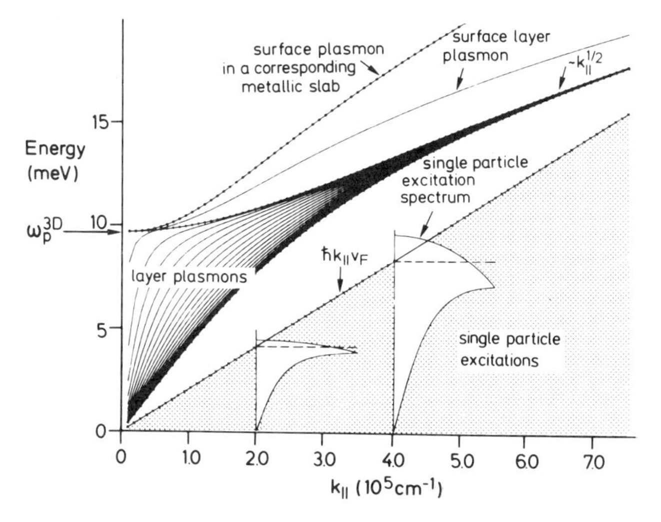

Fig. 3 Schematic view of the electronic excitations of a layered electron gas (here with N = 20 layers), showing the single particle continuum and the coupled layer plasmon eigenmodes. Also shown are the limits of the plasmon band (see text).

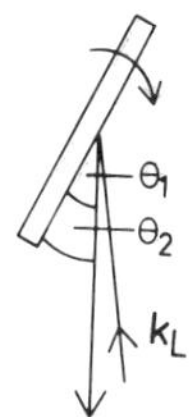

Fig. 4. Scattering geometry permitting the measurement of Raman spectra of a layered electron system as a function of $k_{||}$.

2D and layered 2D carrier systems. This technique has been well developed for three dimensional semiconductor systems[18,19] and has been applied by several workers recently to two dimensional carrier systems in quantum wells and heterojunctions (for a review see Ref. 20). For incoming and scattered light polarised parallel to each other, the Raman spectra are caused by charge density fluctuations, i.e. proportional to the dynamical structure factor of the electrons

$$S(\omega,k) \sim \text{Im}\,[\chi(\omega,k)/\varepsilon(\omega,k)]. \qquad (4)$$

Although scattering from single particle excitations may in principle occur also in this configuration, it is normally unobservable due to heavy screening, which is described by ε in the denominator of Equation (3). In crossed polarisation the Raman signal is proportional to the spin-density fluctuations. The fluctuation dissipation theorem relates the spin density fluctuation spectrum to Im $\chi_0(\omega,k)$. The Raman scattering process in this case is due to spin-flip processes, where the spin-flip can occur because of the spin-orbit interaction, which couples spin up and spin down states in the valence band. This work has been applied to study the electronic excitations in quantum wells and heterojunctions perpendicular to the layers, i.e. the intersubband excitations, where electrons are lifted from the lower to higher subbands in the wells[20]. Raman scattering has been first applied to in-plane plasmon excitations by Olego et al.[12] and Sooryakumar et al.[13]

Plasmon excitations in single layer two-dimensional electron systems have been studied in great detail using infrared spectroscopy and grating coupler techniques. This work has recently been reviewed by Heitmann[21]. A grating is necessary, because in a 2D electron system the plasmon dispersion does not intersect with light dispersion $\omega_{light} = c\,k$. In the Raman process plasmons and single particle excitations are absorbed or emitted by one of the partners of a virtual electron hole pair excited by the laser photon - light does not couple directly to the excitations in Raman scattering. Therefore the in-plane excitations can be studied without the need to use grating structures on the sample surface. The dispersion of a variety of different coupled in-plane excitations and their interactions can therefore be studied in modulation doped superlattices by Raman scattering.

4. EXPERIMENTAL DETAILS

We measure the plasmon eigenmodes and the single particle excitation spectra as a function of the in-plane wavevector transfer $k_{||}$. This technique takes specific advantage of the possibility of varying $k_{||}$ in a light scattering experiment of a 2D system. $|k|$ cannot be changed significantly in 3D systems except by changing the laser frequency (which changes resonance conditions) or the scattering geometry from forward to backscattering. It is not usually possible to change k continuously. In a light scattering experiment from a 2D system or a layered 2D system the wavevector transfer perpendicular and parallel to the plane of the layers has to be considered separately. We use a near back scattering geometry as shown in Figure 4. The wave vector transfer parallel to the layers in this configuration is $k_{||} = k_{Laser}(\cos\Theta_1 + \cos\Theta_2)$. This configuration is similar to the one used by Olego et al.[12], but allows to vary $k_{||}$ over nearly twice the range. The $k_\perp$ transfer is approximately independent on the scattering angle of the experiment because of the high refractive index of the materials studied ($n_{GaAs} \approx 3.6$). An additional factor is, that the perpendicular component of the wave vector transfer is not conserved during the scattering process, because of the lack of translational symmetry perpendicular to the layers in a layered sample with a few layers only. The $k_\perp$ transfer of course still influences the relative Raman scattering intensities of different plasmon eigenmodes.

Most of the excitations studied here lie in the energy range below about 10 meV and the single particle excitation spectra lie in the range below 5 meV for our conditions. Since the thermal energy at a typical experimental temperature of 10K is $kT \approx 0.9$ meV, and heating of the sample by the laser illumination may be important, accurate control of the temperature and low laser power are crucial to interpret the measurements properly. It is important to measure the temperature of the *electron system*. Under the non-equilibrium conditions of laser illumination, there will be a temperature gradient between the electrons in the illuminated volume of the sample, the lattice of the sample in the illuminated volume, the rest of the sample and the cold finger of the cryostat. For all the single particle measurements, we determined the cold finger temperature with a calibrated Ge resistor (Lake-Shore Cryotronics) and measured the temperature of the *electron system* by measuring the exponential tail on the high energy side of the band gap luminescence spectrum, and by measuring the Stokes/Antistokes ration of appropriate Raman signals. We found this procedure absolutely crucial. In all experiments the laser power was adjusted to achieve minimal heating, as monitored by the difference of the temperature of the coldfinger and the temperature of the electron gas.

5. MODULATION DOPED GaAlAs/GaAs/AlAs/GaAlAs QUANTUM WELLS

The samples used in the present study are modulation doped quantum wells which were grown such that the assymmetry of the quantum wells was deliberately enhanced.The layer sequence for our sample is shown schematically in the insert of Figure 6. For sample No. 4849, for which we present the results in this paper, *initial* Shubnikov de Haas (SdH) experiments showed more than 30 oscillations of ρ_{xx} with a single period of oscillation corresponding to a carrier density of n = 6.85×10^{11} cm $^{-2}$, and there was an indication of higher filled subbands. The total carrier density determined from low field Hall effect measurement under saturating illumination conditions (as corresponding to the conditions of the Raman experiments) is $n_{tot} = 3.4 \times 10^{12}$ cm^{-2}. Assuming that five layers contribute, this yields a density of n = 6.8×10^{11} cm $^{-2}$ per layer in agreement with the Shubnikov de Haas oscillation period. In addition, there is good agreement between the measured and the calculated plasmon dispersion for this sample and of the position of the single particle excitation spectra with the theoretical position, when assuming a carrier density of n = 6.8 $\times 10^{11}$ cm^{-2}. Since the mobility (measured under saturating illumination conditions at 4K) is 177 000 cm^2/Vs we assume for the present analysis that there is a single layer of electrons per well, localised at the "normal" interface. We have recently done Shubnikov de Haas measurements with higher accuracy and therefore increased sensitivity for additional electron sheets. We will show in a future paper that more accurate determination of the distribution of electrons in the various subbands of the wells allows the determination of further detail in the plasmon spectra (G. Fasol, D. King-Smith, N. Mestres and K. Ploog, to be published).

The complete characterisation of the electronic structure (determination of the wavefunctions of all electrons levels populated, determination of carrier densities and mobilities of all carriers, at the "normal", the "inverted" interface and the population of higher subbands) needs a large set of experimental data combined with selfconsistent electronic calculations, which we are undertaking at the moment. Preliminary results of our calculations show that the two lowest electron levels in the present quantum wells are closely spaced in energy and have two maxima of the charge density, one spaced close to the "normal" and the other close to the "inverted" interface. We are at present calculating the plasmon dispersion taking the wave functions determined by the selfconsistent band structure calculations into account. Thus we are presently developing a complete picture of the electronic properties of our samples. We have recently developed a method based on resonant Raman scattering which allows the determination of the Fourier components of the carrier wavefunction in a modulation doped quantum well . This technique will eventually be useful in interpreting the details of the plasmon spectra[22], since it will allow to take experimentally determined wave functions into account. Conversely, determination of the plasmon dispersion will allow characterisation of the electron distribution among subbands in modulation doped quantum wells.

6. RAMAN MEASUREMENTS OF DISCRETE LAYER PLASMONS

Olego et al.[12] and Sooryakumar et al.[13] have reported Raman measurements on in-plane plasmons in layered electron gas systems in modulation doped quantum wells with a large number of layers. In these samples the number of layers was so large, that the periodicity of the layers restricts the scattered Raman intensity to those plasmon eigenmodes, for which $k_\perp$ is close to the wavevector transfer of the light scattering process perpendicular to the wells. In this case the Raman signal is expected to consist of the superposition of the Raman signals from several discrete layer plasmon modes close to fulfilling the condition on $k_\perp$. This work showed excellent agreement of the experimental dispersion with the plasmon dispersion of infinite systems[7,8,9,23,24]. Raman spectra from discrete plasmon eigenmodes in systems with N of the order of 15 show several modes for which $k_\perp$ is close to the optical wavevector transfer[14].

In modulation doped multi quantum well systems with few numbers of layers (e.g. N = 2, ..., 5) there is no translational symmetry perpendicular to the layer planes and therefore discrete Raman scattering peaks can be seen, corresponding to most or all of the discrete layer plasmon modes[15]. Figure 5 shows Raman measurements of sample No. 4849. The left hand side (Figure 5a) shows the Raman spectra for parallel polarisation. Due to the selection rules the collective modes appear in this configuration. The spectra on the right hand side of the figure (Figure 5b) are measured in crossed polarisation and show the Raman signals due to spin density fluctuations. These spectra will be discussed in the next section. The spectra of the collective excitations in Figure 5a show several peaks, which all depend on the in-plane wavevector transfer. We have studied more than ten samples in the course of this project sofar, results on these and results of detailed calculations taking the wavefunctions from selfconsistent calculations into account will be published separately.

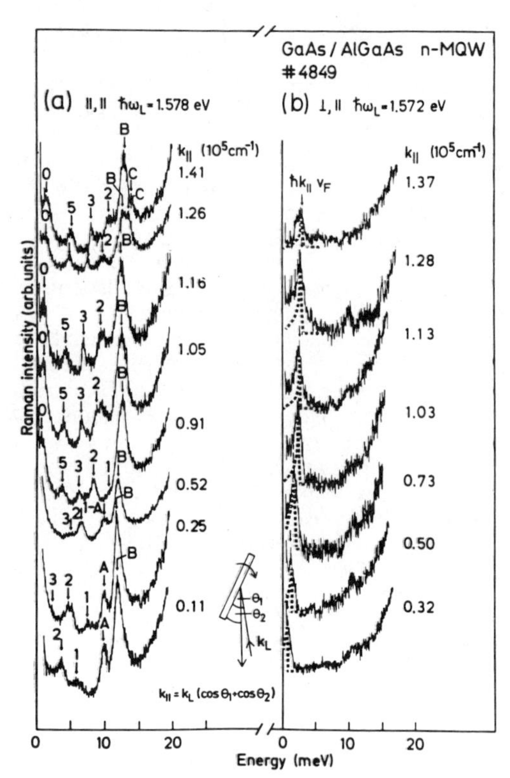

Fig. 5. Electronic Raman spectra of sample No. 4849 (5 layers) as a function of $k_{||}$.
a) in parallel polarisation showing collective modes
b) in crossed polarisation showing single particle excitation spectra. Dashed lines show the calculated imaginary part of the Lindhard response function (calculated here for T=0K, and neglecting scattering).

There are three groups of Raman peaks appearing in parallel polarisation: there is a mode labelled "0", modes 1 to 5 and modes A, B, and C. The modes 1 to 5 are due to the discrete layer plasmon modes and their splitting is due to the *inter-well Coulomb interaction*.. This assignment can be made by comparing the position of the Raman peaks as a function of $k_{||}$, as shown in Figure 6, to the theoretical layer plasmon dispersion curves. These theoretical curves are shown as the thin solid lines in Figure 6. We have computed them using a recent calculation by Jain and Allen[25]. We have programmed their calculation, but our treatment differs in that we found it numerically more convenient to diagonalise Equation (5) of Ref. 25. This calculation assumes that there is one single infinitely thin electron gas layer inside each quantum well, i.e. that all electrons within each well move in phase. The calculation neglects intrawell degrees of freedom - thus it does not take account of the possible filling of higher subbands. King-Smith[26] has recently calculated the plasmon dispersion for our parameters including the effect of the finite perpendicular extent of the electron wave function, and of the filling of higher subbands, improving the agreement with our experimental results, particularly for the lowest energy mode (n = 5). It should be stressed, that the present calculation only uses as input parameters the layer thickness, which is accurately known from the growth parameters, and the electron density, which we have determined both by Shubnikov de Haas and by Hall effect measurements. There are no adjustable parameters. We did not find a mode in the measurement corresponding to mode n = 4. This may be a consequence of the strong dependence of the Raman intensity on the laser energy, i.e. mode n = 4 could be visible at a different laser energy. Modes A, B, and C are attributed to mixed modes of layer plasmons coupled to intersubband excitations. The coupling can be clearly seen in the fact that mode C clearly splits off from mode B. This coupling is seen even more clearly in other sample structures we have investigated. Preliminary calculations[26] show that mode "0" is likely to be due to a mode where the electrons on the two lowest energy levels in the well - the "symmetric" and the "anti-symmetric" state - oscillate out of phase, while they oscillate in phase for modes n = 1,....5.

The insert on the right hand side of Figure 6 shows the eigenvectors corresponding to some layer plasmon modes. In each case we have plotted the deviation of the electron density from it's equilibrium value as a function of the coordinate parallel to the layer. The top mode shown, is a "surface plasmon"[27], where the amplitude associated with the electron density oscillation decreases exponentially from the surface into the interior of the structure. Such a surface mode would only appear if the uppermost electron layer is very close to the vacuum interface of the sample. The mode shown in the middle is the one with the highest energy (n = 1) for our five layer sample. It can be clearly seen that the maximum of the carrier density in each layer occurs for the same lateral position. This mode therefore has the highest Coulomb energy associated with it's motion. The bottom insert shows the layer plasmon mode with the lowest energy. In this case the electron density oscillations on different layers are exactly 180° out of phase - the electrons "avoid" each other as much as they can and therefore the Coulomb energy is lowest. Similar pictures can be drawn of the eigenmodes for various number of layers N. As N increases the phase relationship of the electron movement on different layers, becomes more and more "wave-like". Translational symmetry perpendicular to the layer planes emerges, Bloch's theorem and wavevector conservation of $k_\perp$ during the scattering process start to become applicable as N increases.

Further measurements which will be published separately explore the interaction of the layer plasmon modes with intersubband excitations. The agreement between theory and the experimental plasmon dispersion will be improved by work under way to include the effects of higher subbands, of the presence electron layers with different charge densities and of the presence and possible occupation of higher subbands. Thus measurements of the plasmon dispersion of modulation doped multiple quantum wells is a method of characterising the carrier density and the carrier configuration. Indeed optical measurement of the 3D bulk plasmon resonances is a standard method to determine the carrier density in bulk doped semiconductors and work such as presented here may extend this method of characterisation to 2D and layered 2D systems.

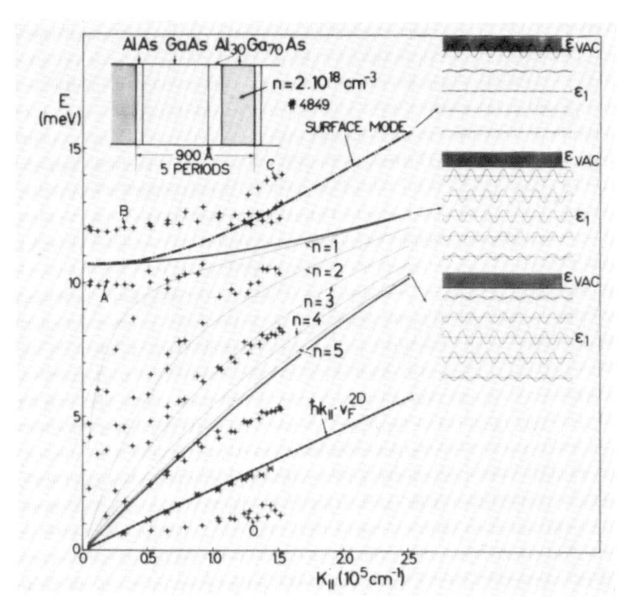

Fig. 6. Sample No. 4849 (5 layers): experimental energies of electronic excitations as a function of in-plane wavevector component $k_{||}$ from measurements shown in Figure 5. Layer plasmon modes (n=0,1,...5), mixed plasmon-intersubband modes (A, B, C) and single particle excitations(*). Plasmons energies are compared with a calculation following Jain and Allen (see text). Positions of single particle excitations are compared with $hk_{||}v_F$ (dashed line), which is computed using the carrier density but no fitting parameter.

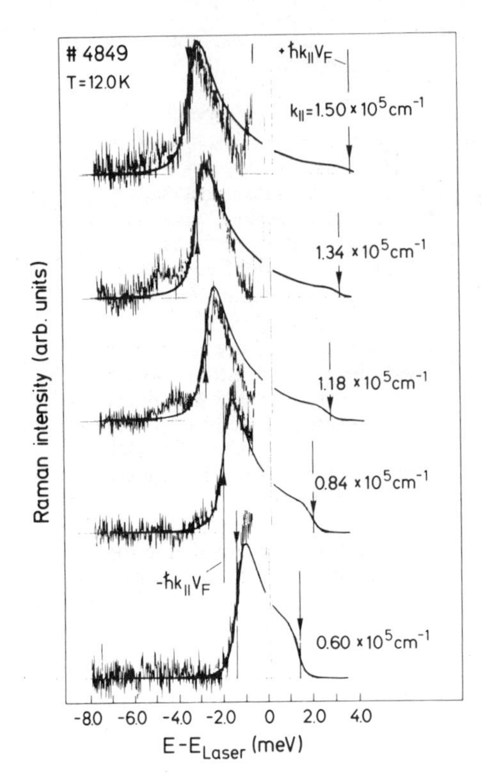

Fig. 7. Single particle excitation Raman spectra measured for different in-plane wavevector components $k_{||}$. Solid line is a fit using the Lindhard-Mermin dielectric function at the appropriate electron temperature as determined from the exponential tail of the luminescence. Only fit parameter is the phenomenological scattering time τ_{sp}.

Fig. 8. Raman spectra from single particle excitations for different *electronic temperatures*. The *electronic temperatures* are determined from the exponential tail of the luminescence spectra.

7. SINGLE PARTICLE EXCITATION SPECTRA

Figure 5b shows the Raman spectra of the five layer modulation doped quantum well sample No. 4849 in crossed polarisation for different values of the in-plane wavevector transfer. Due to the selection rules these spectra are from spin density fluctuations and represent the single particle excitation spectra[18,19,20]. They show a sharp peak slightly below h $k_{||}$ v_F. In Figure 6 the position of the experimental single particle peaks are compared with $hk_{||}v_F$ (dashed line). v_F is determined from the carrier density, known from Shubnikov de Haas and Hall measurements. These spectra are proportional to Im $\chi_o(\omega,k_{||})$ - they measure directly the imaginary part of the dielectric response function. The dashed lines in Figure 5b show Im $\chi_o(\omega,k_{||})$ as calculated by Stern[4] for T = 0 K and in the absence of broadening due to the finite lifetime of the single particle excitations. The agreement is quite good. Necessarily the experimental spectra are considerably broader than the imaginary part of the Lindhard function due to the finite temperature of the measurement and the broadening due to the scattering.

Single particle Raman spectra have been measured in doped bulk GaAs[19]. The scattering process has been shown to be due to spin flip scattering[18]. This work has later been extended by Pinczuk et. al.[28], who have made a very careful lineshape analysis of the 3D single particle Raman spectra in terms of the Lindhard-Mermin response function[29]. In a 2D system, as mentioned above, the wavevector $k_{||}$ can be varied and therefore the wavevector dependence of Im $\chi_o(\omega,k_{||})$ can be studied. We have recently reported the measurement of single particle excitations of the 2D electron gas[15,16,17].

As calculated by Hamilton and McWhorter for GaAs[18], the spectra in this case are proportional to Im($\chi_o(\omega,k_{||})$). To interpret our results, we have calculated

$$\chi_{\text{Mermin}}(\omega,k_{||}) = \frac{(1 + i/\omega\tau_{sp})(\chi_0(\omega + i/\tau_{sp},k_{||}))}{1 + (i/\omega\tau_{sp})(\chi_o(\omega + i/\tau_{sp}, k_{||})/\chi_o(0,k_{||}))} \qquad (5)$$

from Ref. 29, where $\chi_o(\omega, k_{||})$ is the non-interacting Lindhard dielectric response function[3].

Figure 7 shows the Raman spectra of the single particle excitations for sample 4849 as a function of in-plane wavevector transfer $k_{||}$, achieved by changing the angle of measurement as explained above. The solid lines show a fit of the Raman cross section proportional to the imaginary part of the Lindhard-Mermin response function:

$$\frac{d^2\sigma}{d\Omega d\omega} \sim (1 - e^{-h\omega/kT})^{-1} \text{Im}\,\chi_{\text{Mermin}}(\omega, k_{||}) \qquad (6)$$

This is Equation 2.92 from Ref. 20. We calculate $\chi_{\text{Mermin}}(\omega,k_{||})$ numerically for the appropriate temperature of the electron gas by computing Equations (5) and (6) varying the value of the phenomenological scattering time τ_{sp} to achieve the best fit. In this way a measurement of the single particle relaxation time can be obtained from the Raman spectra. Figure 8 shows the temperature dependence of the single particle spectra as a function of temperature for a particular value of in-plane wave vector transfer ($k_{||} = 1.33 \times 10^5$ cm^{-1}). Again fits of the spectra using the Lindhard-Mermin response function are shown, where the only fitting parameter is a phenomenological scattering time $\tau_{s.p.}$. The only other input parameter is the carrier density determined consistently both from Shubnikov de Haas and Hall effect measurements. In this way we determined the temperature dependence of the single particle relaxation time. Since the energies involved here are in the range of thermal energies, it is crucial for the significance of these experiments to have accurate temperature control as explained in the experimental section above.

We determined the single particle relaxation times τ_{sp} determined in this way as a function of electron temperature. The single particle relaxation times τ_{sp} are compared to the mobility scattering time τ_μ, determined from the Hall measurements. Extrapolating to T = 0K, where only scattering due to impurities and inhomogeneities contributes, we find τ_{sp}(T = 0K) = (1.8$\pm$0.3) ps and τ_μ(T = 0K) = (7.1 $\pm$ 0.1) ps and therefore τ_μ/τ_{sp} = (4.0 $\pm$ 0.8). The single particle relaxation time determined by the light scattering experiments is considerably larger than the mobility scattering time - this is a common geometrical effect and shows that small angle scattering contributes to a large extent to the scattering.Thus the single particle Raman spectra provide a way to obtain both carrier density and information on the scattering processes in the structure. To compare the scattering times obtained by the various techniques, a microscopic theory is necessary.

8. SUMMARY

The dielectric response of a layered electron gas system, as occuring in modulation doped quantum wells, is characterised by plasmons and single particle excitations. These excitations exist both for the degrees of freedom of the quantum well perpendicular and parallel to the layers. The work presented here is a study of the excitations parallel to the layers, i.e. the excitations associated with the directions in which the electrons are free to move with high mobility.

For a system with N layers of carriers the plasmon dispersion has N discrete layer plasmon modes. These N modes are split in energy, due to the *interlayer Coulomb interaction* between the different layers. The interlayer Coulomb interaction (the separation of peaks n = 1, ... ,5 in Figure 6 is a direct measure of this Coulomb interaction) between layers is considerable. At the same time there is no overlap of the wavefunctions in different wells in the present samples.

We have measured the spin density fluctuation spectra, corresponding to the single particle excitations by spin-flip Raman scattering in crossed polarisation. The single particle excitation spectra are directly proportional to the imaginary part of the dielectric response function Im $\chi_o(\omega, k_{||})$. Taking account of the appropriate temperature we can describe the Raman spectra well. We determine a phenomenological scattering time $\tau_{s.p.}$ from the fit.

ACKNOWLEDGEMENTS

The authors would like to express their deep gratitude to M. Cardona for the continuing support of this work. They are very grateful to M. Dobers, D. Heitmann, D. King-Smith, K. von Klitzing,

G. Lonzarich and U. Ekenberg for many helpful discussions and help. The work would have been impossible without the technical assistance by H. Hirt, M. Siemers and P. Wurster.

REFERENCES

1. R. Dingle, H. Störmer, A. C. Gossard, and W. Wiegmann, Appl. Phys. Lett. 33, 665 (1978)
2. D. Pines and P. Nozières, "The Theory of Fermi Liquids", W. A. Benjamin Co., Inc., New York, (1966)
3. J. Lindhard, Kgl. Danske Videnskap, Mat. Fys. Medd. 28, No. 8. (1954)
4. F. Stern, Phys. Rev. Lett. 18, 546 (1967)
5. R. H. Ritchie, Phys. Rev. 106, 874 (1957)
6. R. A. Ferrell, Phys. Rev. 111, 1214 (1958)
7. A. L. Fetter, Ann. Phys. (N. Y.) 81, 367 (1973)
8. A. L. Fetter, Ann. Phys. (N. Y.) 88, 1 (1974)
9. D. Grecu, Phys. Rev. B 8, 1958 (1973)
10. J. K. Jain and P. B. Allen, Phys. Rev. B 32, 997 (1985)
11. A. C. Tselis and J. J. Quinn, Phys. Rev. B 29, 3318 (1984)
12. D. Olego, A. Pinczuk, A. C. Gossard and W. Wiegmann, Phys. Rev. B 25, 7867 (1982)
13. R. Sooryakumar, A. Pinczuk, A. C. Gossard and W. Wiegmann, Phys. Rev. B 31, 2578 (1985)
14. A. Pinczuk, M. G. Lamont and A. C. Gossard, Phys. Rev. Lett. 26, 2092 (1986)
15. G. Fasol, N. Mestres, H. P. Hughes, A. Fischer and K. Ploog, Phys. Rev. Lett. 56, 2517 (1986)
16. G. Fasol, H. P. Hughes and K. Ploog, Surface Science 170, 497 (1986)
17. G. Fasol, N. Mestres, M. Dobers, A. Fischer and K. Ploog, Phy. Rev. B 36, 1565 (1987)
18. D. C. Hamilton and A. L. McWhorter in Light Scattering Spectra of Solids, p. 309 (ed. G. B. Wright), Springer Verlag, Berlin
19. A. Mooradian in Light Scattering Spectra of Solids, p. 285 (ed. G. Wright), Springer Verlag, Berlin
    A. Mooradian and A. L. McWhorter in Light Scattering Spectra of Solids, p. 297 (ed. G. Wright), Springer Verlag, Berlin
20. G. Abstreiter, M. Cardona and A. Pinczuk in Light Scattering in Solids IV, p. 5, (ed. by M. Cardona and G. Güntherodt), Springer Verlag, Heidelberg, (1984)
21. D. Heitmann, Surface Science, 170, 332 (1986)
22. T. Suemoto, G. Fasol and K. Ploog, Phys. Rev. B 34, 6034 (1986)
    T. Suemoto, G. Fasol and K. Ploog, in Proceedings of the 18th International Conference on the Physics of Semiconductors, Stockholm 1986 (ed. O. Engström), World Scientific Publ. Co. (Singapore), p. 683
    and T. Suemoto, G. Fasol and K. Ploog, Phys. Rev. B 37, (1988) in print.
23. S. Das Sarma and J. J. Quinn, Phys. Rev. B 25, 7603 (1982)
24. W. L. Bloss and E. M. Brody, Solid State Commun. 43, 523 (1982)
25. J. K. Jain, P. B. Allen, Phys. Rev. Lett. 54, 2437 (1985)
26. D. King-Smith and G. Fasol, to be published.
27. G. F. Giuliani and J. J. Quinn, Phys. Rev. Lett. 51, 919 (1983)
28. A. Pinczuk, G. Abstreiter, R. Trommer and M. Cardona, Solid State Commun. 30, 429 (1979)
29. N. D. Mermin, Phys. Rev. B 1, 2362 (1970)

RESONANT TUNNELING IN DOUBLE BARRIER HETEROSTRUCTURES

Mark A. Reed

Central Research Laboratories
Texas Instruments Incorporated
Dallas, TX 75265

INTRODUCTION

Semiconductor quantum wells are the subject of considerable interest due to the ability to confine carriers in ultrathin (<100Å) semiconductor layers bounded by lower electron affinity (higher bandgap) potential barriers. When the dimension of the thin lower bandgap layer approaches the carrier mean free path, the restriction of the carrier motion in the direction perpendicular to the layer interface causes electric quantization that govern the electronic and optical properties. The formation of these quantum states (subbands) in the thin layer ("quantum well") confined by thick barriers allows for the experimental investigation of quasi-two dimensional carriers. In the case that the barriers are sufficiently thin such that the quantum well states are quasi-bound (due to quantum mechanical tunneling through the barrier), then it is possible to directly probe the electronic structure of the quantum well by carrier transport through the structure. This phenomena of resonantly tunneling through these quantum well states was first experimentally realized by Chang, Esaki, and Tsu[1] in a GaAs quantum well / AlGaAs double barrier heterostructure. Recent advances in advanced semiconductor epitaxy, typified by molecular beam epitaxy (MBE) in the GaAs/AlGaAs system, has generated remarkable interest and success in these investigations.[2-7] A wide range of intriguing physical phenomena can be investigated, since the phenomena is not material specific; resonant tunneling has been observed in such systems as silicon[8] and in a semimetal/semiconductor system.[9]

This chapter presents some of the recent investigations in resonant tunneling in double barrier heterostructures, highlighting the spectroscopic ability that these structures have for determining quantum well states. It is demonstrated that the quantum well states and the resulting electronic characteristics can be "tuned" to a high precision, not only by dimensional control but by the use of multicomponent systems. The effects of impurities in the double barrier structure on resonant tunneling and the resulting temperature dependences are also discussed.

In addition to the size quantization imposed by the epitaxial layer structure, recent advances in microfabrication technology makes possible lateral confinement on a dimension approaching that of the vertical dimensions. Presented are the first observations of telegraph noise switching phenomena seen in quantum wires containing resonant tunneling structures.

RESONANT TUNNELING

The prototype resonant tunneling structure is a one-dimensional, double barrier - single quantum well heterostructure, schematically shown in Figure 1(a). The center quantum well has a spectrum of discrete allowed eigenstates, higher in energy relative to the contact conduction band edge. At low device bias, the quasistationary state (band) in the central quantum well is too high in energy to allow resonant tunneling of electrons through the eigenstate. Tunneling through the entire structure is allowed, but is exponentially small. As the device bias increases, the quantum well state is lowered with respect to the Fermi level of the input contact, and carriers are allowed to tunnel through the quasistationary state in the quantum well (Figure 1(b)). When the device bias is increased to the point that the allowed state in the quantum well is lower in energy than the conduction band edge of the input contact (Figure 1(c)), elastic tunneling is no longer allowed due to the conflicting requirements of energy and momentum conservation. This decreases the tunneling current, producing negative differential resistance (Figure 1(d)). The tunneling current eventually increases at high bias due to Fowler-Nordheim tunneling through the top of the barriers.

The positions of the quantum well states are determined by the height of the potential barriers and by the width of the quantum well. For a given barrier height, widening the quantum well will cause the quantum well states to move downward in energy (i.e., to lower bias), and previously unbound excited states may appear. These higher energy eigenstates will produce the same resonance phenomena as they are biased to the Fermi level. By dimensional and/or alloy content control, one has the ability to continuously tune the resonant voltage positions of the quantum well and, with sufficient ingenuity, to create multiple resonances with an arbitrary spacing.

The typical experimental embodiments of this structure have been realized in the direct gap GaAs III-V compounds fabricated by molecular beam epitaxy (MBE). In this embodiment, the contacts to the double barrier structure are n+ GaAs, while the barriers are $Al_xGa_{1-x}As$ and the central quantum well GaAs. In the initial experiments on these structures, the central quantum well was lightly doped. However, transport occurs via resonant transmission of carriers through the structure, and thus the region need not be doped at all. Indeed, the donors in this region can only serve as scattering centers, degrading the resonant transmission of the structure.

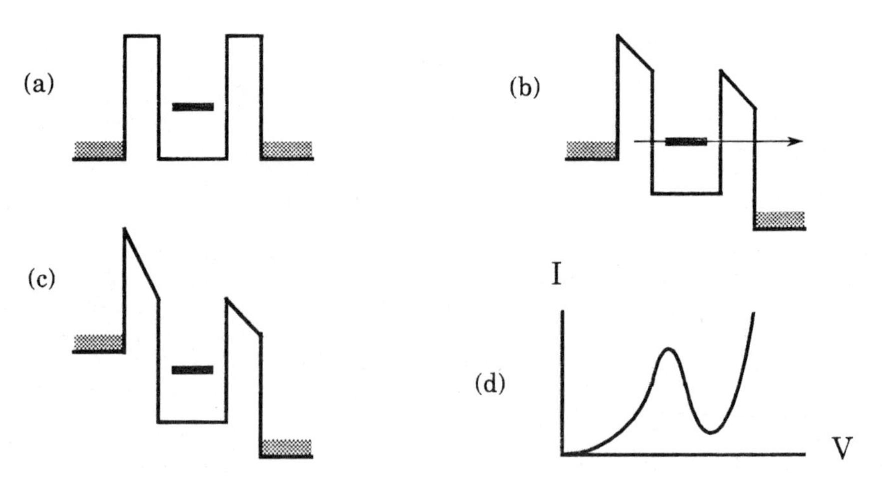

Figure 1. Conduction band diagram in real space of a resonant tunneling structure (a) at zero applied bias, (b) at resonant bias, and (c) at greater than resonant bias. (d) Schematic I-V characteristic.

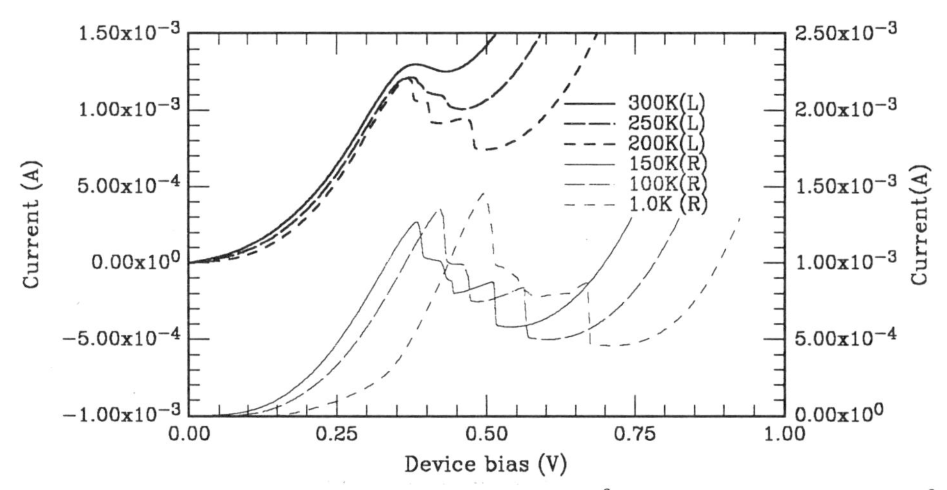

Figure 2. Current - voltage characteristics of a 50Å GaAs quantum well / 50Å $Al_{.3}Ga_{.7}As$ double barrier structure as a function of temperature. For clarity, the current scale is shifted (L and R denote the left hand or right hand current scales, respectively) for half the curves. The device mesa size = 25 μm^2.

Figure 2 shows typical current-voltage characteristics of an undoped 50Å GaAs quantum well / 50Å $Al_{.3}Ga_{.7}As$ double barrier structure grown by MBE. Standard fabrication processes were used to define mesa diode devices. The structure exhibits weak negative differential resistance (NDR) at room temperature due to tunneling through the ground state of the quantum well. The NDR increases as the sample is cooled to lower temperatures, giving a peak current - to - valley current ratio (an often used figure of merit) of 3:1 at 1.0K. A number of effects are evident here; 1) a shifting of the resonant voltage to higher bias with decreasing temperature due to the increasing resistivity of the GaAs contact regions; 2) a slight increase of the peak resonant current with decreasing temperature, and; 3) a dramatic decrease of the valley current with decreasing temperature. The complex structure at biases between the peak and valley is due to self-bias effects of the oscillating negative resistance device.

The enhanced resonant tunneling effect in this structure over previous initial works[1,2] of a nominally identical structure was suspected to be the reduction of impurity-assisted scattering by to the elimination of donor impurities from the double barrier structure. This implies that the double barrier structure should be as impurity-free as possible. However, this imposes strict tolerance on the dopant control in the contact regions. In addition, the dopant (usually Si) incorporated into the n^+ GaAs regions can migrate into the nominally undoped double barrier region during growth. To enhance the purity of the double barrier structure, the doping of the GaAs contacts was reduced to zero prior to the growth of the double barrier structure, and likewise symmetrically resumed after the structure, to minimize the migration of Si impurities into the structure during growth. The undoped GaAs layer between the barrier and the n^+ GaAs contact is called a "spacer layer", and the current-voltage characteristics of a structure incorporating such spacers is shown in Figure 3. This structure is identical to the structure shown in Figure 2, except for the insertion of 100Å spacers. This structure exhibits a dramatic increase in the magnitude of NDR; a 1.75:1 peak-to-valley ratio is achieved at 300K, increasing to almost 8:1 at low temperature. Variants on this technique have produced peak-to-valley ratios of 3.9:1 at 300K, and 21.7:1 at low temperatures in state-of-the-art structures.[3]

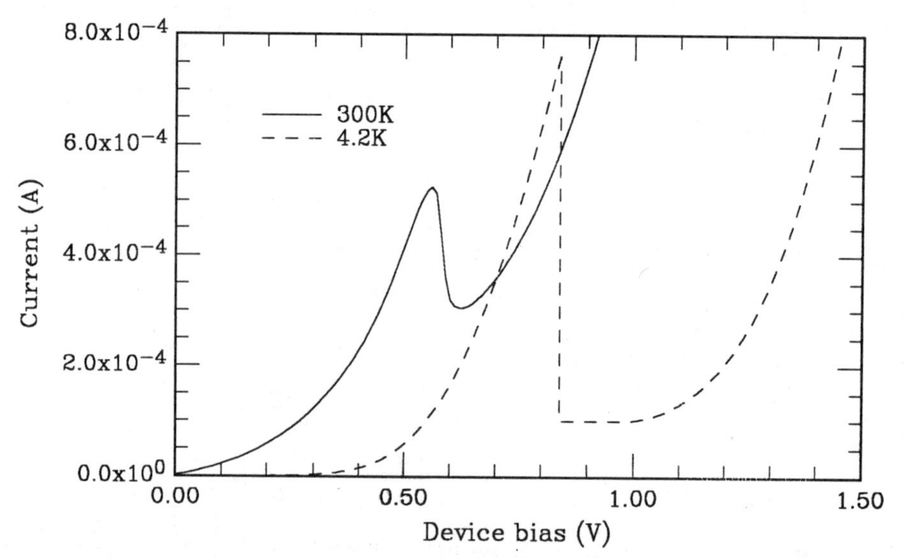

Figure 3. Current - voltage characteristics of a 50Å GaAs quantum well / 50Å $Al_{.3}Ga_{.7}As$ double barrier structure, with 100Å spacers, at 300K and 4.2K. The device mesa size = 25 μm^2.

The mechanisms that limit the valley current are elucidated in Figure 4, where the valley current for the structures detailed above are plotted as a function of temperature. The temperature dependence of the inelastic tunneling current for both structures follow the same functional form, except for a scale factor. The valley current is linear in T above a certain critical temperature (approximately 130K) for both structures, though the magnitude of the effect is

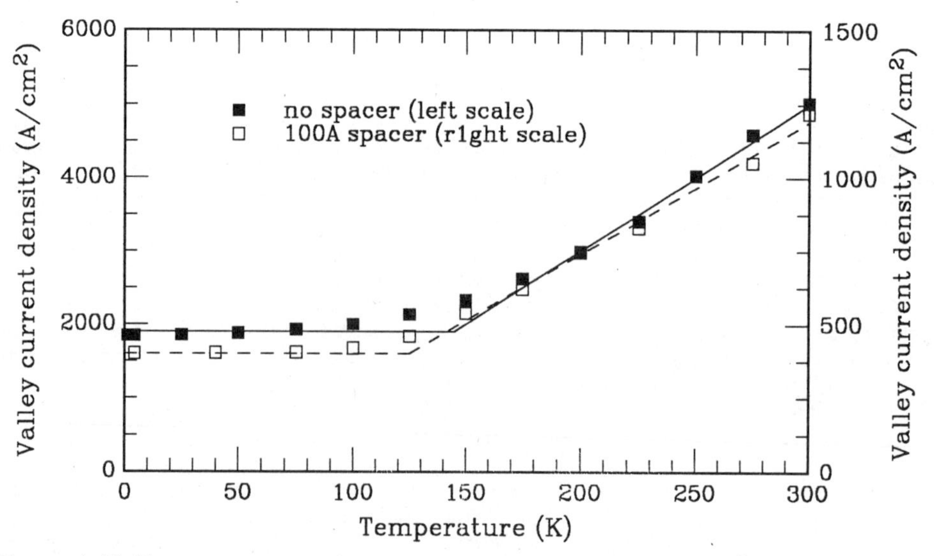

Figure 4. Valley current as a function of temperature for the 50Å GaAs quantum well / 50Å $Al_{.3}Ga_{.7}As$ double barrier structures with zero and 100Å spacers.

much smaller for the 100Å spacer layer sample (20 A/cm^2-K versus 7.2 A/cm^2-K). Below the critical temperature, the valley current is "frozen out", again at a lower value for the 100Å spacer layer sample. This data implies impurity-assisted tunneling for the excess component in both temperature regimes; for $T > T_{critical}$, the process is phonon-assisted, as the excess current is roughly proportional to the phonon density. The density of residual impurities in the structure determine the minimum "frozen-out" excess current and the magnitude of the T linear term. The same impurities in the double barrier structure are responsible for both temperature regimes, since the curves for the two different structures scale.

The ability to tune the position,and even the number of resonant states in these system leads to some interesting quantum well spectroscopy. The examples considered above exhibited a single resonant peak; as was previously mentioned, enlarging the quantum well will result in lowering the energy of the quantum well states, and previously unbound excited states may appear. Figure 5 shows the current-voltage characteristics for a structure with the identical dimensions and barrier height as the structures discussed above, with the exception that the quantum well has been widened to 100Å. The two well-defined peaks in the characteristics correspond to resonant tunneling through the ground state (resonance at 100 mV) and the first excited state (resonance at 900 mV) of the quantum well. The calculated peak positions of the structure are 83 mV for the ground state and 350 mV for the first excited state. This apparent discrepancy is caused by assuming that the device bias is the same as the potential drop across the double barrier structure. However, there is a voltage drop across the contact regions to the double barrier structure, which causes the bias across the double barrier structure to be less than the entire device bias. This can be treated as a contact resistance in series with the heterojunction. Assuming a series contact resistance of 40Ω, the experimental resonant peaks occur at 84 mV for the ground state and 380 mV, in excellent agreement with the predicted values.

This series resistance is also responsible for the apparent "hysteresis", or bistability (double valued current for a fixed bias) in the current - voltage characteristic seen for the excited state resonances. This is because the current is decreasing with increasing device bias in the NDR bias region. If the impedance of the (current limiting) double barrier structure is sufficiently low with respect to the series resistance, then hysteresis will appear.

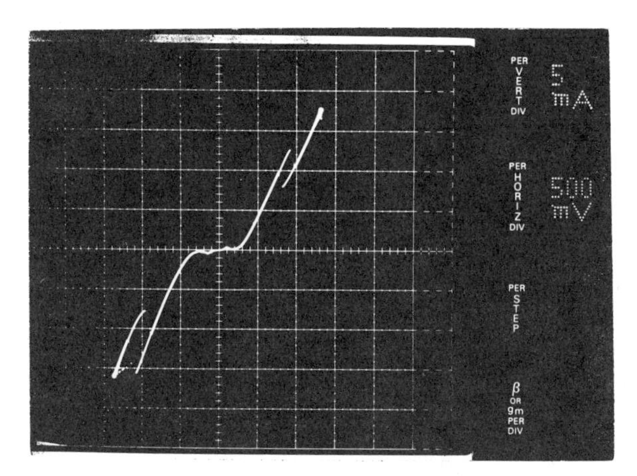

Figure 5. Current - voltage characteristic (positive and negative bias) of a 100Å GaAs quantum well / double $Al_{.3}Ga_{.7}As$ barrier structure at 77K, demonstrating resonant tunneling through the ground (100 mV) and first excited state (900 mV) of the quantum well.

The contact resistance of these structures can often be conveniently treated as a simple resistance; indeed, the observation of multiple resonant states can provide an accurate determination of this resistance. However, more complex structures demand a detailed modeling of the accumulation and depletion regions in the contact regions. This is especially important in view of the trend to incorporate spacer layers between the doped GaAs contact and the AlGaAs barrier.

MULTICOMPONENT RESONANT TUNNELING STRUCTURES

The design of resonant tunneling heterostructures need not be limited to the simple "double square barrier" case; in addition to being able to substitute complex barrier shapes,[6] it has been demonstrated[10] that the central quantum well can be replaced by a material with higher electron affinity than the surrounding GaAs contact regions; specifically, a strained-layer InGaAs quantum well. Figure 6 shows a conduction band diagram of such a structure. The intriguing aspect of these structures is that the InGaAs quantum well states can be lowered with respect to the GaAs contact region simply by the incorporation of In into the well, keeping all other dimensional parameters of the structure fixed. Indeed, sufficient In can be added such that quantum well states can "disappear" below the GaAs contact conduction band edge, making these states invisible to transport.

Figure 7 illustrates the gradual shifting of the ground state resonance to lower voltages in a GaAs contact / $Al_xGa_{1-x}As$ double barrier / $In_yGa_{1-y}As$ quantum well strained layer structure as In is incorporated into the well of nominally identical resonant tunneling structures. It should be noted that the lowest voltage structure ($y=0.08$) has a finite zero-bias conductance since the quantum well state has been lowered below the Fermi level of the GaAs contacts.

If one continues to increase the In content in the quantum well, the ground state of the quantum well will not only be lowered through the Fermi level, but will eventually be lower than the GaAs contact conduction band edge. In this case, the ground state is hidden from transport and resonant tunneling proceeds only through the excited states. Figure 8 shows the current-voltage characteristics of such a structure, along with a nominally identical double barrier structure with a GaAs quantum well for comparison. Notice here that the conduction band edge of the InGaAs was lowered sufficiently to observe the $n=3$ state of the quantum well, which was virtual in the GaAs quantum well case.

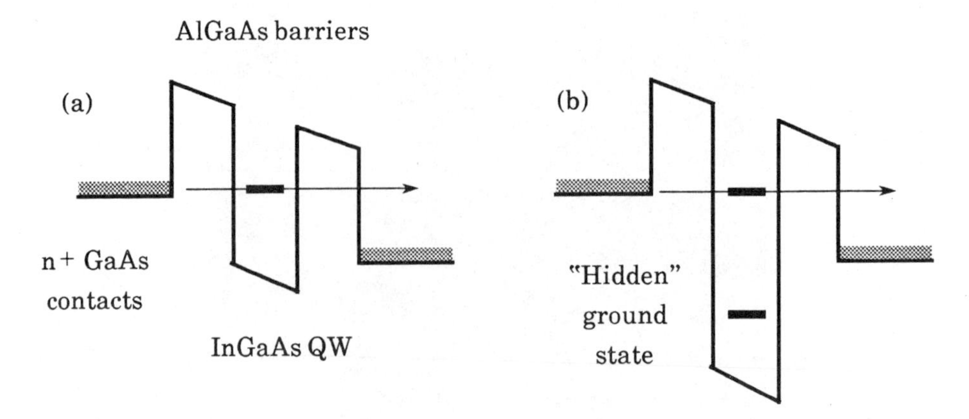

Figure 6. (a) Conduction band diagram in real space of an InGaAs quantum well resonant tunneling structure, shown at resonant bias. (b) Increasing the In content sufficiently can cause the ground state (at any bias) to be "hidden", and tunneling occurs only through excited states.

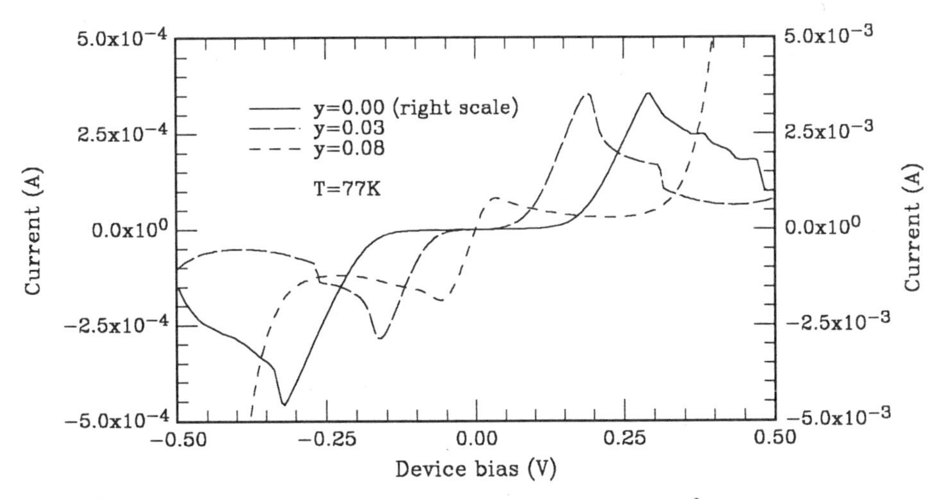

Figure 7. Current-voltage characteristics at 77K of 50Å $In_yGa_{1-y}As$ quantum well/40Å $Al_{.25}Ga_{.75}As$ barrier structures, with $y=0$, $y\sim0.03$, and $y\sim0.08$.

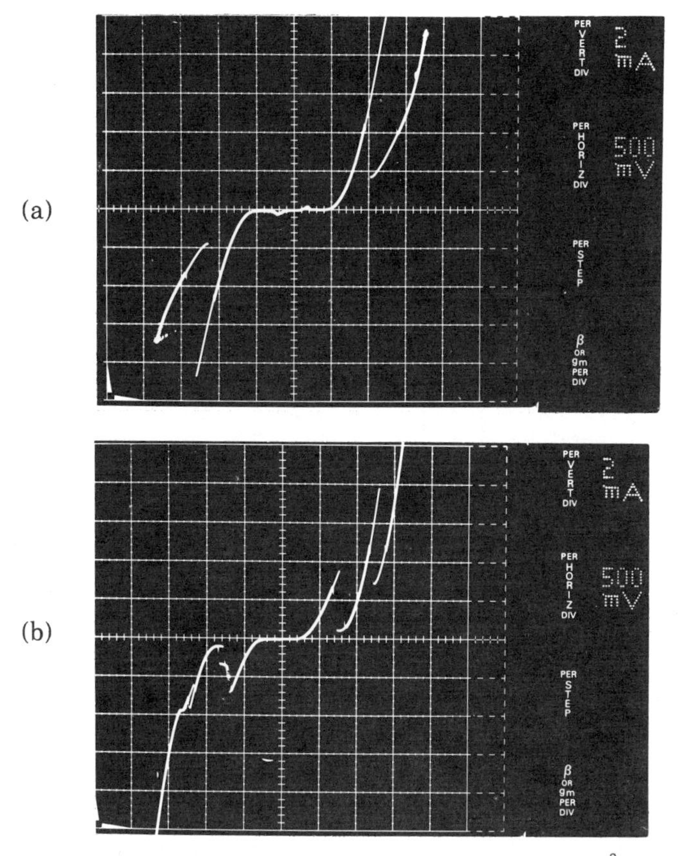

Figure 8. Current-voltage characteristics at 77K of (a) a 85Å GaAs quantum well / 35Å AlAs barrier structure (the ground and first excited state resonances are visible), and (b) an identical structure as in (a), with the replacement of the GaAs quantum well with a 85Å $In_{.1}Ga_{.9}As$ quantum well. The ground state is hidden, and the first and second excited state resonances are visible.

TUNNELING FROM QUANTIZED REGIONS

The ability to engineer peak positions in two component and in multicomponent systems now allows us, in general, to use double barrier structures as tools to understand tunneling phenomena. One obvious use is as a spectrometer on suitably designed quantum well or superlattice structures. Specifically, we demonstrate here that we can perform spectroscopy on large quantum wells placed between different GaAs (or InGaAs) quantum well, double barrier structures. The different resonant positions of the peaks, tunable by In content instead of changing the quantum well width (which introduces other complexities), discriminates which region of the epitaxial structure is being examined. This insertion of these resonant tunneling regions throughout a complex epitaxial structure gives us a microscopic spectrometer to examine the energy states in the structure.

Figure 9 illustrates an experimental embodiment of such a structure. The structure consists of a 40Å $Al_{.25}Ga_{.75}As$ double barrier / 90Å $In_{.05}Ga_{.95}As$ quantum well / 750Å GaAs quantum well / 40Å $Al_{.25}Ga_{.75}As$ double barrier / 90Å $In_{.05}Ga_{.95}As$ quantum well structure, with a n+ GaAs contact on the other side of the GaAs quantum well resonant tunneling structure. The large GaAs quantum well (750Å) will have a small energy splitting in comparison to the quantum wells of the double barrier structures (90Å). In this specific case, the GaAs quantum well resonant tunneling structure has a n+ contact region so as to provide a "source" or "sink" of available carriers. Thus, when the structure is biased such that carriers tunnel from the large quantum well into the double barrier structure, they will be injected from a ladder of states in the large quantum well. For simplicity, we will discuss results using the GaAs quantum well resonant tunneling structure as the spectrometer, though similar results are also obtained from the InGaAs resonant tunneling structure (at a different bias position).

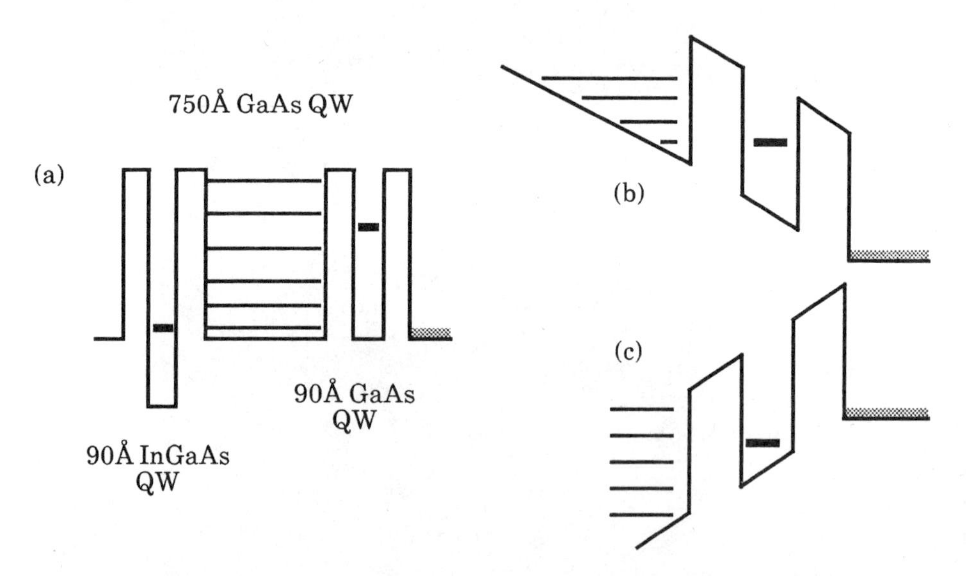

Figure 9. (a) Schematic of a large GaAs quantum well bounded by two resonant tunneling structures which have different resonant voltages by varying the In content. These structures act as spectrometers to probe the states of the large central quantum well region. (b) Bias condition for electron injection from the quantized states in the large GaAs quantum well into the n+ GaAs contact. (c) Bias condition for electron injection from the n+ GaAs contact into the large GaAs quantum well.

Figure 10 shows a current-voltage characteristic at T=4.2K corresponding to electron injection from the large GaAs quantum well. A series of peaks corresponding to electron injection from the states in the 750Å quantum well through the state in the 90Å quantum well are clearly observable. The structure appears on the low bias side of the major peak only. The experimentally observed splittings are 59 meV, 103 meV, 143 meV, and 238 meV. No peaks corresponding to n>4 in the large quantum well are observed, indicative of the position of the Fermi level in this region. The ratios of the splittings to the ground state splitting (1:1.74:2.42:4.03) are in excellent agreement with calculated values (1:1.69:2.36:3.01) except for the n=4 level, presumably due to band bending. Clearly this technique can be generalized to more complex regions, such as parabolic injectors to verify equal splitting in such structures.

Now consider electron injection from the n+ GaAs region, through the 90Å GaAs quantum well double barrier structure, into the 750Å GaAs quantum well region. Below the resonant peak, the excess tunneling current is a sum of elastic scattering to available states on the other side of the structure, and inelastic tunneling through the entire structure or through the intermediate quantum well state (which should be negligible at these temperatures). Thus, no structure in the I-V characteristic below the resonant peak should be seen due to the large quantum well. However, when the structure is biased beyond the major resonant peak position, elastic scattering can then occur via the 90Å quantum well state; i.e., elastic scattering to the 90Å quantum well ground state subband, relaxation to the bottom of this band, then tunneling out of the structure. Thus, peaks in the I-V characteristic should occur at biases greater than the major resonant peak bias.

Figure 11 shows the I-V characteristic of this structure at T=4.2K corresponding to electron injection from the n+ GaAs region. There are no observable peaks at biases less than the resonant bias. The absence of any peaks on the low bias side verify that tunneling through the entire structure (instead of through the intermediate quantum well state) is asymptotically small. However, a series of oscillations appear at biases greater than the resonant bias in accordance with the scattering mechanism described above. The spacings of

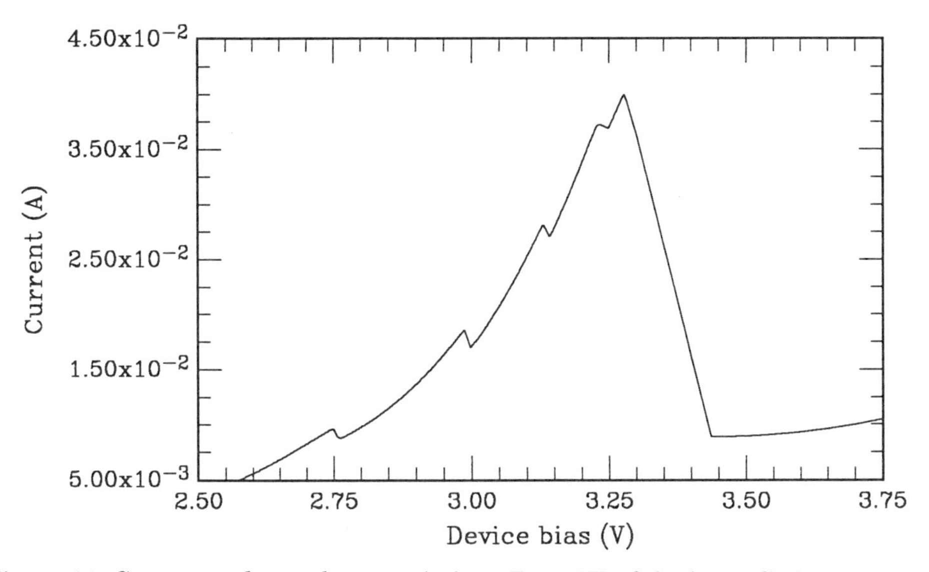

Figure 10. Current-voltage characteristic at T=4.2K of the large GaAs quantum well / double barrier structure, corresponding to electron injection from the 750Å GaAs quantum well. The structure on the low bias side of the major peak is due to tunneling from the levels in the 750Å GaAs quantum well.

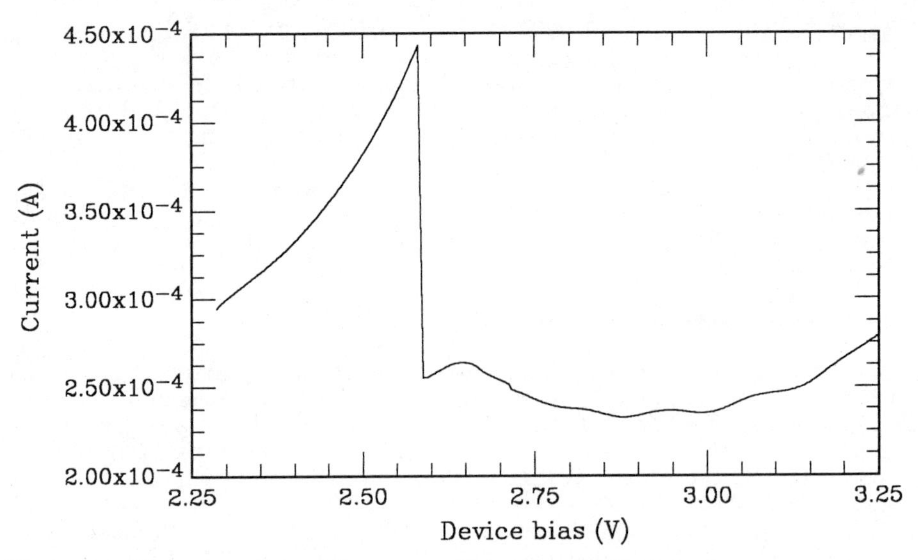

Figure 11. Current-voltage characteristic at T=4.2K of the large GaAs quantum well / double barrier structure, corresponding to electron injection from the n+ GaAs contact. The structure on the high bias side of the major peak is due to tunneling into the levels in the 750Å GaAs quantum well.

these oscillations (~130meV) are approximately constant, corresponding to the asymptotic energy level spacing of a large quantum well, and quantitatively are in good agreement with the size of the splittings of these levels when the asymmetry of the resonant peak positions is taken into account. However, there is insufficient resolution of the splitting to allow an exact determination of the quantum numbers of these levels.

Figure 12 shows the same series of oscillations for a similar sample, as a function of temperature. The oscillations tend to "wash out" with increasing

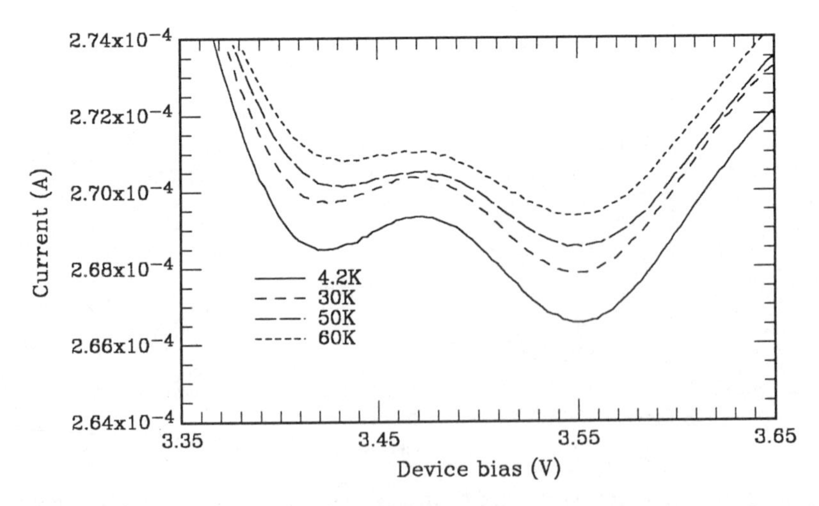

Figure 12. High resolution current-voltage characteristics as a function of temperature for the large GaAs quantum well / double barrier structure, corresponding to electron injection from the n+ GaAs contact.

temperature. This is because the thermal distribution of carriers in the Γ-minimum of the GaAs quantum well is approaching the level seperation of the large GaAs quantum well into which the electrons are tunneling. This implies that the thermalization time of the carriers in the well is less than the tunneling time out of this particular structure. Application of this technique to a series of structures may prove useful in studying carrier dynamics as well as analysing the scattering mechanisms in tunneling structures.

TUNNELING IN QUANTUM WIRES

The resonant tunneling devices discussed so far operate by virtue of the quantum size effect in the quantum well imposed by the lower electron affinity tunnel barriers. The dimension of the structure in the plane of the quantum well is essentially infinite on this scale. However, microfabrication techniques have advanced to the degree that lateral dimensions can approach the dimensions determined by the epitaxial growth layers. In this case, effects due to the confinement potential imposed by the microfabrication may be observable. By extrapolation, structures one dimension below quantum wells are defined as "quantum wires" (likewise, totally confined structures are "quantum dots").

The microfabrication approach used to produce these microstructures is summarized in Figure 13. The initial structure (substrate) is a resonant tunneling diode structure, grown by MBE on a n+ GaAs substrate and consisted of a 0.5 micrometer-thick Si-doped ($2x10^{18}$ cm^{-3}) GaAs buffer layer graded to less than 10^{16} cm^{-3}, a 50Å undoped GaAs spacer layer, a 50Å $Al_{0.27}Ga_{0.73}As$ tunnel barrier, and a 50Å undoped GaAs quantum well. The structure was grown to be nominally symmetric about a plane through the center of the quantum well. Large area (2 micrometers x 2 micrometers) devices fabricated in a conventional manner exhibit a 1.6:1 peak to valley tunnel current ratio and a current density at resonance of 1.6×10^4 A/cm^{-2}.

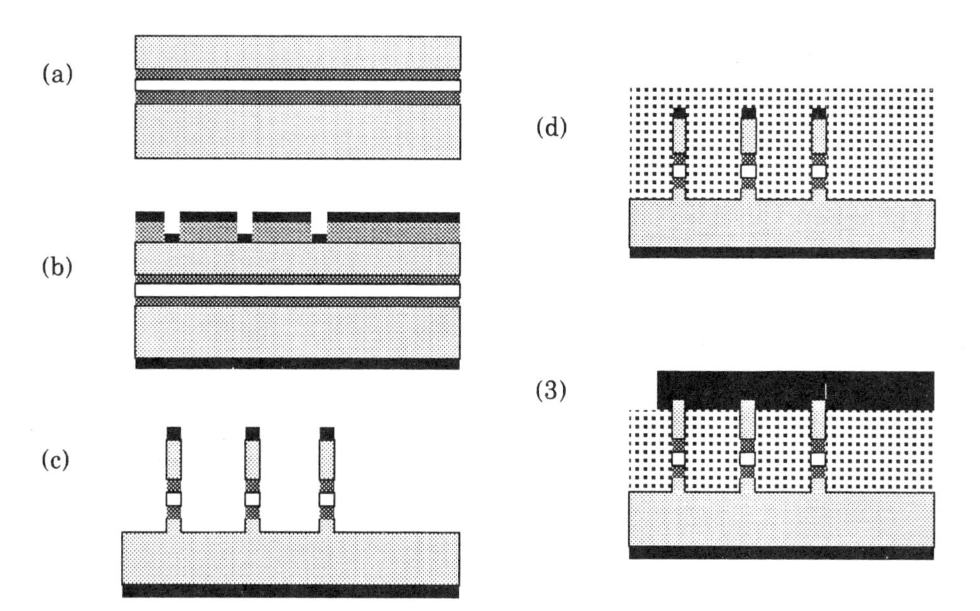

Figure 13. Schematic of fabrication sequence of quantum wire devices. (a) GaAs/AlGaAs double barrier structure starting material, with n+ GaAs contacts on top and bottom. (b) E-beam definition in PMMA of dots (singular or multiple; 3 shown here). Evaporation of AuGe/Ni/Au Ohmic contacts. (c) Liftoff. BCl_3 reactive ion etch to define pillers; i.e., "wires". (d) Planerization with polyimide. (e) Etchback to Ohmic top contacts. Evaporation of Au bonding pads.

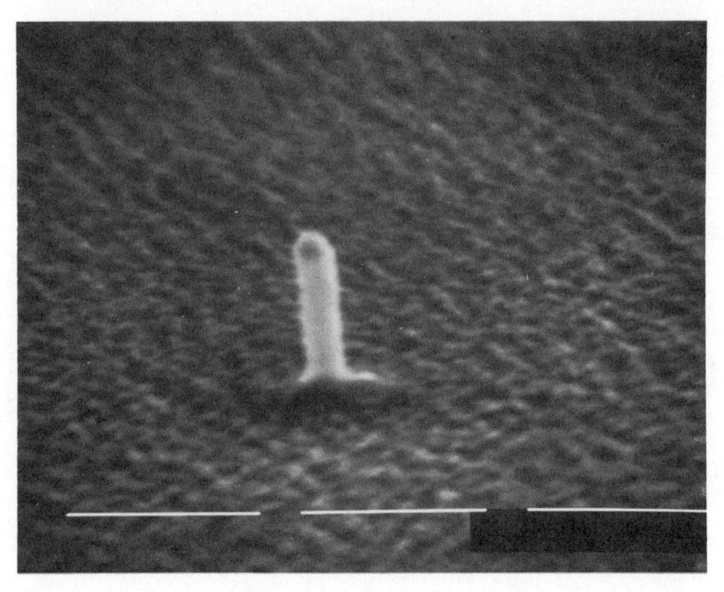

Figure 14. Scanning electron micrograph of an anisotropically etched quantum wire. The width of the wire is approximately 1000Å

E-beam lithography was used to define an ensemble of columns (including single and multiple column regions) nominally 0.25, 0.15, and 0.1 micrometer in diameter, in a bi-layer PMMA resist spun onto the structure. This pattern was then transferred to a AuGe/Ni/Au (500Å / 150Å / 600Å) dual-purpose Ohmic top contact and etch mask by lift-off. Highly anisotropic reactive ion etching (RIE) using BCl_3 as an etch gas defined columns in the structure. A SEM of these etched structures is seen in Figure 14. To make contact to the tops of the columns, an insulating polyimide was spun on the wafer, cured, and then etched back by oxygen RIE until the tops of the columns were exposed. A gold contact layer was lifted off which provided bonding pads for contact to either single wires or arrays of them.

Figure 15 shows a current-voltage characteristic of a single quantum wire resonant tunneling structure at 100K. The lateral dimensions of this single structure is 0.15 micrometers x 0.25 micrometers. The structure clearly shows NDR, though the peak to valley is degraded from the large area structure, probably due to process damage. The characteristic clearly exhibits a "noise" that is far above the system background noise. The origin of this noise is the so-called "single electron switching" phenomena[11] that has been observed in narrow Si MOSFET wires. Traps in or near the narrow conduction channel and are near the Fermi level can emit or capture electrons with a temperature-dependent characteristic time. The lowering of specific traps through the Fermi level is clearly evident (at .6V, .8V through .85V, 1.0V through 1.1V). This phenomena is seen up to room temperature (though for a different set of traps), and is usually "frozen out" by 4.2K.

Figure 16 shows a time-dependent trace of the resistance of a similar device at fixed biases voltage, exhibiting the switching between two discrete resistance states, implying the trapping and detrapping of single electrons onto the same

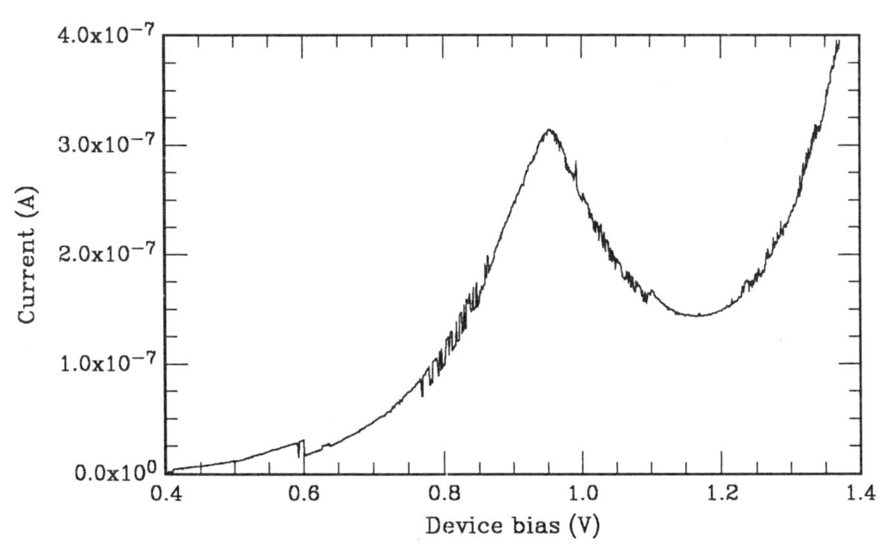

Figure 15. Current-voltage characteristics of a single quantum wire which has lateral dimensions 0.25 x 0.15 micrometer. T=100K.

trap. This switching rate is a function of temperature as well as bias. If the mechanism for the telegraph noise is scattering from traps of varying occupancy as suggested, then the trap(s) should have a well-defined activation energy measurable by the switching rate between discrete values as a function of temperature. Shown in Figure 17 is a measured activation energy of 280 meV (at fixed bias) for the trap producing the telegraph noise shown in Figure 16.

It is curious that telegraph noise can be seen for the physical dimensions (0.15 micrometer x 0.25 micrometer) of the structure. However, the effects of depletion at the etched mesa side surfaces due to pinning of the Fermi level has not been taken into account. Taking the observed current at resonance and

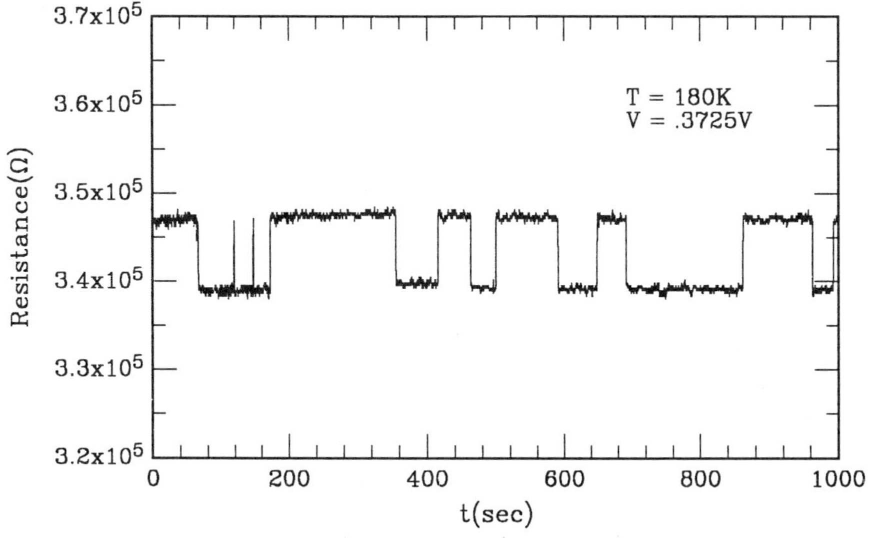

Figure 16. Time dependent resistance fluctuations due to single electron trapping in a single quantum wire. Fixed bias, T=180K.

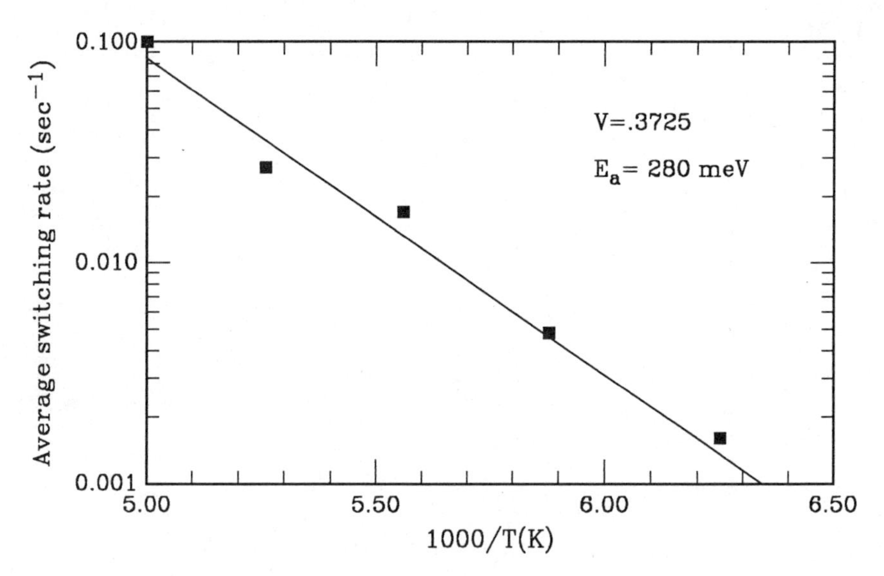

Figure 18. Activation energy of a telegraph noise trap in a single quantum wire.

assuming that the current density must be the same as in the large area device (assuming that the switching is a perturbation), we calculate that the effective (circular) conduction path diameter is ~500Å, consistent with the observation of the switching phenomena. This implies a depletion layer of approximately 500Å. However, transport and switching phenomena have also been observed in an array of wires 2000Å by 1000Å suggesting a depletion layer smaller than 500Å.

SUMMARY

Investigations into resonant tunneling phenomena has intensified in the last few years due to the precise dimensional and compositional control that now exists for ultrathin epitaxial structures. It is now possible to use such structures for the investigation of novel physical phenomena and as spectroscopic tools. With the advent of microfabrication techniques, we are now approaching a realm where lateral dimensions are comparable to the vertical epitaxial dimensions. Investigations into systems that combine both should provide a laboratory for the investigation of fascinating localization and quantum size effect phenomena.

ACKNOWLEDGEMENTS

I am indebted to J. W. Lee and R. J. Matyi for MBE growth, to T. M. Moore and J. N. Randall for microfabrication, to H.-L. Tsai for cross-section TEMs, to R. T. Bate, W. R. Frensley and A. E. Wetsel for analysis and discussions, and to R. K. Aldert, D. A. Schultz, P. F. Stickney, J.R. Thomason, A. E. Wetsel, and J. A. Williams for technical assistance. This work was sponsored by the Office of Naval Research, and the Army Research Office

REFERENCES

1. L. L. Chang, L. Esaki, and R. Tsu, *Appl. Phys. Lett.* **24**, 593 (1974).
2. T.C.L.G. Sollner, W. D. Goodhue, P. E. Tannenwald, C. D. Parker, and D. D. Peck, *Appl. Phys. Lett.* **43**, 588 (1983).
3. C. I. Huang, M. J. Paulus, C. A. Bozada, S. C. Dudley, K. R. Evans, C. E. Stutz, R. L. Jones, and M. E. Cheney, *Appl. Phys. Lett.* **51**, 121 (1987).
4. E. E. Mendez, W. I. Wang, B. Ricco, and L. Esaki,*Appl. Phys. Lett.* **47**, 415 (1985).
5. F. Capasso, K. Mohammed, and A. Y. Cho, *Appl. Phys. Lett.* **48**, 478 (1986).

6. M. A. Reed, J. W. Lee, and H-L. Tsai, *Appl. Phys. Lett.* **49**, 158 (1986).
7. T. Nakagawa, T. Fujita, Y. Matsumoto, T. Kojima, and K. Ohta, *Appl. Phys. Lett.* **51**, 445 (1987).
8. S. Miyazaki, Y. Ihara, and M. Hirose, *Phys. Rev. Lett.* **59**, 125 (1987).
9. M. A. Reed, R. J. Koestner, and M. W. Goodwin, *Appl. Phys. Lett.* **49**, 1293 (1986).
10. M. A. Reed and J. W. Lee, *Appl. Phys. Lett.* **50**, 845 (1987).
11. K. S. Ralls, W. J. Skocpol, L. D. Jackel, R. E. Howard, L. A. Fetter, R. W. Epworth, and D. M. Tennant, *Phys. Rev. Lett.* **52**, 228 (1984).

EXTRINSIC PHOTOLUMINESCENCE IN UNINTENTIONALLY AND MAGNESIUM DOPED GaInAs/GaAs STRAINED QUANTUM WELLS

A.P. Roth, R. Masut,* D. Morris and C. Lacelle

Laboratory of Microstructural Sciences
Division of Physics
National Research Council Canada
Ottawa, Ontario, Canada K1A 0R6

ABSTRACT

Low temperature photoluminescence spectra of various $In_xGa_{1-x}As/GaAs$ strained quantum wells and superlattices have been analyzed to study the contribution of residual and magnesium acceptors to the photoluminescence emissions. Sharp and intense structures due to free exciton recombinations dominate the spectra of undoped samples. These structures broaden, and weaker emissions on their low energy side become more prominent when intentional p-type doping is larger than 10^{16} cm^{-3}. The broadening of the excitonic structure is interpreted as being due to additional emission from excitons bound either to neutral acceptors in the wells or to impurities or defects at the well-barrier interfaces. The small structures are due to recombinations between electrons and acceptors confined in the wells. The binding energy of the acceptors deduced from the spectra is compared to that calculated taking into account the strain induced valence band splitting and the large heavy hole anistropy, together with the binding energy enhancement due to confinement.

INTRODUCTION

High quality strained-layer superlattices (SLS) or quantum wells (QW) of semiconductors with various degree of mismatch are routinely grown by MBE or MOVPE. Among them, structures of $In_xGa_{1-x}As/GaAs$ are interesting because of their fundamental properties and their potential for electronic or optoelectronic applications. For instance, high mobility field effect transistors using $In_xGa_{1-x}As$ quantum wells with $Al_xGa_{1-x}As$ and GaAs barriers have received a lot of attention recently.[1-3] However, despite this technological interest, little work has been published so far on the fundamental properties of $In_xGa_{1-x}As/GaAs$ structures. Most of the work has been concentrated on transport properties[4-9] and on intrinsic optical properties.[10-16]

---

*Now at GCM, Ecole Polytechnique, Montréal

The particular properties of strained quantum wells and superlattices arise from the removal of the valence band degeneracy at the Γ point under biaxial strain. In these structures, the individual layer thickness must be small enough that the mismatch between adjacent layers of different lattice constants is accommodated elastically, therefore minimizing the density of misfit dislocations. In this case, biaxial compression or tension in the plane of growth (x,y) generates a tetragonal distorsion of the lattice which removes the valence band degeneracy at the Γ point. The m_J=3/2 and m_J=1/2 hole bands are not coupled by strain whereas the m_J=1/2 band is coupled with the split-off band at k=0. Also, the two hole masses become highly anisotropic with in plane (x,y) and out-of-plane (z) masses very different within each band. For instance in $In_xGa_{1-x}As/GaAs$ structures the alloy layers are under compressive (x,y) strain and the upper hole energy band at the Γ point is the m_J=3/2 state. The (k_x,k_y) mass for that band is much smaller than the out-of-plane (k_z) mass, it becomes comparable to that of electrons in the conduction band. This reduction of the in-plane mass results in high hole mobility for transport confined in the (x,y) plane.[5-8] A second consequence of the valence band splitting is the creation of two sets of acceptors and excitons each of them associated with either the m_J=3/2 or m_J=1/2 hole band. On the other hand, in any quasi two-dimensional structure the spatial anisotropy due to the reduced layer thickness splits the two hole bands such that the upper energy band is that with m_J=3/2 at k=0. Therefore, in strained quantum wells and superlattices the strain induced valence band splitting can be increased or reduced by spatial confinement depending on the direction of the strain in the wells. If they are under biaxial compression the splitting is enhanced, (as in $In_xGa_{1-x}As/GaAs$) whereas it is reduced in layers under tension since in that case the strain induced splitting reverses the order of the two hole bands. When both strain and confinement act in the same direction, band mixing is smaller than in lattice matched structures. Then, optical transitions involving either kind of hole band should reflect the light or heavy mass character specific to each band. It has been shown both experimentally[17,18] and theoretically[19-22] that in lattice matched structures the binding energies of excitons and shallow impurities confined in quantum wells increase from their 3D values upon reducing the well width. In strained layer structures the same effect is present although the initial 3D binding energy of acceptors, and to a lesser extent of excitons, is quite different from that in unstrained bulk. In particular, acceptor states derived from the m_J=3/2 hole band have a binding energy smaller in the strained layer than in unstrained bulk, which offsets partly the effect of confinement.

In this paper we present experimental evidence for the reduction of acceptor binding energy in strained layers and we study the effect of well width on this energy. Samples with various strain, well width, alloy composition and doping concentration are studied by low temperature photoluminescence. The experimental results are compared to values calculated for large strain[23] and assuming infinite barrier heights.[19]

EXPERIMENTS

All the samples were grown by low pressure MOVPE using trimethylgallium, trimethylindium, and arsine. Uniform p-type doping was carried out using bis-cyclopentadienyl magnesium.[24] The sample parameters (alloy composition, layer thickness and uniformity) given in Table 1 are nominal values deduced from measurements made on thick layers.[25] Low temperature photoluminescence (PL) was measured with samples in a continuous flow

Table 1. Characteristics of the $In_xGa_{1-x}As/GaAs$ SLS and Strained SQW

Sample	In %	nb. of periods	W/B (Å)	Doping p(cm^{-3})	Strain %	FWHM (meV)
1	12	10	20/100	und.	-0.86	3.9
2	12	10	20/100	10^{15}	-0.86	3.2
3	12	10	25/100	5×10^{16}	-0.86	11.4
4	12	10	30/100	und.	-0.86	4.7
5	12	10	50/100	10^{16}	-0.86	5.2
6	12	20	50/ 50	und.	-0.86	3.5
7	12 on buffer with 6	10	50/ 50	und.	-0.43	8.0
8	12	10	75/100	10^{17}	-0.86	12.0
9	12	SQW	75	und.	-0.86	6.0
10	12	20	100/100	und.	-0.86	5.0
11	16	10	30/100	und.	-1.14	4.0
12	16	10	35/100	10^{17}	-1.14	11.0
13	16 on buffer with 8	10	50/ 50	und.	-0.57	10
14	19	SQW	75	und.	-1.35	4.4
15	19	SQW	75	10^{17}	-1.35	15.6

W : well width
B : barrier width
FWHM : Full width at half maximum of excitonic luminescence at 20 mW/cm^2 excitation power and 4.5K.

liquid Helium cryostat. The sample temperature, measured with a silicon diode, could be varied between 4.5 and 300K. The luminescence was excited by an argon ion laser (λ=5145Å) focussed on the sample surface. The detection was performed with a 3/4 meter spectrometer and a cooled S1 photomultiplier.

RESULTS

Figure 1 shows the 4.5K PL spectra of 2 sets of SLS samples with $In_{.12}Ga_{.88}As$ wells of width varying between 20 and 100Å grown directly on GaAs. In this case only the wells are strained, and the strain is the same in all the structures as long as both the well thickness and the total SLS thickness are smaller than their respective critical value. Otherwise the SLS would be partly relaxed and the GaAs barriers would be under biaxial tension. This is not the case for the samples of Table 1 grown on GaAs. On the contrary in those grown on lattice matched buffers (samples 7 and 13) the misfit strain is shared almost equally between the wells and the barriers. The first set of samples (Fig. 1a) is undoped with a residual carrier concentration less than 5×10^{-14} cm^3. In the second set (Fig. 1b) the samples have been uniformly doped with magnesium. The concentrations indicated in the figures are estimates based on Hall measurements made on thick layers (>2.5 µm) doped under the same conditions. The spectrum of an undoped (n~2×10^{14} cm^{-3}) $In_{.12}Ga_{.88}As$ thick layer is shown in Figure 2. In this alloy, the random distribution of In and Ga atoms creates local variations of potential which result in a broadening of excitons and other optical transitions.[26] Furthermore, free and bound excitons are separated by less than 3 meV so that with an alloy broadening of ~2 meV they cannot be resolved even in high purity samples grown on lattice matched substrates.[27] The large exciton peak of Figure 2 is thus the enveloppe of free and bound exciton recombinations. The weak structure on the low energy side is due to e-A recombinations. The residual acceptors are mainly carbon.[24] A similar spectrum is obtained

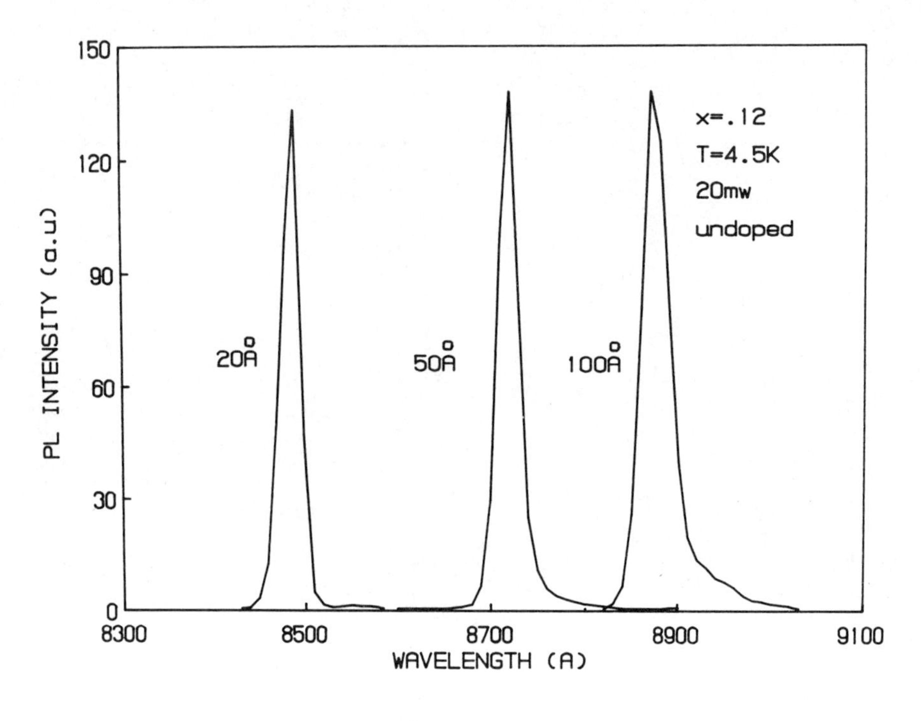

(a) Undoped

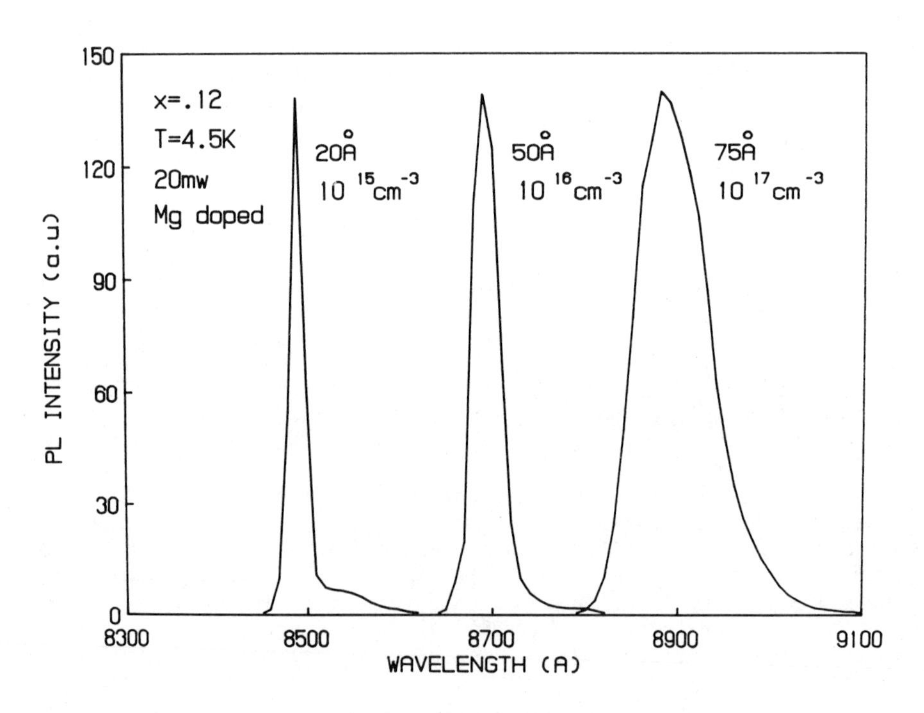

(b) Magnesium Doped

Fig. 1. Photoluminescence spectra of 2 sets of $In_{.12}Ga_{.88}As/GaAs$ multiple quantum wells.

from Mg doped layers with a small shift of the e-A emission (≈1.5 meV) due to the different central cell potential of magnesium.

In undoped quantum wells (Fig. 1a) low temperature PL is dominated by excitons made of electrons and $m_J=3/2$ holes from the n=1 levels (e_1-h_1). The large splitting of the valence band due to biaxial compression and confinement moves the light ($m_J=½$) hole level towards the GaAs valence band so that usually (e_1-ℓ_1) excitons are not observed in contrast with $A\ell_xGa_{1-x}As$/GaAs structures. The width of this exciton emission is due partly to alloy broadening in the well and partly to interface fluctu-

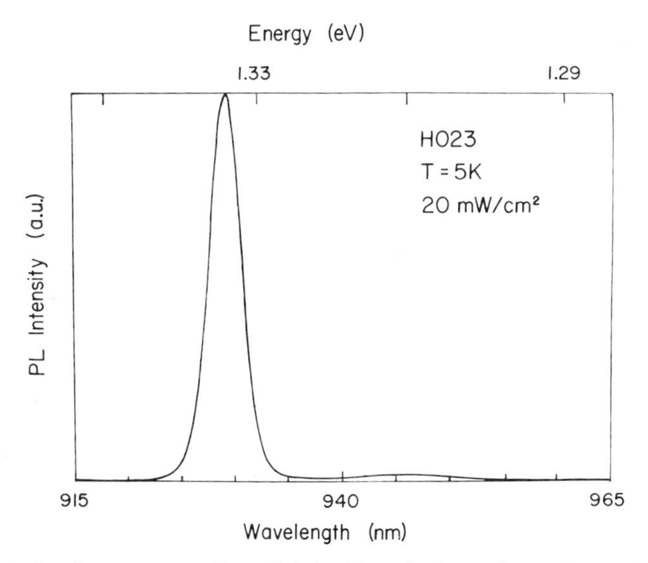

Fig. 2. Photoluminescence of a thick (5 μm) $In_{.12}Ga_{.88}As$ undoped single layer.

ations. Interface induced broadening is unavoidable when at least one of the two semiconductors is an alloy. Even on an atomically flat surface, the distribution of the Gallium and Indium atoms is random. When it is covered with GaAs the local composition varies along the interface so that both barrier height and strain vary locally. This produces an interface broadening which is therefore an intrinsic feature of quantum well structures which involve alloy semiconductors. The full width at half maximum (FWHM) of the excitonic peak in undoped structures is small even for narrow well width (see Table 1). The FWHM is somewhat larger for structures grown on ternary buffer layers (Table 1). The buffer layers are thicker than the critical thickness above which misfit dislocations multiply. Therefore their in-plane lattice constant is that of the alloy and is equal to that of the SLS. However, the surface roughness of the buffer layer is important because of the large density of misfit dislocations and this influences the quality of the SLS grown on it. The fact

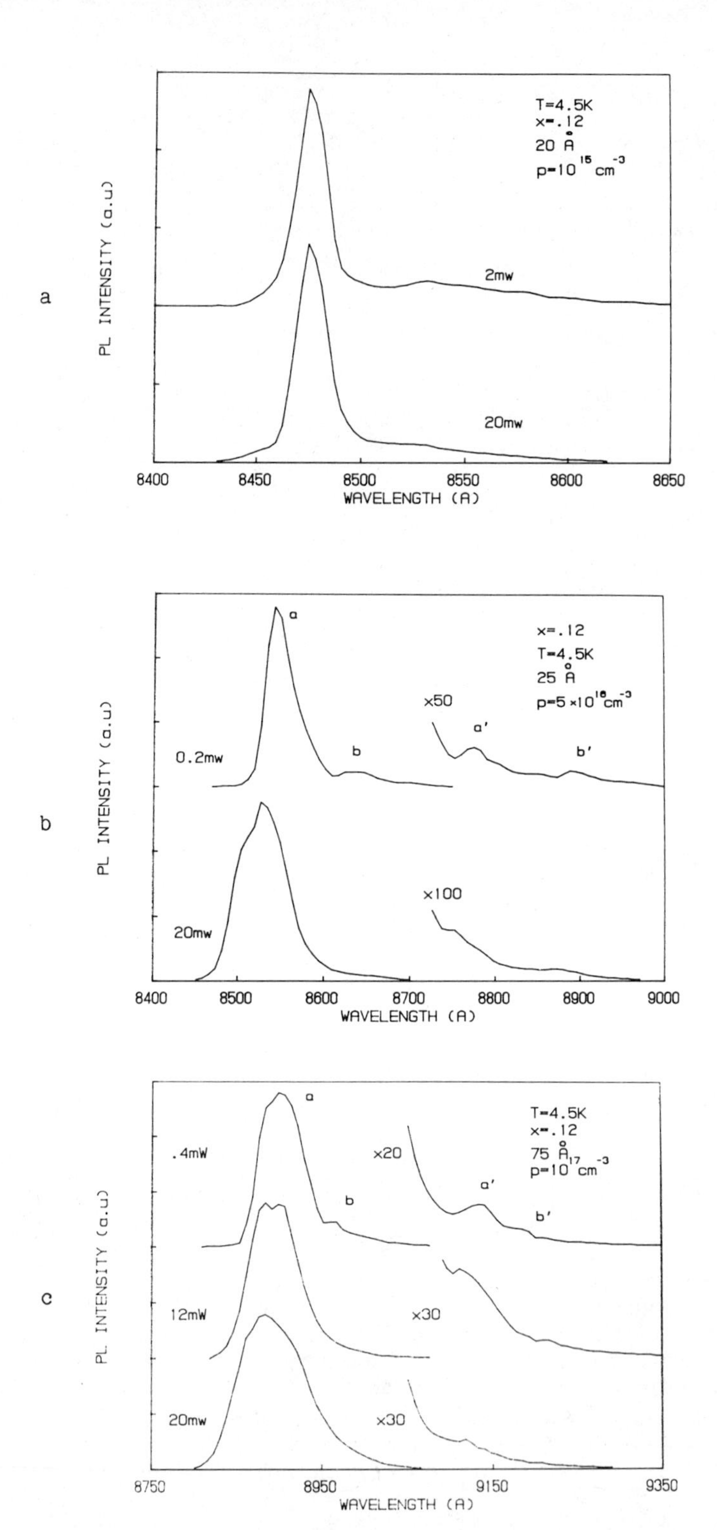

Fig. 3. Photoluminescence spectra as a function of excitation power for 3 samples of different well width and doping levels, (a) 20Å, p=10^{15} cm^{-3}; (b) 25Å, p=5×10^{16} cm^{-3}; (c) 75Å, p=10^{17} cm^{-3}.

that the FWHM does not increase at narrow well width proves that it is not controlled by step-like interface roughness due to nonoptimized growth conditions but by intrinsic alloy fluctuations. In doped structures the FWHM increases with doping concentration to reach 12 meV at $p \approx 10^{17}$ cm^{-3}. Although the large increase of the FWHM could be due to interface disordering created by the larger impurity concentration, this is most unlikely as usually this arises only at higher concentration and after annealing. The origin of the broadening can be found by studying the PL spectra variations as a function of excitation power or temperature. The excitation power dependence of the PL spectra of several samples is shown in Figure 3. Increasing the excitation power does not affect the FWHM in undoped or lightly doped ($p \approx 10^{15}$ cm^{-3}) samples, nor does it affect the energy of the luminescence maximum. However the intensity follows the excitation power linearly. When $p \gtrsim 10^{16}$ cm^{-3} the FWHM increases with excitation power while the peak maximum shifts to higher energy. At the highest doping a double structure is resolved sometimes at intermediate powers (Fig. 3c). When the sample temperature is increased up to 40K, the energy maximum either does not shift or increases by up to 3.5 meV. This shift is the same as that observed when the excitation power increases. Above 40K the luminescence maximum follows the normal band gap variation with temperature. These experiments suggest that the broadening is due to the presence of bound excitons which contribute together with free excitons to the luminescence peak. When the temperature or the excitation power increases, recombination through free excitons is more favourable than through bound excitons and the energy maximum shifts to higher energy while the FWHM increases due to the presence of both free and bound excitons. At low excitation power or temperature, bound excitons are not saturated and do not thermalize into free excitons. As the doping concentration increases the contribution from bound excitons increases as well suggesting that these excitons are bound to neutral acceptors in the well as it has been observed in $A\ell_xGa_{1-x}As/GaAs$ quantum wells.[28-29] However it is also possible that the excitons are bound to impurities or defects located at the well-barrier interfaces.[30] This interpretation would suggest that in doped structures the concentration of acceptors trapped at the interfaces increases with doping. In the absence of photoluminescence excitation spectra we cannot distinguish between the two possibilities.

In all the samples, a small structure (b in Figs. 3,4) appears at an energy lower than the exciton peak (a). This structure saturates rapidly at high excitation power or when temperature is raised. The energy difference between (a) and (b) varies with the sample parameters (well width, alloy composition and strain). The intensity of structure (b) increases with p-type doping as shown for instance in Figures 1a and 1b for wells of 20Å with x=0.12. Finally it does not shift in energy when the excitation power changes. We conclude therefore that this structure (b) is due to $e\text{-}A^0$ recombinations between electrons and neutral acceptors confined in the ternary layers.

Smaller spectral structures (labeled a',b',a" in Figs. 3,4) are also observed a lower energies. Their intensity follows closely that of (a) and (b) emissions and the energy differences a-a', a'-a", b-b' is 35.5±0.3 meV in all the samples regardless of the well width or alloy composition. This value agrees rather well with that of LO phonons in the alloy which is 35.9 meV at x=0.12 and decreases slightly at large x values.[31] Structures a' and a", are therefore LO phonon replica of the excitonic structure whereas b' is a replica of the $e\text{-}A^0$ emission. The existence of these phonon replica which are not observed in AℓGaAs-GaAs structures can be explained by the alloy fluctuations enhancement of the localization of excitons in ternary alloy wells which increases coupling to LO phonons.[32]

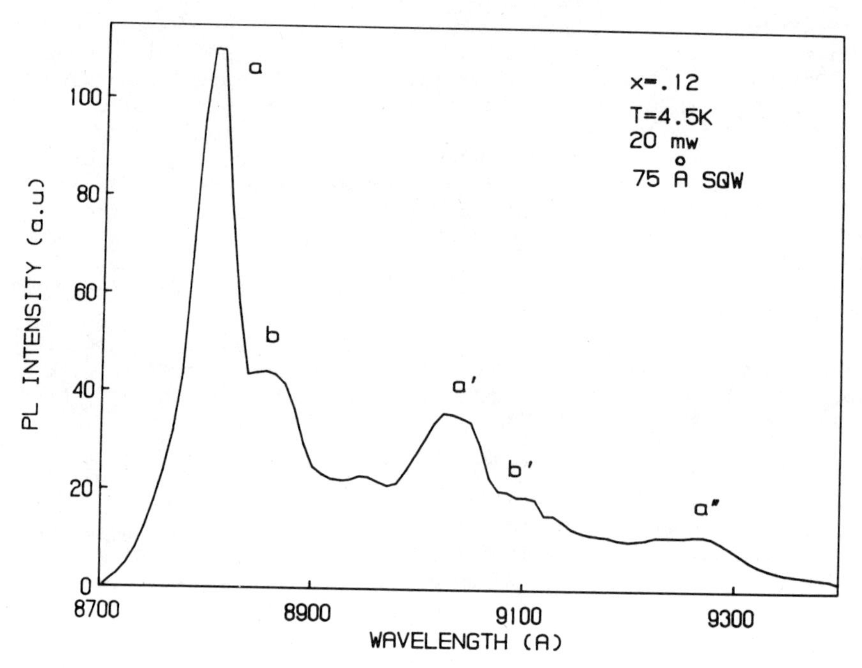

Fig. 4. Photoluminescence spectrum of a single quantum well of 75Å. Structures a', a" and b' are LO phonon replica of a and b respectively.

In strained quantum wells or superlattices the transition energy between confined electrons and holes (e-h and e-ℓ) can be calculated as in lattice matched structures provided the strain induced band gap variation and valence band splitting are taken into account. This can be done using an isolated square well model for single wells or multiple wells with thick barriers.[16] For samples with thin barriers the envelope function approximation including strain effects is more appropriate.[33] The conduction band and valence band discontinuities are not known in the $In_xGa_{1-x}As/GaAs$ system. However, several experiments[10,12] have suggested that the conduction band discontinuity is about 70% of the difference between the band gaps of the two materials, once strain is taken into account. Because in $In_xGa_{1-x}As/GaAs$ structures the ternary layers are always under biaxial compressive strain, the upper most valence band in the well is that with m_J=3/2 (heavy holes in the bulk). The strain induced splitting between the m_J=3/2 and m_J=1/2 bands is 20 to 65 meV in the samples discussed here depending on the alloy composition and on the buffer layer lattice constant. The GaAs barriers are not strained when the buffer is also GaAs. Otherwise the valence band ordering in the barrier is opposite to that in the wells. In all cases the choice of band discontinuity mentioned above results in m_J=1/2 hole being confined in the barrier. Following that approach and using a standard square well model we obtain e_1-h_1 transition energies which are always slightly larger than the energy of the exciton recombinations observed in the PL spectra. Part of the difference can be accounted for by the binding energy of the exciton which increases with decreasing well width from 8 meV at 100Å to 10 meV at 20Å. Another part is due to the uncertainty in the actual well width (~1 monolayer) and composition ($\Delta x \sim 0.005$) of the wells. Finally, in multiple well structures the envelope function approach would be more accurate because even with 100Å barrier, coupling between shallow wells is likely to be significant.

Nevertheless, these results show that the shift of excitonic photoluminescence in strained quantum wells is due to a combination of strain and confinement. One expects therefore that the e-A° transition energy and the acceptor binding energy are affected by these two factors as well. A precise calculation of the binding energy of acceptors in the wells as a function of strain and well wi'th requires the solution of the usual Luttinger hamiltonian including the impurity potential,[34] to which the strain hamiltonian[23] and confinement potential terms[21] are added. The binding energy E_A^u of acceptors in thick unstrained $In_xGa_{1-x}As$ layers is smaller than the strain induced valence band splitting in strained layers. For instance in a fully strained layer with x=0.12, $E_A^u \approx 60\% \ \Delta E_V$ where ΔE_V is the splitting between the light and heavy hole bands at the Γ point; at x=0.19 $E_A^u \approx 40\% \ \Delta E_V$. However, in strained layers the acceptor binding energy decreases so that in first approximation $E_A^S \ll \Delta E_V$. Therefore we calculate first the binding energy of acceptor states derived from the $m_J=3/2$ band in strained $In_xGa_{1-x}As$ neglecting the effects of the $m_J=1/2$ band.

Table 2. Measured and Calculated Acceptor Binding Energies

Sample	In%	$m_{//}/m_{\perp}$	E_A^S (meV)	$m_{//}^*$	W/B (Å)	E_A (meV) Exp.	E_A (meV) Theory Center	E_A (meV) Theory Edge
1					20/100	23	35	22
2					20/100	25	34	22
3					25/100	33	32	20
4					30/100	32	31	19
5	12	0.084/0.483	11.5	0.14	50/100	23	25	13
6					50/50	20	25	13
7				0.16⁺	50/50	28	27	
8					75/100	22	22	
9					75	20	22	
10					100/100	20	20	
11					30/100	21	30	19
12	16	0.079/0.470	10.2	0.13	35/100	20	29	18
13				0.16⁺	50/50	32	27	
14	19	0.076/0.464	9.6	0.12	75	24	19	9
15					75	21	19	9

*equivalent effective mass for hydrogenic acceptor.
⁺larger equivalent mass to take into account the reduced valence band splitting (see text).

The effect of strain on shallow impurities has been discussed in the past.[23] With our approximation, the problem is that of a single anisotropic band which can be treated by a variational method.[23,35] The binding energy of the acceptor depends on the anisotropy factor $m_{//}/m_{\perp}$ where $m_{//}$ and $m_{\perp}$ are the $m_J=3/2$ hole effective masses in the x (or y) and z directions respectively. Table 2 gives the values of the anisotropy factors and the resulting binding energies for the 3 values of x studied. The values of the effective masses were calculated[23] using interpolated Luttinger parameters.[16]

The experimental values of the binding energy of carbon and magnesium acceptors in thick unstrained $In_{.12}Ga_{.88}As$ are (24±1) and (27±1) meV respectively.[36] The value calculated for a layer of the same composition strained on GaAs is only 11.5 meV neglecting chemical shifts (Table 2).

The second step in the analysis is to calculate the increase of binding energy due to confinement. However, the situation is somewhat different from that of the 3D biaxially strained layer discussed above. The approximation $E_A^S \ll \Delta E_V$ should become $E_A^S \ll \Delta E_{h_1-h_2}$ and $\Delta E_{h_1-\ell_1}$ where $\Delta E_{h_1-h_2}$ and $\Delta E_{h_1-\ell_1}$ are the differences of energy between the first two heavy hole levels and between the first heavy and light hole levels respectively. In very narrow wells there is only one confined heavy hole level but at w=100Å and x=0.12 for instance $\Delta E_{h_1-h_2}$ is only ~14 meV. On the other hand, in all samples the light holes ($m_J=½$) are confined in the GaAs barriers. Therefore even in narrow wells where $\Delta E_{h_1-\ell_1}$ is small (at

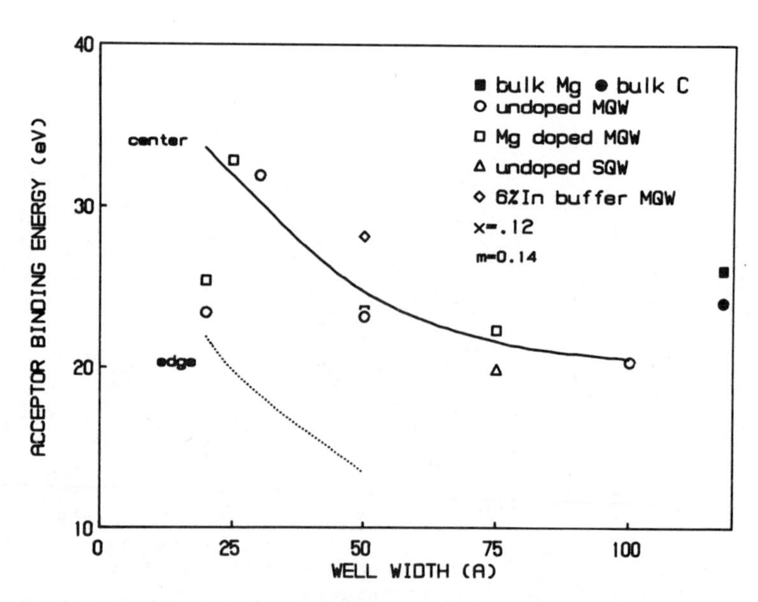

Fig. 5. Variation of acceptor binding energy with well width for x=0.12. The solid line has been calculated for acceptors at the center of an infinite well, the dotted line corresponds to acceptors at the edge of the well.

x=0.12 and w=20Å, $\Delta E_{h_1 - \ell_1}$=28 meV), the two kinds of holes do not mix because they are spatially separated and the acceptor states are thus derived only from the m_J=3/2 hole states.

Thus the acceptor wavefunction is always built from the heavy hole states in the well and to that extent the one band approximation is still acceptable. However, this applies strictly to center well acceptors as the wavefunction of edge acceptors depends also on the light hole states because the overlap with these barrier states is not negligible. To study the effect of confinement, we use the infinite barrier height approximation[19] taking as 3D binding energies those of Table 2 with the corresponding effective Bohr radius. This is equivalent to consider hydrogenic acceptors with the effective masses given in Table 2. The results of the calculation are shown in Figure 5 for x=0.12 and in Table 2 for all the samples. We have considered acceptors both at the center and at the edge of the wells. The experimental points are obtained from the relation $E_A = E_{Bx} + (E_{e-A^\circ} - E_x)$ where E_{Bx} is the binding energy of the free exciton, E_{e-A° and E_x are the energies of the electron-acceptor and exciton emissions measured on the experimental spectra. The binding energy of the exciton E_{Bx} in the well has been calculated using the same approximations as for the acceptors.[19] The effect of strain on the binding energy is much smaller for excitons than for acceptors since the exciton reduced mass is dominated by that of the electron which is not affected much by strain. However the relative binding energy increase due to confinement is more important for excitons than for acceptors since the Bohr radius of excitons is about 3 times that of acceptors. The infinite barrier height model overestimates the binding energy increase especially when the well width is smaller than the Bohr radius because it does not take into account the extension of the wavefunction in the barrier. Thus, the exciton binding energies we have calculated are overestimated so that the values of E_A deduced from emissions observed in the spectra and E_{Bx} are overestimated as well. The apparent good agreement between the points and the calculated curves is therefore somewhat fortuitous in view of all the approximations used. Nevertheless, the results of Figure 5 and Table 2 suggest that at small well width the electron-neutral acceptor recombinations involving acceptors at the edge of the well dominate those due to acceptors in the well, similar to what has been observed in AℓGaAs-GaAs[17] and predicted theoretically.[19]

The effect of strain is clearly seen by comparing the acceptor binding energies in structures of same well width and composition but grown on lattice matched and mismatched (GaAs) buffers. The experimental binding energies (Figure 5 and Table 2 for x=0.12 and 0.16 and w=50Å) are much larger for structures grown on lattice matched buffers. In this situation the compression in the quantum wells is smaller (~50%) so that the splitting of the valence band is smaller. Then the acceptor binding energy is larger than in a fully strained layer. In Table 2 we have shown that increasing the equivalent hydrogenic effective mass from 0.14 to 0.16 yields a better agreement. Finally, a series of transport[5-8] and optical[11] experiments on several SLS samples with x≈0.2 and well width larger than 90Å has shown that the hole mass in the (x,y) plane is light, with values varying between 0.13 and 0.17. However no relation between the value of the mass and that of the strain has been observed.

CONCLUSION

The analysis of the PL spectra of undoped and uniformly Mg doped

$In_xGa_{1-x}As/GaAs$ strained layer superlattices and quantum wells shows that excitonic recombinations which dominate the PL spectra at low temperature become broader at high doping levels. The increased broadening and the energy shift of the excitonic peak at high excitation power suggest that bound excitons contribute to the PL in doped layers. Smaller structures assigned to recombinations between electrons and acceptors confined in the wells have been studied using a single anisotropic band model with infinite barrier wells. The results show that the binding energy of acceptors is reduced by strain and enhanced by confinement so that in the fully strained wells studied, the acceptor binding energy is never much larger than in bulk unstrained crystals. In structures grown on lattice matched buffers however, the effect of strain is less important and the confinement enhancement of the binding energy yields an acceptor binding energy larger than in the bulk.

ACKNOWLEDGEMENTS

We wish to thank W. Trzeciakowski for many valuable comments on the manuscript.

REFERENCES

1. T. E. Zipperian and T. J. Drummond, Electron. Lett. 21:823 (1985).
2. W. T. Masselink, A. Ketterson, J. Klem, W. Kopp and H. Morkoç, Electron. Lett. 21:939 (1985).
3. J. J. Rosenberg, M. Benlamri, P. D. Kirchner, J. M. Woodall and G. D. Peltet, IEEE Electron Device Lett. EDL 6:491 (1985).
4. J. E. Schirber, I. J. Fritz, L. R. Dawson and E. C. Osbourn, Phys. Rev. B28:2229 (1983).
5. J. E. Schirber, I. J. Fritz and L. R. Dawson, Appl. Phys. Lett. 46:187 (1985).
6. I. J. Fritz, L. R. Dawson, T. J. Drummond, J. E. Schirber and R. M. Biefield, Appl. Phys. Lett. 48:139 (1986).
7. I. J. Fritz, T. J. Drummond, G. C. Osbourn, J. E. Schirber and E. D. Jones, Appl. Phys. Lett. 48:1678 (1986).
8. I. J. Fritz, B. L. Doyle, J. E. Schirber, E. D. Jones, L. R. Dawson and T. J. Drummond, Appl. Phys. Lett. 49:581 (1986).
9. I. J. Fritz, J. E. Schirber, E. D. Jones, T. J. Drummond and L. R. Dawson, Appl. Phys. Lett. 50:1370 (1987).
10. J. Y. Marzin, M. N. Charasse and B. Sermage, Phys. Rev. B31:8298 (1985).
11. E. D. Jones, H. Ackermann, J. E. Schirber, T. J. Drummond, L. R. Dawson and I. J. Fritz, Solid State Commun. 55:525 (1985).
12. I. J. Fritz, B. L. Doyle, T. J. Drummond, R. M. Biefield and G. C. Osbourn, Appl. Phys. Lett. 48:1606 (1986).
13. J. Y. Marzin and E. V. K. Rao, Appl. Phys. Lett. 43:560 (1983).
14. N. G. Anderson, W. D. Laidig, R. M. Kolbas and Y. C. Lo, J. Appl. Phys. 60:2361 (1986.
15. A. P. Roth, M. Sacilotti, R. A. Masut, P. J. D'Arcy, B. Watt, G. I. Sproule and D. F. Mitchell, Appl. Phys. Lett. 48:1452 (1986).
16. A. P. Roth, R. A. Masut, M. Sacilotti, P. J. D'Arcy, Y. LePage, G. I. Sproule and D. F. Mitchell, Superlattics and Microstructures 2:507 (1986).
17. R. C. Miller, A. C. Gossard, W. T. Tsang and O. Munteanu, Phys. Rev. B25:3871 (1982).
18. D. C. Reynolds, K. K. Bajaj, C. W. Litton, P. W. Yee, W. T. Masselink, R. Fischer and H. Morkoç, Phys. Rev. B29:7038 (1984).
19. G. Bastard, Phys. Rev. B24:4714 (1981).

20. R. L. Greene and K. K. Bajaj, Solid State Commun. 9:831 (1983).
21. W. T. Masselink, Y. C. Chang and H. Morkoç, Phys. Rev. B32:5190 (1985).
22. U. Ekenberg and M. Altarelli, Phys. Rev. B35:7585 (1987).
23. G. L. Bir and G. E. Pikus, "Symmetry and Strain Induced Effects in Semiconductors," John Wiley & Sons, N.Y. (1974).
24. A. P. Roth, M. A. Sacilotti, R. A. Masut, A. Machado and P. J. D'Arcy, Jour. Appl. Phys. 60:2003 (1986).
25. A. P. Roth, R. A. Masut, M. Sacilotti, P. J. D'Arcy, B. Watt, G.I. Sproule and D. F. Mitchell, Jour. Crystal. Growth 77:571 (1986).
26. E. F. Schubert, E. O. Göbel, Y. Horikoshi, K. Ploog and H. J. Queisser, Phys. Rev. B30:813 (1984).
27. E. F. Schubert and W. T. Tsang, Phys. Rev. B34:2991 (1986).
28. R. C. Miller, A. C. Gossard, W. T. Tsang and O. Munteanu, Solid State Commun. 43:519 (1982).
29. A. N. Balkan, B. K. Ridley and J. Goodridge, Semicond. Sci. Technol. 1:338 (1986).
30. M. H. Meynadier, J. A. Brun, C. Delalande, M. Voos, F. Alexandre and J. L. Liévin, J. Appl. Phys. 58:4307 (1985).
31. M. Nakayama, K. Kubota, H. Kato and N. Sano, Solid State Commun. 51:343 (1984).
32. M. S. Skolnick, K. J. Nash, P. R. Tapster, D. J. Mowbray, S. J. Bass, and A. D. Pitt, Phys. Rev. B35:5925 (1987).
33. J. Y. Marzin, "Strained Superlattices" in: "Heterojunctions and Semiconductor Superlattices," G. Allan, G. Bastard, N. Boccara, M. Lannoo and M. Voos, ed, Springer Verlag, Berlin (1986).
34. J. M. Luttinger and W. Kohn, Phys. Rev. 97:869 (1955).
35. W. Kohn and J.M. Luttinger, Phys. Rev. 98:915 (1955).
36. A. Roth, R. Masut, M. Sacilotti and B. Watt, to be published.

MAGNETO-OPTICS OF EXCITONS IN GaAs-(GaAl)As QUANTUM WELLS

W. Ossau, B. Jäkel, E. Bangert and G. Weimann*

Physikalisches Institut der Universität Würzburg
* Forschungslabor der Deutschen Bundespost Darmstadt

ABSTRACT

We report low temperature luminescence measurements of the energy shift and splitting of excitons in GaAs-(GaAl)As quantum wells as a function of well width and magnetic field strength. The maximum strength of the field was 9.5 T and its direction was perpendicular to the interfaces. We have performed variational calculations of the exciton binding energy to explain the experimental data of the diamagnetic shift. The present calculations incorporate the valence-band nonparabolicity resulting from the mixing of heavy- and light-hole states.

1. INTRODUCTION

In recent years the properties of the quasi-two dimensional nature of Wannier excitons confined in superlattices and quantum wells have attracted considerable interest. Due to the quantum size effect excitons have characteristics different from that in bulk material[1]. The exciton becomes quasi-two dimensional, its binding energy is enhanced, the degeneracy of heavy- and light-hole band is removed, resulting in two exciton systems. High magnetic fields, by providing discrete optical spectra, have proved to be a powerful means to study exciton states in quantum wells. Recently several groups have investigated both, heavy- and light-hole states in the presence of a magnetic field[2-4]. These studies have revealed discrepancies between theory and experiment suggesting that a more rigorous model is required to describe exciton states in quantum wells. In a previous publication[5] we took into account the coupling of heavy- and light-hole states due to the off-diagonal terms of the Luttinger-Hamiltonian used to describe the complex valence-band structure of GaAs near the Γ-point. The agreement between experiment and theoretical calculations, however, was not satisfying. In that publication we used the spherical approximation to incorporate the nonparabolicity of the valence band due to the mixing of heavy- and light-hole states. The spherical approximation is rather drastic, it neglects warping of the valence bands altogether and produces substantial errors in the $k_{||}=0$ subband energies.

In this study we apply a refined exciton wavefunction and consider the warping of the valence band states. In section 2 we report our experimental results of the diamagnetic shift and the splitting of the exciton states as a function of the well width. Section 3 presents the variational calculations performed to obtain the exciton binding energies. Section 4 deals with the calculation of the subband dispersion and the incorporation of the numerical results into the variational calculations.

2. EXPERIMENTAL RESULTS

We have investigated several GaAs multi-quantum wells (MQWs) of thicknesses L_z between 2.7 and 30 nm with low temperature photoluminescence experiments in high magnetic fields up to 9.5 T. The MQWs were grown by molecular beam epitaxy and consist of up to 110 GaAs layers, embedded in (GaAl)As barriers. The width of the barriers is large, in order to ensure decoupling of the individual GaAs wells. The high quality of the samples made it possible to observe excitonic lines with line width less than 0.2 meV.

Fig. 1 shows a typical luminescence spectrum of a quantum well with 18 nm thickness. We observe excitons formed between the lowest electron subband and the heavy-, light- and excited-hole subbands. In addition we identify recombinations of excited exciton or interband transitions as well as excitons bound to defects[6]. All these lines have an individual magnetic field dependence which can be recognized in fig. 2. The excitonic lines E(11h), E(11l) and E(13h) show a nonlinear, quadratic energy shift. The other lines, varying strongly and approximately linear with magnetic field, can be explained by excited exciton or Landau level transitions. In a magnetic field all ground-state excitonic lines split into two components with circular polarization σ+ and σ-. As we observe in Faraday configuration the luminescence is emitted by dipole allowed m = ± 1 recombinations of an electron (J = 1/2) with a light-hole (J_z = ± 1/2) or a heavy-hole (J_z = ± 3/2) respectively. Figure 3 shows the difference of the

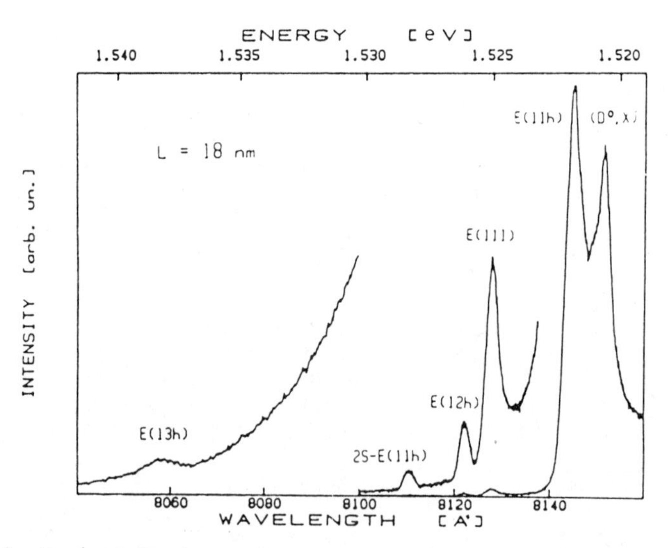

Fig. 1. Typical luminescence spectrum for a quantum well with L_z = 18 nm. The sample temperature was 1.8 K.

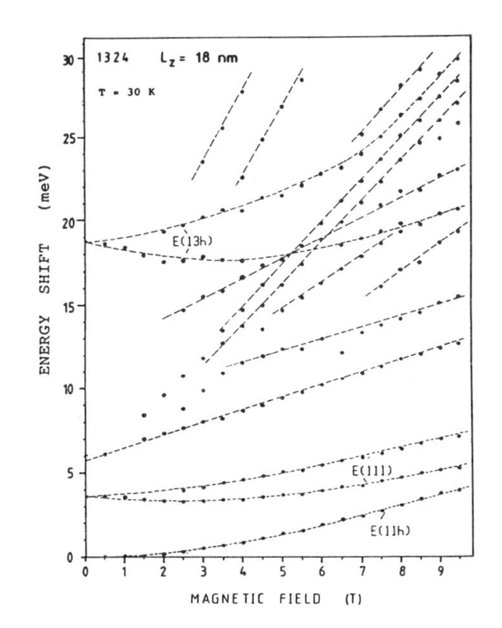

Fig. 2. Relative shift of photon energy as a function of the magnetic field. The sample temperature is 30 K and L_z = 18 nm.

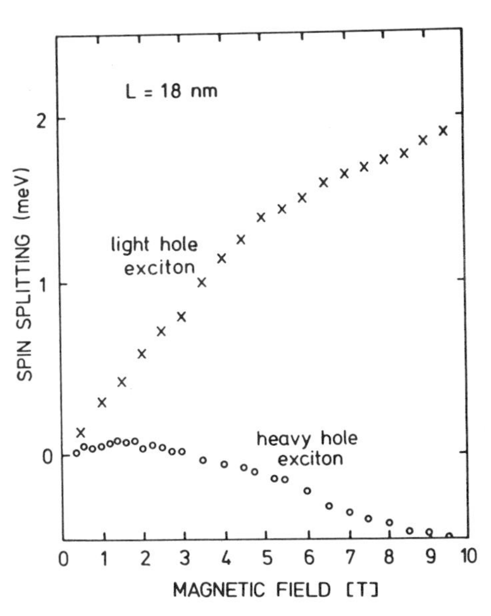

Fig. 3. Spin splitting of the heavy- and light-hole exciton as a function of the magnetic field.

photon energies of the two components. We observe a nonlinear splitting, which is small for the heavy-hole exciton and observable only at temperatures below 2 K. Furthermore the nonlinear splitting behaviour leads to a sign reversal of the difference between the heavy-hole $\sigma+$ and $\sigma-$ lines at field strength of about 3 T. In contrast the splitting of the lines attributed to the light-hole exciton is more pronounced and shows no sign reversal. We, however, observe a change in the slope of the splitting at field strengths equal to those where the heavy-hole exciton shows a vanishing splitting. These nonlinear splittings are a consequence of the mixing of heavy- and light-hole states and make a determination of effective g-factors very uncertain. To compare our data for different quantum wells we used the slope of the spin-splitting for weak fields. Due to the small slope and the sign reversal of the splitting the obtained data are uncertain for the heavy-hole exciton. Nevertheless the splitting parameters show a distinct dependence on the width as displayed in fig. 4. With decreasing well thickness the splitting increases for the light-hole- and decreases for the heavy-hole exciton. This is qualitatively the same result as observed by Bajaj et al.[7] with reflection measurements for magnetic field direction parallel to the interfaces. The maximum field strength applied by these authors was 3.6 T only, what explains that they may haven't found the pronounced nonlinearity observed in our study. Again the dependence of the spin-splitting on the well width is a consequence of the mixing of heavy- and light-hole states. For wide quantum wells the eigenstates of heavy- and light-hole are closer in energy, which leads to an enhanced mixing compared to narrow wells where the eigenstates are more separated. This different degree of coupling is important for the interpretation of the diamagnetic shift measurements, too, and will be discussed in detail later.

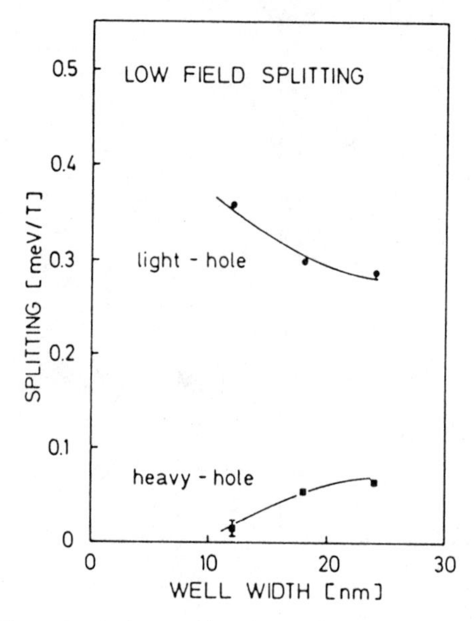

Fig. 4. Spin splitting of heavy- and light-hole exciton for low magnetic field strength as a function of the well width.

The diamagnetic shift, this means the energy shift of the exciton's ground state in a magnetic field, is an excellent measure of the quasi-two dimensional character of excitons confined in quantum wells. By a change from three to two dimensions the diamagnetic shift is reduced as can be shown for the low field case.

In general the change in energy of the ground state is given by:

$$E_{hh}(1S) = \frac{e^2 B^2}{8\mu} \langle \Phi \mid x^2 + y^2 \mid \Phi \rangle \qquad (1)$$

The term in brackets indicates the extension of the exciton parallel to the interface. In three dimensions this value is $2 * a_B^2$, where a_B is the effective Bohr radius of the exciton. For a completely two dimensional exciton the expectation value is $3/8 * a_B^2$. Therefore the diamagnetic shift of a two dimensional exciton is 16/3 times smaller than that of a 3D one. This means by measuring the diamagnetic shift it is possible to determine the degree of quasi-two dimensional character of excitons confined in quantum wells.

In fig 5 we have plotted the diamagnetic shift of the heavy hole exciton as a function of well width. As expected, due to a more pronounced two-dimensional character the diamagnetic shift becomes smaller by narrowing the well width. However for L_z about 7 nm the data reach a minimum and then increase rapidly with further decreasing L_z. We have explained this increase of the diamagnetic shift by the penetration of the exciton wavefunction into the adjacent barriers[4]. This means that the exciton progressively loses its quasi-two dimensional character. To compare these experimental data with the theory we have performed variational calculations for the exciton binding energy.

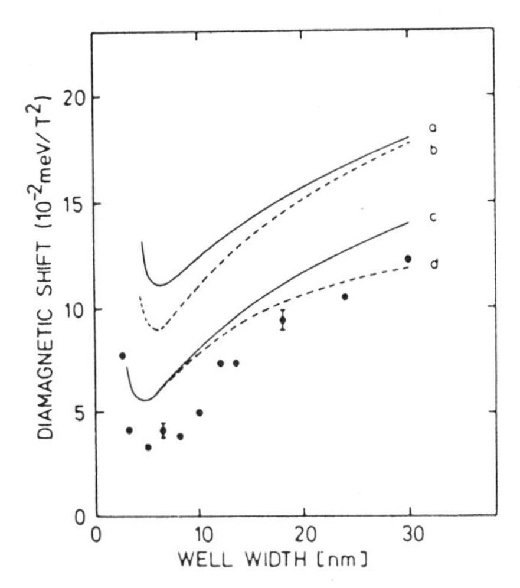

Fig. 5. Diamagnetic shift of the heavy hole exciton in dependence of the well width. The curves are calculated (See text).

3. VARIATIONAL CALCULATIONS

Neglecting spin the upper valence-band in zinkblende-structures is threefold degenerated at the Γ-point. By spin-orbit interaction the valence-band states are splitted into a upper fourfold (J = 3/2) and a lower twofold (J = 1/2) state separated by spin-orbit splitting (0.34 eV in GaAs). As this energy difference is larger than the relevant exciton binding energies, we consider the upper fourfold degenerated valence band only. Away from Γ, the bands split further into a twofold degenerated light-hole and a twofold degenerated heavy-hole band. In the neighbourhood of the Γ-point this is described by the well known Luttinger-Hamiltonian[8]. In the basis of the four Γ_8 states with m_j = +3/2, +1/2, -1/2, -3/2 it reads:

$$H = \begin{bmatrix} a_+ & b & c & 0 \\ b^* & a_- & 0 & c \\ c^* & 0 & a_- & -b \\ 0 & c^* & -b^* & a_+ \end{bmatrix} \tag{2}$$

where

$$a_\pm = (-\hbar^2/2m_o)[(-\gamma_1 \pm 2\gamma_2)k_z^2 - (\gamma_1 \pm \gamma_2)(k_x^2 + k_y^2)]$$

$$b = (-\hbar^2/2m_o)\sqrt{3}[\gamma_3(k_x - i k_y)k_z]$$

$$c = (-\hbar^2/2m_o)\sqrt{3}/2[\gamma_2(k_x^2 - k_y^2) - 2i\gamma_3 k_x k_y]$$

The γ_i are material parameters describing the hole effective masses different for well and barrier material as can be seen in table I. The values for (GaAl)As are obtained by linear interpolation.

TABLE 1. Luttinger parameters for GaAs and AlAs /9/

	GaAs	AlAs
γ_1	6.85	3.45
γ_2	2.10	0.68
γ_3	2.90	1.29

The presence of the valence band discontinuity at the interface of quantum wells removes the band degeneracy due to the reduction in symmetry along the axis of growth. This separates the heavy- and light-hole band even at k = 0. If one neglects the off-diagonal elements - i.e. b = c = 0 in equ. (2), both bands can be treated independently, what finally leads to two separated exciton systems commonly referred to as the heavy-hole exciton ($J_z = \pm 3/2$) and the light-hole exciton ($J_z = \pm 1/2$). This often used treatment is very simplifying and leads to discrepancies between theory and experiment. For the moment we will follow the approach of two separate exciton systems and include the coupling of light- and heavy-hole bands later.

For nondegenerate bands the exciton Hamilton operator reads:

$$H = \frac{p_{ze}^2}{2\,m_e} + \frac{p_{zh}^2}{2\,m_{hz}} + \frac{p_x^2 + p_y^2}{2\,\mu} - \frac{e^2}{\varepsilon[\,x^2 + y^2 + (\,z_e - z_h\,)^2]^{1/2}}$$

$$+ V_{conf}(z_e) + V_{conf}(z_h) \qquad (3)$$

where m_e, m_h, $\vec{r}_e$, $\vec{r}_h$ are the effective masses and positions of electrons and holes respectively, ε is the relative dielectric constant and μ the reduced exciton mass determined by the electron and hole mass value parallel to the interfaces. We take into account the penetration of the wavefunctions into the adjacent barriers, therefore these parameters have to be changed for well and barrier material respectively. To obtain energies of the ground state excitons we apply a variational approach with the following trial functions for the well:

$$\Phi = N \cos(k_e z_e) \cos(k_h z_h) \exp(-(\,\alpha(x^2+y^2) + \beta(z_e - z_h)^2)^{1/2}) \qquad (4)$$

and the barrier:

$$\Phi = C \exp(-æ_e z_e) \exp(-æ_h z_h) \exp(-(\,\alpha(x^2+y^2) + \beta(z_e - z_h)^2)^{1/2}) \qquad (5)$$

where α and β are trial parameters. In the wavefunction of the barrier the cosines are replaced by exponentials. The parameters k_e, k_h, $æ_e$, $æ_h$ and C are determined by the subband energy and the condinuity conditions. We minimized the energy as a function of α and β and calculated the diamagnetic shift with equation (1) making use of the obtained wavefunction. As can be seen from this equation the reduced mass of the exciton is a crucial input parameter. The values for the reduced mass μ and the heavy hole mass m_h used for the calculations are expressed in terms of the Luttinger parameters γ_1 and γ_2. The expressions are deduced from the diagonal terms of the Luttinger matrix for the limit of small k_z and $k_{||}$.

$$m_{hz} = \left[\gamma_1 - 2\gamma_2 \right]^{-1} m_o = 0.377\ m_o$$

$$\mu = \left[\frac{1}{m_e} + \gamma_1 + \gamma_2 \right]^{-1} m_o = \left[\frac{1}{0.0665} + \frac{1}{0.11} \right]^{-1} m_o = 0.041\ m_o \tag{6}$$

Using these parameters we calculated the binding energy of the (1S)-state of the heavy-hole exciton. We assumed an aluminium content of 40 % and a ratio of 60/40 for bandgap discontinuities. The numerical results are drawn as full line in figure 5 (curve a). The qualitative agreement between calculations and experiment is fairly good. The experimental findings - decreasing diamagnetic shift with decreasing well width L_z, a minimum near 7 nm and subsequently increasing diamagnetic shift with further decreasing L_z due to the penetration of the wavefunction into the barriers - are well described. The quantitative agreement, however, is poor. The inclusion of the nonparabolicity of the conduction band electron[10] produces a small improvement for narrow wells only, as can be seen by the dashed line b in fig. 5. Some groups[3,11] try to overcome the displayed discrepancies between experimental data and numerical results by using the reduced mass as a fit parameter. For example to get agreement between experiment and theory a reduced mass of 0.0501 m_o is needed for a quantum well with 18 nm width. This requires an enhanced parallel hole mass of 0.191 m_o instead of 0.11 m_o which is obtained from the diagonal terms of the Luttinger matrix. Such drastically increased hole masses are employed in a variety of studies to fit the experimental data. Our aim is not to give parameters but to understand the mechanisms leading to heavier heavy-hole masses. In the case of completely decoupled valence bands ($E_{hh}-E_{lh}=\infty$) the parallel heavy hole mass is $m_{h||} = 0.11\ m_o$ as described above. In the case of degenerated bands ($E_{hh}-E_{lh} = 0$) one has $m_{h||} \approx m_o/(\gamma_1 - 2\gamma_2) = 0.377\ m_o$. Thus we have energy differences of some meV between the heavy- and light-hole subbands, we expect for $m_{h||}$ a value in the range between

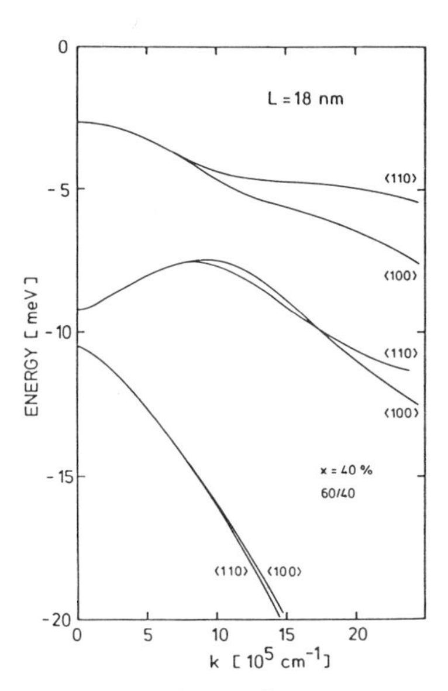

Fig. 6. Subband dispersion of a quantum well with 18 nm width and x = 0.4. The solid curves are the dispersion in ‹100› and ‹110› direction

$0.11\ m_o \leq m_{h||} \leq 0.377\ m_o$. To deduce reasonable $m_{h||}$ values from the complete Luttinger Hamiltonian, the parameters of well and barriers we have performed in addition calculations of the hole subband dispersion with inclusion of the degeneracy and warping of the bulk valence bands described by the Luttinger Hamiltonian.

Fig. 6 shows the result for a quantum well with 18 nm embedded in (GaAl)As barriers with 40 % aluminium content for two different directions in the plane perpendicular to the ⟨001⟩ axis of growth. For small $k_{||} = 0$ the subbands are decoupled completely and thus can be denoted as heavy- or light-hole subbands. The coupling between heavy- and light-hole states brings about interesting features of the subband structures. One can observe electron like behaviour of the light hole for small $k_{||}$ values and a reversal of the mass sign for $k_{||} \approx 10^5\ cm^{-1}$. Furthermore the lowest heavy-hole subband shows a strong nonparabolicity. The influence of warping of the valence bands is reflected by the directional dependence of the subband dispersion in the $k_{||}$ plane. The maximum deviation in the plane parallel to the interfaces is represented in fig.6 by the dispersion in the ⟨100⟩ and ⟨110⟩ directions. As can be seen from this plot, too, the influence of warping is effective for k-vectors greater than $10^6\ cm^{-1}$ only. We are mainly interested in the values for the heavy-hole mass parallel to the interface. To obtain numerical results for $m_{h||}$ from the subband calculations we used the relation

$$m_{h||} = \frac{h^2}{2\,\Pi^2} \left[dA(E)/dE \right] \qquad (7)$$

where A(E) is the area in the 2D-space enclosed by a contour of constant energy E. We find strong nonparabolicities which are more pronounced when the well width is large and the fraction of the bang-gap discontinuity for the valence band is small. This dependence is caused by the energy diffe-

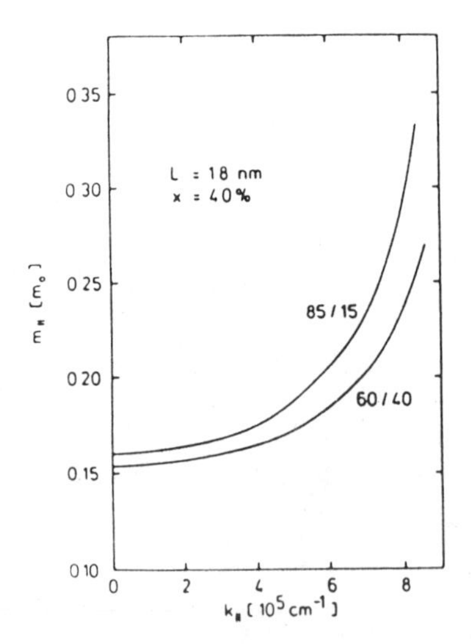

Fig. 7. Calculated heavy hole mass $m_{h||}$ as a function of $k_{||}$ for a quantum well with L_z = 18 nm. The aluminium content is 40% .

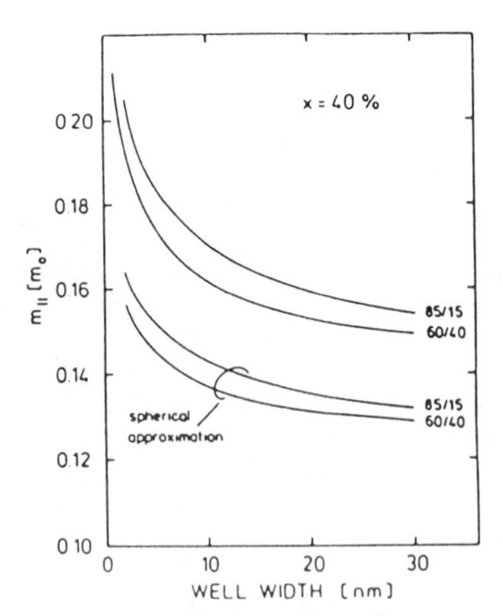

Fig. 8. Calculated heavy hole mass $m_{h||}$ for $k_{||}$=0 as a function of the well width. The two lower curves are obtained with the spherical approximation[5].

rence between heavy- and light-hole subbands, which is small for the mentioned conditions and therefore results in a strong coupling between the hole states.

As can be seen in fig. 7 the parallel hole mass for small $k_{||}$ values is greater than that obtained in the usual way discussed above from the Luttinger diagonal elements only. In fig. 8 we have plotted the dependence of $m_{h||}$ for small $k_{||}$ as a function of the well width. The parallel hole mass is increasing with decreasing well width. For very wide quantum wells only the parallel hole mass approaches the parameter deduced from the diagonal elements. Fig. 8 also shows that the consideration of warping leads to heavier mass values than we have obtained in a previous publication[5] using the spherical approximation.

Since the calculated subband dispersion results from the 4*4 Hamiltonian, effects of heavy-hole light-hole coupling are incorporated into the exciton states if we insert the hole mass $m_{h||}$ of the 2D-subbands into equation (6) to determine the reduced mass of the exciton.

We calculated the diamagnetic shift of the heavy-hole exciton (1S)-state using for $m_{h||}$ the zone-centre hole mass. The result can be seen in fig. 5 as full curve denoted with c. Comparing this result with curve b and the experimental data it is obvious that the consideration of hybridization of heavy- and light-hole states leads to a better agreement between experimental and numerical data.

It has already been shown that the parallel hole mass has a strong dependence on the wavevector $k_{||}$. Taking into account these nonparabolicities the calculation of the diamagnetic shift will be more difficult. In the case of k dependent heavy-hole mass values it is not allowed to write the reduced mass μ as a constant factor in front of the expectation value in equation (1). This means that equation (1) has to be written as follows:

$$E_{hh}(1S) = \frac{e^2B^2}{8} \left\langle \Phi \left| \frac{x^2 + y^2}{\mu} \right| \Phi \right\rangle \tag{1a}$$

In order to obtain numerical results for the diamagnetic shift we have expanded the heavy-hole mass in a power series of $k_{||}$.

$$\frac{1}{m_{h||}} = \frac{1}{m_{h||}(k_{||}=0)} + D\cdot k_{||}^{2} \tag{8}$$

D is a factor describing the nonparabolicity of the heavy-hole dispersion. As can be seen from fig. 6 $m_{h||}$ shows only a smooth dependence on the wavevector, therefore it is sufficient to neglect powers higher than $k_{||}^2$. Then equation (1a) can be written

$$E_{hh}(1S) = \frac{e^2B^2}{8\mu(o)} \langle \Phi|x^2 + y^2|\Phi\rangle + \frac{e^2B^2}{8} D \langle \Phi|k_{||}^2\cdot(x^2+y^2)|\Phi\rangle \tag{1b}$$

The first term is the diamagnetic shift determined by m_e and the zone-centre hole mass. The second term is a small correction resulting from the nonparabolicity of the parallel hole mass. The first contribution to the diamagnetic shift has already been calculated (curve c in fig.5). The dashed curve d in figure 5 considers the corrections influenced by the second term of equation (1b). This corrections reflecting the nonparabolicities of the heavy-hole are nearly unimportant for narrow quantum wells. A fact that is easy to see, because in narrow wells the subband states - especially the first heavy- and light-hole subbands - are separated in energy and therefore the mixing of these states is unimportant. In contrary, for wide quantum wells the states are closer in energy leading to strong nonparabolicities in the dispersion.

SUMMARY

We have shown that the consideration of hybridization of heavy- and light-hole states as well as the inclusion of nonparabolicities of electron and heavy-hole in the calculations for the binding energies of ground state excitons in quantum wells leads to an improved agreement between experimental and calculated data of the diamagnetic shift. The quantitative agreement is fairly good for wells wider than 12 nm. In the narrow well regime the calculated diamagnetic shifts are still greater than those observed experimentally. This behaviour is the same we found with different wavefunctions, mentioned in a former publication[5] and is confirmed by the calculations of G.E.W. Bauer and T. Ando[12]. The measured diamagnetic shift for narrow wells is always smaller than that calculated, independent of the wavefunctions applied. Similar discrepancies are found for the data of the exciton binding energies. The experimental data are always greater than the calculated ones. Exciton binding energy and diamagnetic shift are coupled. In general the diamagnetic shift is small when the binding is strong. Therefore the observed discrepancies between experiment and theory are consistent and can be explained with an additional binding of the exciton. The exciton may be trapped in states induced by interface defects. It is shown by Bastard et al.[13] that exciton trapping on interface defects is an important effect even in good quantum wells smaller than 15 nm.

Acknowledgement: The authors are very much indebted to Prof. Dr. G. Landwehr for his continuous encouragement in the course of this work. We are grateful to Dr. G.E.W. Bauer for communicating results prior to publication. This work has been supported by the Deutsche Forschungsgemeinschaft.

References

1. R. Dingle, Festkörperprobleme, Vol. XV, p.21, (Pergamon), 1975
2. J.C. Maan, G. Belle, A. Fasolino, M. Altarelli and K. Ploog Phys. Rev. B30, p.2253, (1984)
3. N. Miura, Y. Iwasa, S. Tarucha and H. Okamoto, Proc. 17th Int. Conf. on the Physics of Semiconductors, San Francisco, p.360, (1984)
4. W. Ossau, B. Jäkel, E. Bangert, G. Landwehr and G. Weimann Surface Science, 174, p.188, (1986)
5. W. Ossau, B. Jäkel and E. Bangert, Solid State Sci. 71, p.213, (1987)
6. Y. Nomura, K.Shinozaki and M. Ishii, J. Appl. Phys.58, p.1864, (1985)
7. K.K. Bajaj, D.C. Reynolds, C.W. Litton, R.L. Greene, P.W. Yu, C.K. Peng and H. Morkoc, Int. Phys. Conf. Ser. 83, p.325, (1986)
8. J.M. Luttinger, Phys. Rev. 102, p.1030, (1956)
9. Numerical data and Functional Relationships in Science and Technology Ladolt-Börnstein, Vol. 17, Springer, (1982)
10. W. Zawadski and P. Pfeffer, Solid State Sci. 71, p.523, (1987)
11. D.C. Rogers, J. Singleton, R.J. Nicholas, C.F. Foxon and K. Woodbridge Phys. Rev. B 34, p.4002, (1986)
12. G.E.W. Bauer and T. Ando, priv. comm., To be published
13. G. Bastard C. Delalande, M.H. Meynadier, P.M. Frijlink and M. Voos Phys. Rev. B29, p.7042, (1984)

IV. TWO DIMENSIONAL AND OTHER ELECTRONIC PROPERTIES

THE INFLUENCE OF IMPURITIES ON THE SHUBNIKOV-DE HAAS AND HALL RESISTANCE OF TWO-DIMENSIONAL ELECTRON GASES IN $GaAs/Al_xGa_{1-x}As$ HETEROSTRUCTURES INVESTIGATED BY BACK-GATING AND PERSISTENT PHOTOCONDUCTIVITY

J. Wolter, F.A.P. Blom, P. Koenraad and P.F. Fontein

Department of Physics
Eindhoven University of Technology
P.O. Box 513, 5600 MB Eindhoven
The Netherlands

G. Weimann

Forschungsinstitut der Deutschen Bundespost beim FTZ
P.O. Box 50.000
6100 Darmstadt
Germany

ABSTRACT

It is well-known that in the Quantum-Hall (QH) regime the Shubnikov-de Haas (SdH) oscillations and the Hall resistance show a non-ohmic behaviour, which apparently is spin-dependent. We report on new measurements on this effect, which have been carried out with the help of a back-gate and by the use of the effect of persistent photoconductivity induced by light of out-gap energy.

Further, we show that also in the low-magnetic field range ($B < 1.5T$), where the spin-up and the spin-down contributions cannot be observed seperately, the amplitude of the SdH oscillations behaves anomalously as a function of temperature. This temperature dependence, if interpreted in terms of the effective mass, leads to a strong decrease of the mass with decreasing magnetic field. Also this effect can be influenced by a back-gate voltage.

The discovery of the QH Effect [1] has led to an extensive study of the transport properties of two-dimensional electron gases in semiconductors. Despite this enormous effort a quantitative theory of the QH Effect is still missing. Such a theory should explain not just the values and the positions of the quantized Hall plateaus, but should also be able to predict the behaviour in the transition region between the plateaus. Furthermore, a theory should account for the transverse magnetoconductivity σ_{xx}.

In most measurements [2] on the SdH effect of a two- dimensional electron gas in $GaAs-Al_xGa_{1-x}As$ heterojunctions it has been found that the SdH oscillations ρ_{xx} as a function of the magnetic field B usually show a strong asymmetric line shape. This effect appears predominantly in samples with high mobility. Moreover, different spin polarisations belonging to the same Landau level behave quite differently. In a recent paper by Haug et al. [3] this asymmetry in the SdH oscillations was analysed. The authors could explain the asymmetry on the basis of an asymmetric shape of the density of states within one Landau level. Such an asymmetric shape can be expected if either attractive or repulsive scatterers dominate and the number of scatterers within one Landau orbit is small. Haug et al. demonstrated that the relative contributions of the attractive scatterers in the $Al_xGa_{1-x}As$ (ionized silicon donors) and the repulsive scattering centers in the GaAs can be modified by changing the wave functions of the two-dimensional electron gas with a back-gate voltage.

In reference [4] a strong non-ohmic behaviour in the SdH oscillations was reported. Indications of such a non-ohmic behaviour can also be found in [5] and [6]. The shape of the SdH magneto resistivity ρ_{xx} (B) and the Hall resistivity ρ_{xy} (B) depend strongly on the current through the sample. Figure 1 gives an example for a GaAs-$Al_xGa_{1-x}As$ heterojunction with a low-field mobility of 610 000 cm^2/Vs at a temperature T = 4.2 K. The current dependence of the resistance is clearly different for the spin-up and spin-down contributions. Furthermore it is dependent on the location along the Hall-bar.

In this paper we report on new measurements of this effect. In particular, we have studied the dependence of this non-ohmic behaviour by means of a back-gate voltage and by use of the effect of persistent photoconductivity induced by light of out-gap energy. From these measurements we are able to show that not the number of charge carriers in the two-dimensional electron gas, but the confinement of the wave function and thus the interaction of the electrons with the scatterers is responsible for the observed asymmetric behaviour of the line-shape of the SdH oscillations. Moreover, we have found that the spin-dependent non-ohmic behaviour can be reversed, if a back-gate voltage is applied during the cooling-down of the sample. This gives additional evidence that the shape of the wave function and thus the interaction between the electrons and the scatterers is the clue to the non-ohmic behaviour. The phenomenon cannot be explained by an increased electron temperature of the two-dimensional electron gas.

Finally we show that also in the low-magnetic field range (B < 1.5T), where the spin-up and spin-down contributions cannot be observed separately, the amplitude of the SdH oscillations behaves anomalously. This

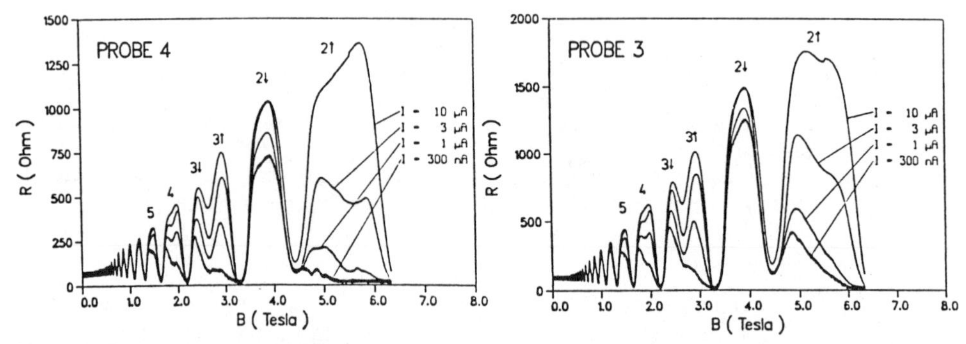

Fig. 1 Potentials on two probes divided by the current through the Hall bar (magneto resistivity) at 4.2 K for a sample with a mobility of 610000 cm^2/Vs. From [4].

temperature dependence, if interpreted in terms of the effective mass, leads to a strong decrease of the mass with decreasing magnetic field. Also this effect can be influenced by a back-gate voltage. Such a decrease of the effective mass at low magnetic fields has previously been reported by Fang et al. [7] for silicon MOSFETS. Further Galchenkov et al. [8] found that in a GaAs-$Al_xCa_{1-x}As$ heterostructure the effective mass, as determined from the temperature dependence of the SdH amplitude, decreased from about 0.080 m_o to 0.063 m_o in the magnetic field range from 2 to 1 T.

The experiments have been carried out on selectively doped n-type (silicon, $2 \cdot 10^{18}/cm^3$) GaAs-$Al_xCa_{1-x}As$ (x = 0.38) heterojunctions. The samples had been grown by Molecular Beam Epitaxy (MBE). The spacer layer has a thickness of 36 nm. The samples had the typical shape of a Hall bar with a channel width of 300 μm. The substrate was thinned down until a total thickness of about 100 μm was achieved. Figure 2 shows a typical experimental curve of the QH resistance (a) and the SdH resistance (b) for two different currents and different back-gate voltages. The samples had been cooled down to liquid helium temperature with the back-gate short-circuited.

For a current of I = 0.05 μA the Hall-plateaus are wider than for a current of 5 μA. Moreover, for a current of 0.05 μA an increase of the back-gate voltage leads to a narrowing of the plateaus. In the SdH oscillations we observe the corresponding behaviour. For a current of 0.05 μA the spin-up contribution increases with increasing back-gate voltage. These observations are consistent with the measurements reported already by Haug [3]. For a current of 5 μA there are almost no changes in ρ_{xx} and ρ_{xy} as a function of the back-gate voltage apart from a small change of the electron density. The current dependence of ρ_{xx} and ρ_{xy} is thus a function

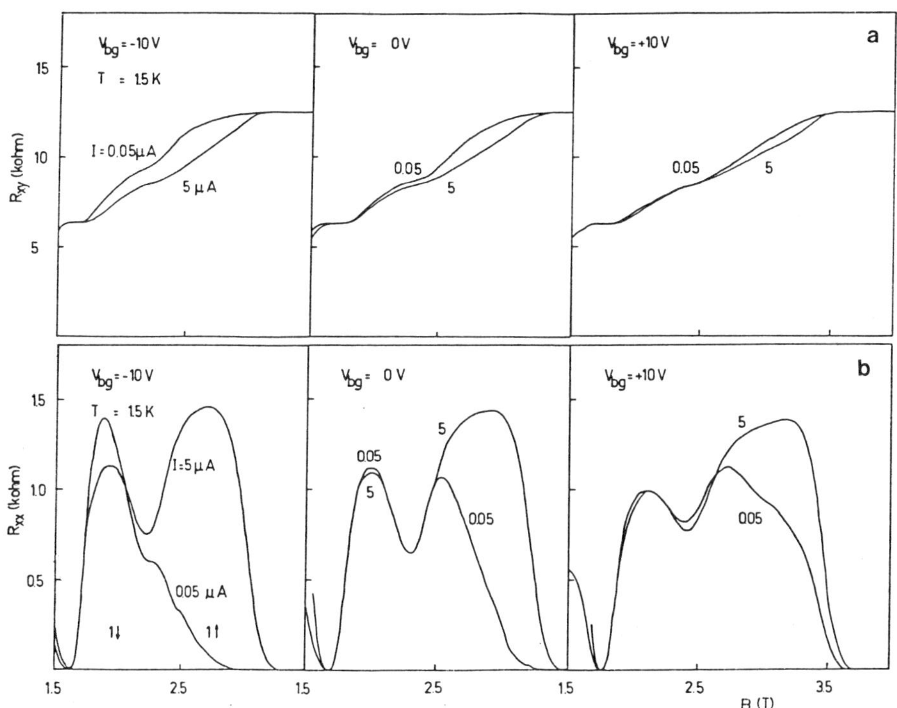

Fig. 2 QH (a) and SdH resistances (b) at 1.5 K for different back-gate voltages V_{bg} and two currents.

of the back-gate voltage. For a back-gate voltage of 30 V we do not observe any current dependence in ρ_{xx} and ρ_{xy}.

We observe an anomalous behaviour in the spin-down peak of the Shubnikov-de Haas oscillations for a back-gate voltage of −10 V: ρ_{xx} increases with decreasing current. A minor indication of this behaviour can already be seen without a back-gate voltage. Since normally the amplitude of the Shubnikov-de Haas oscillations decreases with increasing temperature, independent of the spin polarisation, the observed behaviour cannot be related to a temperature effect. The spin-down peak does not show any current dependence for positive back-gate voltages, while the spin-up peak - depending on the value of the back-gate voltage - depends on the current. The most convincing evidence against the assumption that the phenomenon observed is due to a temperature effect has already been published in [4] where measurements on the current and temperature dependence of the width of a Hall-plateau are reported. In figure 3 we observe a symmetric broadening when the temperature is lowered and a asymmetric broadening when the current is lowered.

When a back-gate voltage is applied, the number of electrons in the two-dimensional electron gas is changed. Also the wave function of the electrons in the potential well and thus the relative contribution of the interaction with the repulsive and attractive scatterers is influenced. In order to investigate whether it is just the electron scattering which is responsible for the effects observed or whether the electron density also plays a major role, we carried out a number of "back-gate" experiments, in which we also used the persistent photoconductivity to keep the electron density the same. In figures 4a, 4b and 4c we plot the Hall resistance and SdH resistance as a function of magnetic field. Figure 4a without a back-gate voltage, figure 4b with a back-gate voltage of +20 V, both in dark, and figure 4c without a back-gate voltage after illumination for 9.2 µs. In this way the experiments of figure 4b and figure 4c were carried out for almost the same electron density. Comparing figure 4a and figure 4b we see that with a positive back-gate voltage the spin-up peak has strongly increases, while the spin-down peak slightly decrease. We observe quite a different behaviour in figure 4c compared to figure 4b. The spin-up peak is strongly decreased, while the spin-down peak remains almost unchanged. From these measurements we conclude that not the electron density but predominantly the shape of the wave function of the electrons and the interaction with the scatterers in the GaAs and $Al_xGa_{1-x}As$ are responsible for the effects observed.

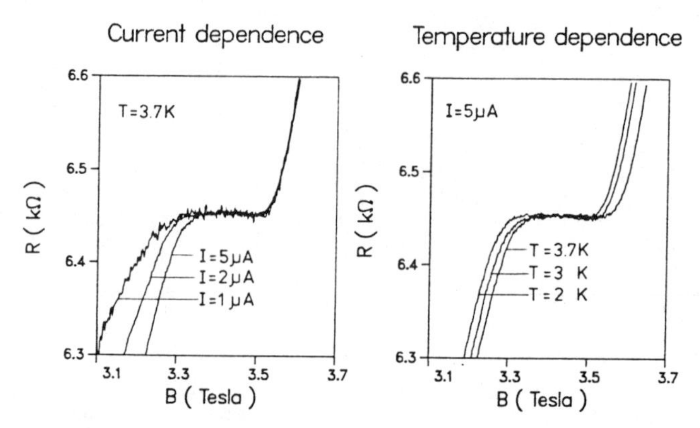

Fig. 3 Broadening of the i = 4 Hall plateau as a function of current and as a function of temperature.

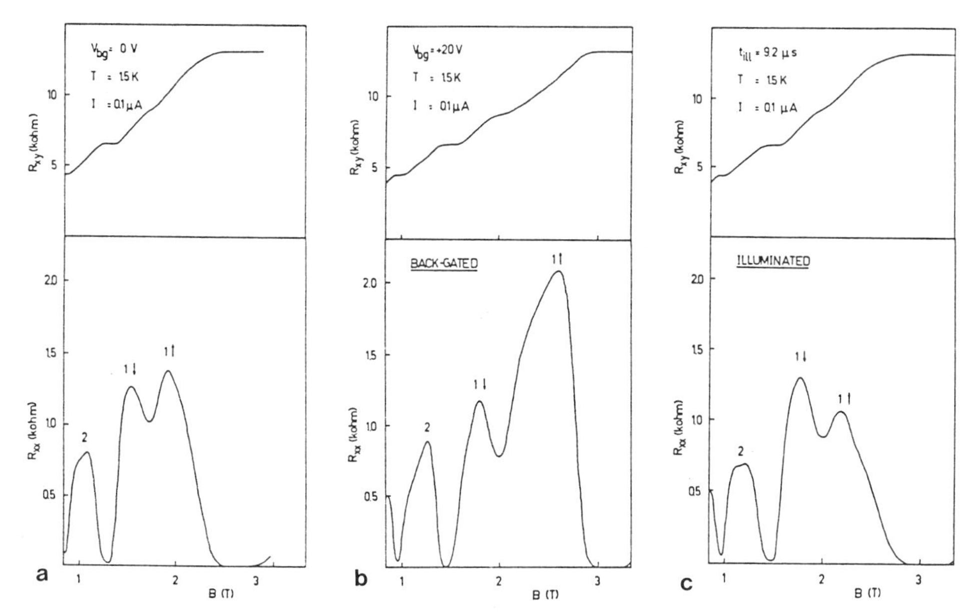

Fig. 4a SdH and QH resistances at 1.5 K as function of magnetic field without a back-gate voltage and illumination.

Fig. 4b With a back-gate voltage V_{bg} = +20 V and no illumination. Electron concentration n = 1.37 * 10^{11} cm^{-2} and mobility μ = 412.000 cm^2/Vs.

Fig. 4c Without a back-gate voltage after illumination during 9.2 μs n = 1.41 * 10^{11} cm^{-2}, μ = 405 000 cm^2/Vs.

In order to investigate this behaviour more quantitatively we determined from Hall measurements at 0.125 T the number of electrons in the two-dimensional electron gas as a function of the back-gate voltage. We observe a linear relationship between the number of electrons and the back-gate voltage. The back-gate behaves predominantly as a condensor with only small effects on the density of the remote ionized silicon scatterers. From a model based on Ando [9], we determined the average distance ⟨z⟩ of the electron wave function from the GaAs-$Al_xCa_{1-x}As$ interface. Table I gives the values for ⟨z⟩ for some back-gate voltages.

Table I

V_{bg} (V)	n_{2D} ($10^{11}/cm^2$)	⟨z⟩ (nm)
- 30	1.31	8.18
0	1.55	8.60
+ 30	1.82	9.11

Thus, for positive back-gate voltages the electrons are shifted away from the interface. For negative back-gate voltages they are shifted towards the interface. Negative back-gate voltages increase the interaction with the ionized silicon donors in the $Al_xCa_{1-x}As$, positive back-gate voltages decrease this interaction.

Comparing again the figures 4a, 4b and 4c we note that <z> is increased with a positive back-gate voltage, while it is decreased after illumination. Further, the back-gate voltage does not significantly change the number of remote ionized scatterers illumination on the other hand increases predominantly the number of ionized scatterers. Both effects lead to a reduced interaction of the electrons with the remote ionized silicon donors in the case of a back-gate voltage. This is also consistent with the observation that with the positive back-gate voltage the same electron density yields a higher mobility. This behaviour supports the assumption that the anomalous behaviour of the spin-up peak originates predominantly from the interaction with the attractive silicon donors.

In a third series of experiments we cooled down the sample to liquid helium temperature, while we applied a positive back-gate voltage V_{af} of + 12.5 V (figure 7) and + 10 V (figure 8). In Table II we present the relevant parameters of these figures. For comparison we also add some numbers from figure 4

Table II

(a)

	V_{af}(V)	V_{bg}(V)	n ($10^{11}/cm^2$)	<z>(nm)
Fig. 4a	0	0	1.55	8.60
Fig. 5	+ 12.5	0	1.72	8.92
Fig. 6	+ 10.0	0	1.84	9.20

(b)

	V_{af}(V)	V_{bg}(V)	n ($10^{11}/cm^2$)	<z>(nm)	μ(cm^2/Vs)
Fig. 4a	0	+ 35		9.14	
Fig. 5	+ 12.5	+ 35	1.96	9.45	500 000
Fig. 6	+ 10.0	+ 10	1.98	9.38	490 000

From Table II we deduce that the electrons are further shifted away from the interface, when a back-gate voltage is applied during the cooling-down of the sample. Further, we deduce that there is no linear relationship between the number of electrons or <z> and the back-gate voltage during cooling-down. This is not surprising, because at room temperature a back-gate voltage leads to a change of the number of ionized silicon donors. The dominant effect of this back-gate voltage is thus no longer just that of a condensor. When we now cool down, this state is frozen. It is essentially different from the situation that no back-gate voltage is applied during cooling-down.

The striking effect of a back-gate voltage during cooling-down can be seen in the figures 5 and 6. The spin-up peak does no longer show a current dependence, while the spin-down peak now displays a rather strong dependence on the current. Also in this case it is possible to reverse the current dependence. Depending on the value of the back-gate voltage a decrease of the current may lead to an increase of the spin-down peak (figure 6). Remember that without a back-gate voltage during cooling down we did not observe any current dependence when we applied a back-gate voltage of + 35 V to the cold sample. We further note that the electron density and the mobility in figure 5 and 6 are almost equal. The current dependence, however, is totally different.

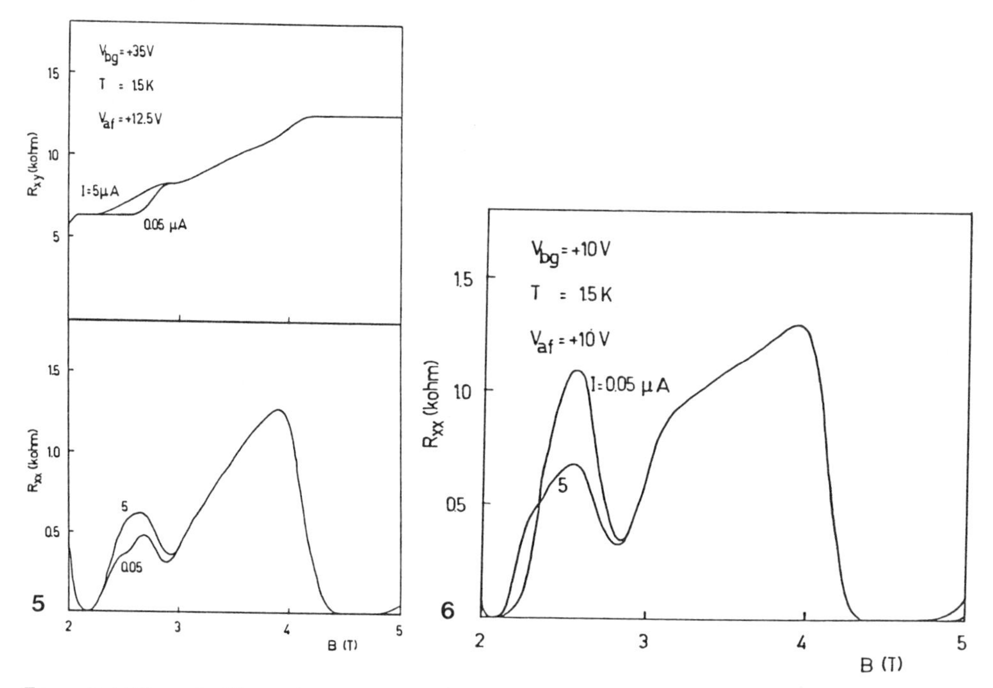

Fig. 5 SdH and QH resistances with V_{bg} = +35 V after cooling down to helium temperature with an applied back-gate voltage V_{af} = 12.5 V.

Fig. 6 As fig. 5 with V_{bg} = + 10 V and V_{af} = +10 V.

When we get a higher density at V_{BG} = 0 with cooling-down the mean distance <z> is increased. This means a smaller interaction with the attractive scatterers in the $Al_xGa_{1-x}As$ so at a lower back-gate voltage we expect a shift of the current dependence from the spin-up peak to the spin-down peak just as we observed. The different current dependence we observe in figure 5 and 6 we cannot interpret at present in terms of attractive and repulsive scatterers.

Also in the low-magnetic field range (B < 1.5T), where the spin-up and the spin-down contributions cannot be observed seperately, we analysed our SdH measurements. From the temperature dependence of the amplitude of the SdH oscillations, we deduce the effective mass in the usual manner [10]. We found (figure 9) that at low magnetic fields the effective mass is smaller than the cyclotron effective mass and apparently decreases with decreasing field. The bending-down at fields below about 0.6 T can be influenced by applying a back-gate voltage in the sense that the curvature decreases (increases) with positive (negative) voltage. At higher fields (0.8-1.5 T) the effective mass is practically independent of the back-gate voltage and approaches the value of the cyclotron effective mass.

Galchenkov et al. [8], who analysed the SdH oscillation of a GaAs-$Al_xGa_{1-x}As$ heterostructure in somewhat higher fields (1-2 T), also found a decrease of the effective mass with decreasing field. However, in their case the value of the effective mass exceeds the cyclotron value at fields above 1.1 T and reaches a value of 0.080 m_o at 2 T. Fang et al. [7] have already reported such a behaviour for a n-channel inversion layer in a silicon MOSFET. No satisfactory explanation of this effect has been given thus far.

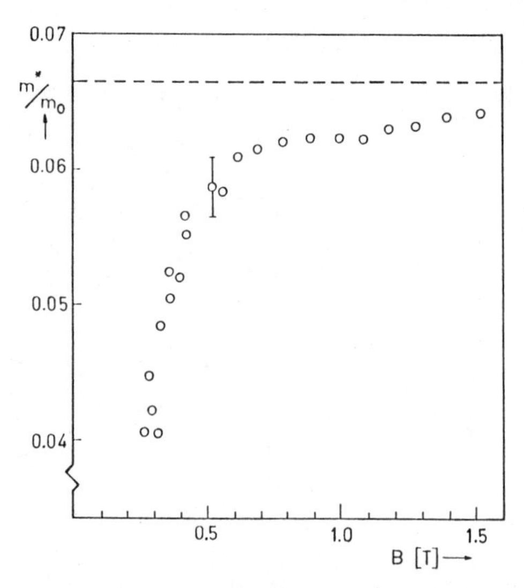

Fig. 7 The "apparent" effective mass m^*/m_o, determined from the temperature dependence of the SdH oscillations, as function of the magnetic field B for a $GaAs-Al_xGa_{1-x}As$ heterostructure with $n = 3.5 * 1011\ cm^{-2}$ and $\mu = 95\ 000\ cm^2/Vs$. The data were taken in the temperature range from 1.3 to 4.2 K.

In measurements of the effective mass with cyclotron resonance [11] such a decrease has never been observed. We thus question whether an analysis of the temperature dependence of the SdH oscillations on the basis of the model given by Ando et al. [10] yields the cyclotron mass. In this model a symmetric semi-elliptic density of states in the Landau levels is assumed. In view of the asymmetric shape of the density of states, which apparently is responsible for the observed anomalous behaviour of the SdH oscillations in the quantized Hall regime, we expect this asymmetry of the density of states also to play a major role in the temperature dependence of the amplitude of the SdH oscillations. Also the discrepancy between the value for the electron densities of two-dimensional electron gases, as determined from Hall or from SdH measurements may originate from this asymmetric shape of the density of states. In a forthcoming publication [12] we will deal with these two problems in detail.

SUMMARY

We have shown that not the number of electrons but the confinement of a wave function and thus the scattering centers is responsible for the observed asymmetric behaviour on the line-shape of the SdH oscillations in the QH regime. Moreover, we were able to reverse the current dependence by applying a back-gate voltage during the cooling of a sample. This gives further evidence that the shape of a wave function and thus the interaction between the electrons and the scattering centers is the clue to the phenomenon observed and that it is not simply related to the electron temperature of the two-dimensional electron gas. Finally we have shown that also for low-magnetic fields, where the spin-up and spin-down contributions cannot be observed separately, the amplitude of the SdH oscillations behaves anomalously.

REFERENCES

[1] K. von Klitzing, G. Dorda, and M. Pepper, Phys. Rev. Lett. 45, 494 (1980).
[2] See for example: M.A. Paalanen, D.C. Tsui, and A.C. Gossard, Phys. Rev. B25, 5566 (1982); G. Ebert, K. von Klitzing, C. Probst, and K. Ploog, Solid State Commun. 44, 95 (1982); G. Ebert, Ph.D. Thesis, Universität München (unpublished 1984).
[3] R.J. Haug, K. von Klitzing, and K. Ploog, Phys. Rev. B35, 5933 (1987) and R.J. Haug, R.R. Gerhardts, K. von Klitzing and K. Ploog, Phys. Rev. Lett. 59, 1349 (1987).
[4] R. Woltjer, J. Mooren, J. Wolter, J.-P. André, and G. Weimann, Physica B134, 352 (1985).
[5] K. von Klitzing, Physica 126B, 242 (1984).
[6] S. Kawaji, Surface Science 73, 46 (1978).
[7] F.F. Fang, A.B. Fowler, and A. Hartstein, Phys. Rev. B16, 4446 (1977).
[8] D.V. Galchenkov, I.M. Grodnenskii, O.R. Matov, T.N. Pinsker, and K.V. Starostin, JETP Lett. 40, 1228 (1984).
[9] T. Ando, J. Phys. Soc. Jpn. 53, 3126 (1984).
[10] T. Ando, A.B. Fowler, and F. Stern, Rev. Mod. Phys. 54, 437 (1982).
[11] M.A. Hopkins, R.J. Nicholas, M.A. Brummell, J.J. Harris, and C.T. Foxon, Superlattices and Microstructures 2, 319 (1986).
[12] P.F. Fontein et al., to be published.

CYCLOTRON RESONANCE OF POLARONS IN TWO DIMENSIONS

J. T. Devreese*, Xiaoguang Wu, and F. M. Peeters°

Department of Physics, University of Antwerp (UIA)
Universiteitsplein 1, B-2610, Antwerpen, Belgium

The magneto-optical absorption of a gas of (quasi) two-dimensional (2D) electrons is calculated using linear response theory as based on a memory function formulation. The theory takes into account the electron LO-phonon interaction (Fröhlich coupling), the non-parabolicity of the electron energy bands, the many-body character of the system (i.e. Fermi-Dirac statistics and screening). The results are compared to recent measurements of the cyclotron mass as a function of magnetic field in quasi 2D electron systems in GaAs-$Al_xGa_{1-x}As$ heterostructures.

INTRODUCTION

Recently there has been growing interest in the study of properties of two-dimensional (2D) electron systems [1]. Those (quasi) 2D electron systems, such as formed in inversion layers and semiconductor heterostructures (e.g. GaAs-$Al_xGa_{1-x}As$ heterosturcturs, InSb inversion layers,...) are of great fundamental and technological interest. In a 2D electron system, the behaviour of electrons is dynamically two-dimensional, i.e. the electrons are free to move parallel to an interface but are bound perpendicular to it by a static potential. In the direction normal to the interface, the motion of the electrons is quantized into a series of subbands.

Those quasi 2D electron systems like GaAs-$Al_xGa_{1-x}As$ heterostructures are made of weakly polar materials. As an electron moves through the crystal it will distort the lattice and consequently a polarization field of the longitudinal optical (LO) phonons is induced which in turn interacts with the electron. The quasi-particle consisting of the electron and its surrounding polarization cloud is called a polaron [2-4]. In 3D systems, polarons have been extensively studied [2-4].

In the present paper the influence of the electron LO-phonon interaction on the cyclotron resonance frequency or the cyclotron resonance mass of electrons in 2D electron systems will be investigated, and the theory will be applied to GaAs-$Al_xGa_{1-x}As$ heterostructures. There exists a considerable amount of studies on the cyclotron resonance of polarons in a 2D system both experimental [5-11] and theoretical [12-20]. Like in all other theoretical studies, the model system under investigation here will be a quasi 2D electron gas interacting with 3D (bulk) LO-phonons and the form of interaction will be assumed to be the so-called Fröhlich coupling [2-4].

The effects of electron LO-phonon interaction (polaron effects) have been observed in a number of cyclotron resonance experiments in GaAs-$Al_xGa_{1-x}As$ heterostructures [5-9] and in InSb inversion layers [10,11]. It is found that in GaAs-$Al_xGa_{1-x}As$ heterostructures the polaron effects are smaller than in corresponding 3D systems.

The reduction of the polaron effect can be caused by two reasons. First, in a realistic 2D electron system, e.g. a GaAs-$Al_xGa_{1-x}As$ heterostructure, the 2D electron layer has a non-zero extent in the direction normal to the 2D plane. In earlier studies on the polaron cyclotron resonance, the 2D electron layer was often approximated to be ideally 2D which results in an enhanced polaron effect compared to a 3D system. It has been realized recently that in order to be able to explain the experimental results it is necessary to take into account the non-zero width of the 2D electron layer [5-12]. Another reason for the reduction of the polaron effect is the many-particle character of the 2D system which will be discussed in detail in the following.

In order to observe a sharp cyclotron resonance peak and to accurately determine the cyclotron resonance frequency, the sample must be very pure [21]. In the case of bulk material, this restricts the electron density to be sufficiently low. In a quasi 2D electron system, in spite of the relatively high electron density (in a GaAs-$Al_xGa_{1-x}As$ heterostructure the electron density may vary from 10^{11} to $10^{12}cm^{-2}$), one still is able to detect a narrow cyclotron resonance peak. Therefore, for such a quasi 2D system, one must take into account the many-particle effects like occupation effects which result from the fact that electrons obey Fermi-Dirac statistics, and the screening of the electron-phonon interaction. In this paper we will show that it is necessary to include those effects in order to get a close agreement between theory and experiment. The occupation effects were also discussed recently by Larsen [15] and Lassnig [18].

To calculate the polaron correction to the cyclotron resonance mass or the cyclotron resonance frequency, there exist several approaches. The most elementary one is to calculate the polaron correction to the energy levels of the electron [13-18] where the electron-phonon interaction is usually treated within a perturbation scheme which may be justified by the fact that the electron-phonon coupling strength is small (for GaAs the electron LO-phonon coupling constant is $\alpha = 0.068$). Then the cyclotron resonance frequency can be obtained from the difference between two energy levels. In the present paper, however, a different approach will be used. Instead of evaluating individual energy levels the dynamical conductivity of the system will be calculated from which the polaron correction to the cyclotron resonance frequency and the cyclotron resonance mass can be directly deduced [19,20]. To calculate the dynamical conductivity the so-called memory function approach will be used [22-25]. This approach allows one to study the many-particle effects in a systematic way. In the present paper, the electron-phonon interaction is treated as a perturbation, i.e. the memory function will be calculated to first-order in the electron-phonon coupling constant.

In a cyclotron resonance experiment, a magnetic field is applied perpendicular to the 2D electron gas which completely quantizes the motion of the electrons into discrete Landau levels and the density of states of the system becomes a series of delta functions if no scattering is present. In a realistic quasi 2D system, however, the Landau levels will be broadened due to the interaction between electrons and impurities, between electrons and phonons, and due to the mutual interaction between the electrons. Theoretically, this requires a self-consistent calculation [26,27]. For simplicity, in the present paper the broadening of Landau levels will be neglected. Because the main concern here is the calculation of the cyclotron resonance mass of polarons, this approximation will not cause any inconveniences and is expected to be valid as long as the broadening of the Landau levels is small.

I. THEORY

The electron LO-phonon interaction is described by the many-particle Fröhlich Hamiltonian

$$\begin{aligned} \mathrm{H} = & \sum_j \frac{1}{2m_b}(\vec{p}_j + \frac{e\vec{A}(\vec{r}_j)}{c})^2 + \sum_{\vec{k}} \hbar\omega_{\mathrm{LO}}\, a^\dagger_{\vec{k}} a_{\vec{k}} \\ & + \sum_{i<j} V(\vec{r}_i - \vec{r}_j) + \sum_j \sum_{\vec{k}} \left(V_{\vec{k}}\, a_{\vec{k}}\, e^{i\vec{k}\cdot\vec{r}_j} + V^*_{\vec{k}}\, a^\dagger_{\vec{k}}\, e^{-i\vec{k}\cdot\vec{r}_j} \right) \quad , \end{aligned} \tag{1}$$

where $\vec{p}_j$ ($\vec{r}_j$) is the 2D momentum (position) operator of the j-th electron. $a^{\dagger}_{\vec{k}}$ ($a_{\vec{k}}$) is the creation (annihilation) operator of a bulk LO-phonon with 3D wave vector $\vec{k}$ and energy $\hbar\omega_{\text{LO}}$. ω_{LO} is assumed to be dispersionless. $\vec{A}$ is the vector potential describing the magnetic field which is taken perpendicular to the 2D electron layer. The z-axis is chosen along the direction of the magnetic field. In Eq.(1) $V_{\vec{k}}$ represents the interaction of the electron with the LO-phonons and is given by

$$V_{\vec{k}} = i\hbar\omega_{\text{LO}} \left(\frac{4\pi\alpha}{Vk^2}\right)^{1/2} \left(\frac{\hbar}{2m_b\omega_{\text{LO}}}\right)^{1/4} \langle\psi_0|e^{ik_z z}|\psi_0\rangle \quad . \tag{2}$$

Here $\psi_0(z) = (b^3/2)^{1/2} z e^{-bz/2}$ is the variational wave function of the electron in the direction normal to the 2D electron layer [1]. The variational parameter b is chosen as $b = (48\pi N m_b e^2/\hbar^2\epsilon_0)^{1/3}$, where $N = n_d + (11/32)n_e$ and n_d and n_e are the depletion and carrier charge densities respectively. In Eq.(2) only the lowest subband is included and all higher subbands are neglected. This approximation should be valid for systems with not too high electron density ($n_e < 10^{12}\text{cm}^{-2}$) and for systems at low temperature, so that there are no electrons in higher electric subbands. $V(\vec{r}-\vec{r}\,')$ represents the electron-electron interaction in the quasi 2D electron layer. The Fourier transform of $V(\vec{r})$ is $v(k_{\|}) = (2\pi e^2/k_{\|}\epsilon_\infty) f(k_{\|}, b)$ with the standard form factor [1] $f(k,b) = (8b^3 + 9b^2k + 3bk^2)/(8(b+k)^3)$ and $k_{\|}^2 = k_x^2 + k_y^2$. In this section the electron is assumed to have a parabolic energy band.

The dynamical conductivity of the system in the presence of a magnetic field can be written as [22-25]

$$\sigma_{\pm}(\omega) = \frac{i n_e e^2/m_b}{\omega \pm \omega_c - \Sigma(\omega)} \quad , \tag{3}$$

where $\Sigma(\omega) = \Sigma(\alpha, \omega_c, b; \omega)$ is the memory function. To first-order in the electron-phonon coupling constant (α) the memory function takes the form

$$\Sigma(\omega) = \frac{1}{\omega}\int_0^\infty dt(1 - e^{i\omega t})\text{Im}F(t) \quad , \tag{4a}$$

with

$$F(t) = -\sum_{\vec{k}} \frac{k_{\|}^2}{n_e m_b \hbar} \left|V_{\vec{k}}\right|^2 \left(i\,D(k_{\|},t) + i\,n(\omega_{\text{LO}}) D^{\text{R}}(k_{\|},t)\right) e^{-i\omega_{\text{LO}} t} \quad , \tag{4b}$$

where $n(\omega) = (e^{\beta\hbar\omega} - 1)^{-1}$ is the occupation number. $D(k_{\|},t)$ ($D^{\text{R}}(k_{\|},t)$) is the electron (retarded) density-density correlation function [28].

At low temperature and in the absence of broadening of the Landau levels, the cyclotron resonance frequency ω_c^* is determined by the following non-linear equation

$$\omega - \omega_c - \text{Re}\Sigma(\omega) = 0 \quad , \tag{5a}$$

and from the solution one defines the cyclotron resonance mass m^* as

$$m^* = \frac{\omega_c}{\omega_c^*} m_b \quad . \tag{5b}$$

The real part of the memory function can be obtained by substituting Eq.(4b) into Eq.(4a)

$$\begin{aligned} \mathrm{Re}\Sigma(\omega) = &\sum_{\vec{k}} \frac{k_{\parallel}^2}{n_e m_b \omega} |V_{\vec{k}}|^2 \frac{\omega^2}{\pi} \int_{-\infty}^{+\infty} dx \frac{(1+n(x))\mathrm{Im}\Pi^{\mathrm{R}}(k_{\parallel}, x)}{((x+\omega_{\mathrm{LO}})^2 - \omega^2)(x+\omega_{\mathrm{LO}})} \\ &+ n(\omega_{\mathrm{LO}}) \sum_{\vec{k}} \frac{k_{\parallel}^2}{n_e m_b \omega} |V_{\vec{k}}|^2 \frac{1}{2} \big[\mathrm{Re}\Pi^{\mathrm{R}}(k_{\parallel}, \omega + \omega_{\mathrm{LO}}) \\ &+ \mathrm{Re}\Pi^{\mathrm{R}}(k_{\parallel}, \omega - \omega_{\mathrm{LO}}) - 2\mathrm{Re}\Pi^{\mathrm{R}}(k_{\parallel}, \omega_{\mathrm{LO}}) \big] \quad , \end{aligned} \tag{6}$$

where $D = \hbar\Pi$ is the polarization function of the 2D electron gas [28] which has been extensively studied [29-31]. Remark that in the calculation of the memory function the frequency dependence of the polarization function of the 2D electron gas is required.

I.A. MAGNETIC FIELD DEPENDENCE

To study the magnetic field dependence of the cyclotron resonance mass, one could simply look into the zero temperature limiting case. At zero temperature, only the imaginary part of the polarization function of the 2D electron gas is needed. The occupation effect or Pauli-blocking effect can be taken into account by approximating $\mathrm{Im}\Pi(k,\omega)$ by $\mathrm{Im}\Pi^0(k,\omega)$ the imaginary part of the polarization function of a non-interacting 2D electron gas [32]. To include the static screening of the electron LO-phonon interaction, $\mathrm{Im}\Pi(k,\omega)$ will be approximated by $\mathrm{Im}\Pi^0(k,\omega)/\epsilon^2(k)$ where $\epsilon(k) = 1 - v(k)\Pi^0(k,\omega = 0)$ is the static dielectric function of the 2D electron gas in the random-phase-approximation (RPA). To consider the screening of the electron LO-phonon interaction in a dynamical way, $\mathrm{Im}\Pi(k,\omega)$ will be approximated by $v^{-1}(k)\mathrm{Im}\epsilon^{-1}(k,\omega)$ with $\epsilon(k,\omega)$ the dieletric function of the 2D electron gas in the RPA approximation [33].

In our numerical calculations, all physical parameters are chosen as corresponding to the GaAs-$\mathrm{Al}_x\mathrm{Ga}_{1-x}$As heterstructures. For completeness, the polaron correction to the cyclotron resonance frequency will first be studied in the one-polaron picture [19]. In Fig.1 the cyclotron resonance mass is plotted as a function of the magnetic field strength. It is clear that, in this one-polaron approximation, an ideal 2D system ($b = \infty$) results in an enhanced polaron effect as compared to the 3D case. The non-zero width of the 2D electron layer reduces the polaron effect and leads to a mass correction which will vary between ideal 2D and 3D corresponding results. In Fig.1 cyclotron resonance masses m^* are plotted for both cases that $\omega_c^* < \omega_{\mathrm{LO}}$ and $\omega_c^* > \omega_{\mathrm{LO}}$. At $\omega_c = \omega_{\mathrm{LO}}$ there is a splitting of the cyclotron resonance frequency due to the splitting of degenerate energy levels [19].

Next, the occupation effect will be studied [32]. In Fig.2 the polaron correction to the cyclotron resonance mass is plotted as a function of the magnetic field for a quasi 2D system with a fixed electron density. The curve indicated with $\nu = 0.4\omega_{\mathrm{LO}}/\omega_c$ is for an electron density $n_e = 4 \times 10^{11}\mathrm{cm}^{-2}$. As seen from the figure, the inclusion of the occupation effect reduces the polaron correction to the cyclotron resonance mass as compared to the one-polaron result which is labeled with $\nu = 0$.

In Fig.3 the polaron correction to the cyclotron resonance mass is plotted as a function of the magnetic field strength for a quasi 2D electron system within different approximations [32,33]. $\nu = 0.4\omega_{\mathrm{LO}}/\omega_c$ corresponds to an electron density of $n_e = 4 \times 10^{11}\mathrm{cm}^{-2}$. The screening of the electron LO-phonon interaction has, for the first time, been treated within a full dynamical way in the presence of a strong magnetic field [33]. Including the dynamical screening the polaron correction to the cyclotron resonance mass is reduced as compared to the calculation where only the occupation effect is considered.

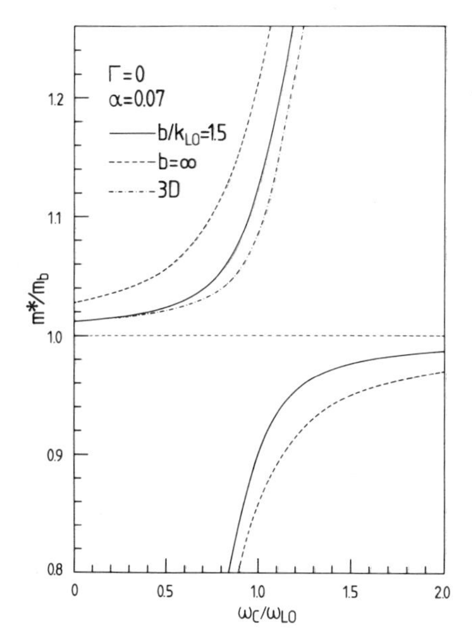

Fig.1 The cyclotron resonance mass is plotted as a function of the magnetic field for an ideal 2D and quasi 2D system. The 3D result is given by the dash-dotted curve.

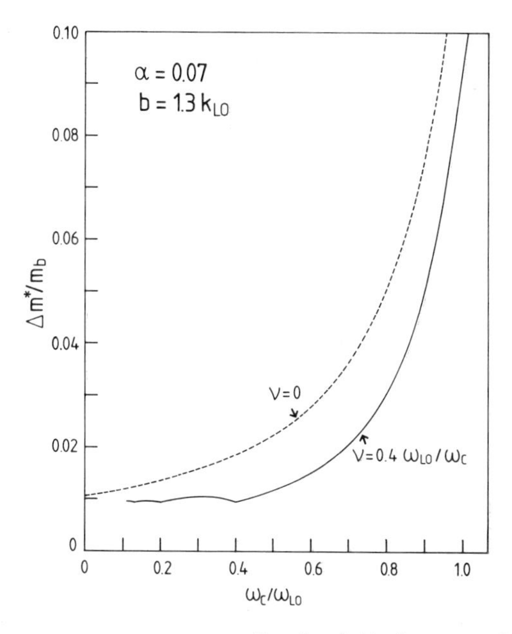

Fig.2 The cyclotron resonance mass correction is plotted versus the magnetic field for a fixed electron density ($n_e = 4 \times 10^{11}\mathrm{cm}^{-2}$) for a quasi 2D system. Only the occupation effect is included. The zero electron density result ($\nu = 0$) is given by the dashed curve.

The correction to the polaron cyclotron resonance mass is reduced as the filling factor ν, which is defined as $\nu = n_e \pi \hbar / m_b \omega_c$, approaches integer values which leads to an oscillating behaviour of the cyclotron resonance mass. For integer values of the filling factor the mass correction is approximately equal to the zero magnetic field mass renormalization. The amplitude of the oscillation of the cyclotron resonance mass decreases as the magnetic field strength decreases. In the zero magnetic field limit the polaron cyclotron resonance mass seems to approach the corresponding result given by Ref.34.

It is found that, in the case of zero magnetic field, the polaron effect on the electron effective mass behaves differently within static as compared to dynamical screening treatments. The static screening is found to overestimate the effect of screening [34]. Thus it is very interesting to look into the case where a strong magnetic field is applied to the quasi 2D electron gas.

As clearly seen from Fig.3, the static screening approach gives almost the same results as the dynamical screening treatment. The difference between the results of both treatments is very small for large magnetic fields ($\omega_c/\omega_{LO} > 0.5$). Only for small magnetic fields the difference becomes noticeable on the scale of our plot. This is because the dominant term contributing to the memory function in the resonant region (i.e. $\omega_c \sim \omega_{LO}$) is the same within the dynamical screening and the static screening approximations [33]. Therefore we may conclude that when the broadening of the Landau levels is neglected the static screening approach gives fairly good results for the polaron cyclotron resonance mass renormalization in the relevant magnetic field region, i.e. $\omega_c/\omega_{LO} > 0.5$ or $H > 10.5$ T for a GaAs-$Al_xGa_{1-x}As$ heterostructure.

Finally, in Fig.4, the cyclotron resonance mass renormalization (at $\omega_c = 0.8\omega_{LO}$) is plotted as a function of the electron density. The electron density is in units of $k_{LO}^2/2\pi$ (with $k_{LO} = (2m_b\omega_{LO}/\hbar)^{1/2}$) which equals 10^{12}cm^{-2} for a GaAs-$Al_xGa_{1-x}As$ heterostructure. $\omega_c = 0.8\omega_{LO}$ is chosen since we are mainly interested in the region where the polaron effects are dominant, i.e. in the region where ω_c is close to ω_{LO}. As expected the polaron effect (hence the cyclotron resonance mass correction) decreases as the electon density increases. Note also that the dynamical screening approximation, as discussed above, gives almost the same result as the static screening treatment over the whole region of the electron densities under study here.

I.B. TEMPERATURE DEPENDENCE

In this section the temperature dependence of the cyclotron resonance mass will be studied [34,35]. To simplify the calculation, the zero magnetic field case will be considered. In this case the polaron correction to the electron effective mass is given by [35-37]

$$\frac{\Delta m}{m_b} = -\lim_{\omega \to 0} \frac{\text{Re}\Sigma(\alpha, \omega_c = 0, \omega)}{\omega} \quad . \tag{7}$$

In Fig.5 the electron-phonon interaction correction to the electron effective mass is plotted as a function of the temperature for two values of the electron density. The calculation is performed for an ideal 2D system, i.e. the 2D electron layer has a zero width. The result corresponding with the zero electron density is the 2D analogue of the FHIP theory for 3D polarons [36,37]. All physical parameters are taken corresponding to a GaAs-$Al_xGa_{1-x}As$ heterostructure. The polaron mass correction is found to increase first with increasing temperature ($T < 80$ K) and to reach a maximum around 90 K. For still higher temperature it starts to decrease. Dynamical screening reduces the polaron mass renormalization over the whole temperature region as expected. It is found that the absolute reduction of the polaron mass correction due to the electron screening is larger at 100 K than at zero temperature. The increase of the mass correction at low temperature is due to the nonparabolicity induced by the electron-phonon interaction as was explained earlier [37] in the case of 3D polarons.

In Fig.6 the polaron mass renormalization is plotted as a function of the temperature for different values of the electron density. But now the calculation is performed

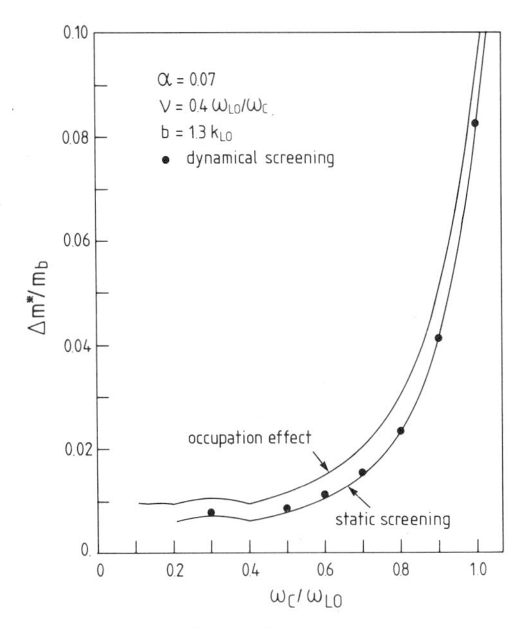

Fig.3 The polaron correction to the cyclotron resonance mass as a function of the magnetic field. The many-particle effects are treated within different approximations and the non-zero width of the 2D electron layer is taken into account. The electron density is $n_e = 4 \times 10^{11}\text{cm}^{-2}$ and all other parameters are taken corresponding to a GaAs-Al_xGa_{1-x}As heterostructure.

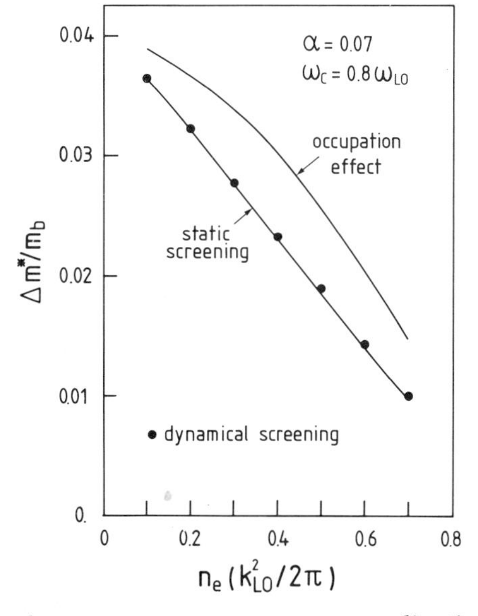

Fig.4 The polaron cyclotron resonance mass renormalization ($\omega_c = 0.8\omega_{LO}$) is plotted as a function of the electron density. The electron density is in units of $k_{LO}^2/2\pi$ which is 10^{12}cm^{-2} for a GaAs-Al_xGa_{1-x}As heterostructure.

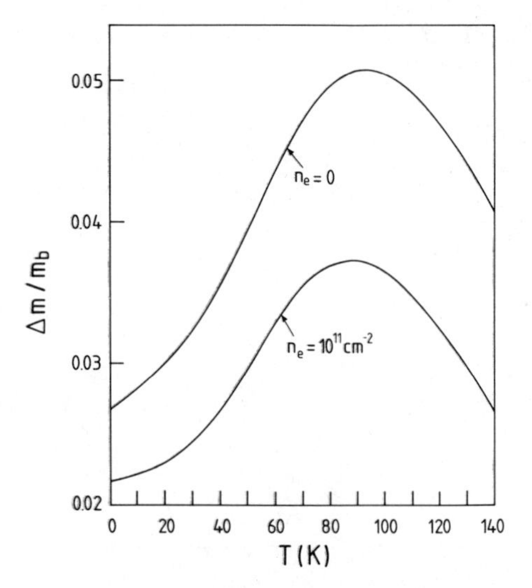

Fig.5 The electron-phonon interaction correction to the electron effective mass is plotted as a function of the temperature for two values of the electron density. The calculation is performed for an ideal 2D system.

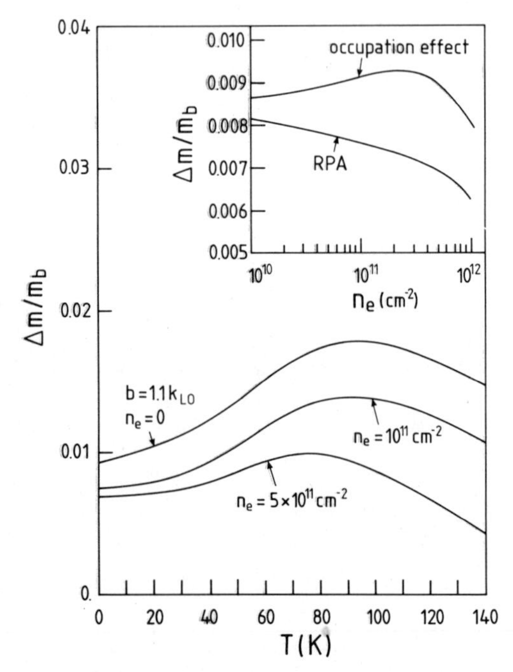

Fig.6 The polaron mass renormalization due to the electron-phonon coupling is plotted as a function of the temperature for different values of the electron density. The inset shows the mass correction as a function of the electron density at zero temperature. The calculation is performed for a quasi 2D electron system where only the lowest subband is taken into account.

for a quasi 2D electron system. Only the lowest subband is taken into account using a Stern-Fang-Howard variational wave function. The inset shows the mass correction as a function of the electron density in the zero temperature limit. Again it is found that the reduction of the polaron mass correction due to the screening is larger around 100 K than at zero temperature. The position of the maximum of the polaron mass renormalization shifts to lower temperature as the electron density increases. The present theoretical results agree qualitatively with a recent experiment [38] but the absolute value of the mass renormalization is not obtained correctly [35].

II. COMPARISON OF THE MAGNETIC FIELD DEPENDENCE WITH EXPERIMENTS

In this section, the theoretical results on the magnetic field dependence of the cyclotron resonance mass will be compared to experimental data.

In the previous section the electron was assumed to have a parabolic energy band. However, to make a realistic comparison between the theoretical results and the experimental data, one has to take into account the fact that in a GaAs-$Al_xGa_{1-x}As$ heterostructure the electron energy band is not perfectly parabolic [5-11]. In this paper the effect of energy band nonparabolicity will be included within a local parabolic energy band approximation [19,20]. This means that in Eq.(5a) ω_c will be replaced by $(\omega_c)_{np}$, the cyclotron resonance frequency which only contains the effect of band nonparabolicity without polaron effects, and hence the cyclotron resonance frequency is given by

$$\omega - (\omega_c)_{np} - \mathrm{Re}\Sigma(\alpha, (\omega_c)_{np}, b; \omega) = 0 \quad . \tag{8}$$

$(\omega_c)_{np}$ is calculated within a $\vec{k} \cdot \vec{p}$ theory [14,19]. The polaron effects then enter via $\Sigma(\omega)$ (see Eqs.(4)).

The validity of this approximation relies on the fact that this local parabolic energy band approximation holds for systems in the limit of zero electron-phonon coupling strength and in the limit of parabolic energy band. Further justification of the approach is provided by calculating the memory function with all energy levels evaluated in the $\vec{k} \cdot \vec{p}$ theory. The latter approach has been proved to be rather accurate [39]. Only a very small difference is found which can hardly be seen in the figures presented below.

In Fig.7 the cyclotron resonance mass is plotted as a function of the magnetic field for a sample with electron density $n_e = 1.4 \times 10^{11}\mathrm{cm}^{-2}$. The dots are the experimental data of Hopkins *et al.* [8]. The theoretical results are given by the full solid curve where the static screening of the electron-phonon interaction is included. The depletion charge density is determined from the subband energy difference [8]. The electron band mass, which is an unknown parameter, is chosen in such a way that an overall fit is obtained and is found to be $m_b = 0.0661 m_e$.

The theoretical results are compared to the experimental data with a higher electron density of $n_e = 3.4 \times 10^{11}\mathrm{cm}^{-2}$ [8] in Fig.8 where the experimental data are given by solid squares. The electron band mass is taken to be the same as obtained from the lower electron density sample (Fig.7), i.e. $m_b = 0.0661 m_e$. For a magnetic field strength $H = 7.03$ T, the filling factor $\nu = 1$ which is indicated by a vertical dashed line. For $\nu < 1$ only the transition of the $n = 0 \rightarrow n = 1$ Landau level is possible, while for $1 < \nu < 2$ both $0 \rightarrow 1$ and $1 \rightarrow 2$ transitions are allowed since higher Landau levels are also occupied. It is expected that the measured cyclotron resonance mass for $1 < \nu < 2$ lies between the two theoretical curves for those two transitions. Because in the calculation of $(\omega_c)_{np}$ the Landau levels are assumed to be perfectly sharp, i.e. no broadening of the energy levels occurs, the theoretical cyclotron resonance mass is discontinuous at integer fillings.

Finally, in Fig.9 the present theoretical result is compared to the experimental data with an even higher electron density of $n_e = 5 \times 10^{11}\mathrm{cm}^{-2}$. In the upper part of

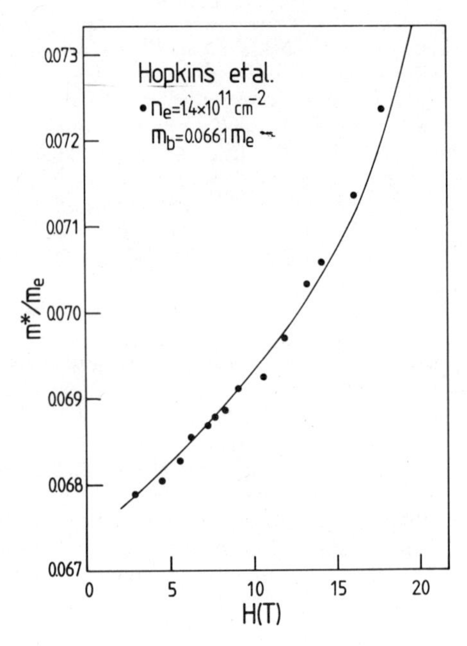

Fig.7 The cyclotron resonance mass as a function of the magnetic field for a sample with electron density $n_e = 1.4 \times 10^{11}\mathrm{cm}^{-2}$. The dots are the experimental data of Hopkins *et al.* [8]. The full curve is the present theoretical result.

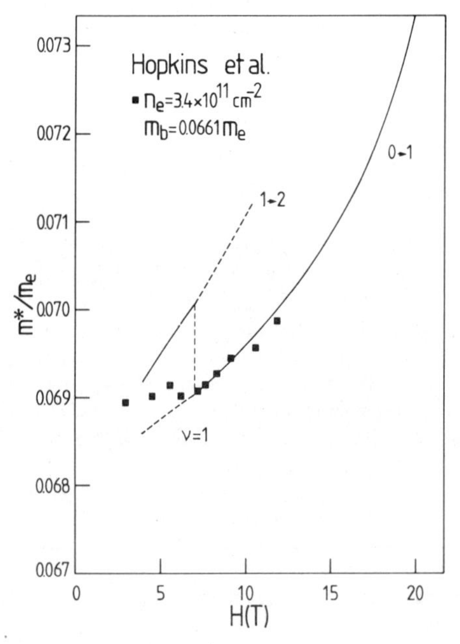

Fig.8 The same as Fig.7 but for a sample with a higher electron density of $n_e = 3.4{\times}10^{11}\mathrm{cm}^{-2}$ [8]. The transitions between higher Landau levels are also included. ν is the filling factor.

Fig.9, the calculated cyclotron resonance mass is plotted as a function of the magnetic field together with the experimental data of Thiele *et al.* [9] which are presented as solid dots. For $\nu > 2$ the experimental data lie above the theoretical results. The theory fits the experimental data well for $\nu < 2$ except at very high magnetic fields ($H > 18$ T) where the experimental data seem to saturate. This saturation is not fully understood yet since even without polaron effects the cyclotron resonance mass should increase linearly as a function of the magnetic field due to the band nonparabolicity. In the lower part of Fig.9, the polaron correction to the cyclotron resonance mass is plotted versus the magnetic field. As found by Thiele *et al.* in Ref.9, the polaron effect is indeed very small for $H < 16$ T and the experimental results are mainly determined by the band nonparabolicity effect in that magnetic field region.

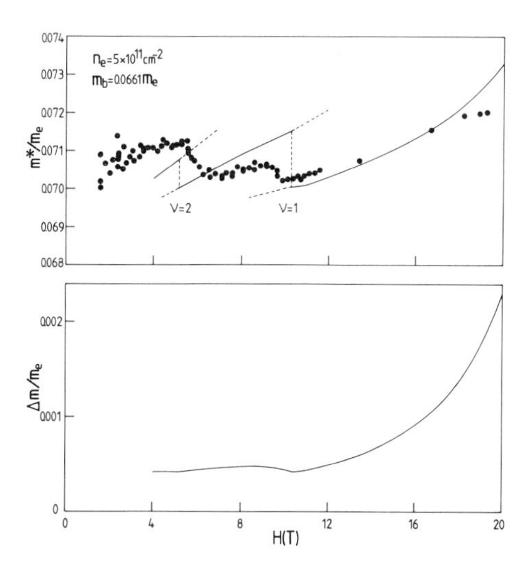

Fig.9 In the upper part of the figure, the cyclotron resonance mass is plotted as a function of the magnetic field. In the lower part of the figure, the polaron correction to the cyclotron resonance mass versus the magnetic field. The dots are the experimental data of Thiele *et al.* [9]. The sample electron density is $n_e = 5 \times 10^{11}\mathrm{cm}^{-2}$.

III. CONCLUSIONS

In conclusion, in the present paper, a detailed analysis has been made of the magnetic field dependence of the polaron cyclotron resonance mass for a 2D electron gas in GaAs-$\mathrm{Al}_x\mathrm{Ga}_{1-x}$As heterostructures. The influence of the electron-phonon interaction on the cyclotron resonance mass is investigated theoretically within different approximations where the role played by the Fermi-Dirac statistics, the static screening and the dynamical screening of the electron-phonon interaction is studied in detail. The present theory accurately describes the experimental data in the magnetic quantum limit, i.e. $\nu < 1$. For samples with electron density $n_e > 4 \times 10^{11}\mathrm{cm}^{-2}$, it is found that the experimental cyclotron resonance mass data are larger than predicted by the theory in the low magnetic field region where $\nu > 2$. The temperature dependence of the polaron correction to the electron effective mass is also studied and is found to agree qualitatively with the experiment.

ACKNOWLEDGMENTS

This work is partially supported by F.K.F.O. (Fonds voor Kollektief Fundamenteel Onderzoek, Belgium), project No. 2.0072.80. One of us (F.M.P.) acknowledges financial support from the Belgium National Science Foundation. X. Wu is supported by a PACER fellowship (CDC, Minneapolis).

REFERENCES

* Also at University of Antwerp (RUCA), B-2020 Antwerpen, Belgium and Eindhoven University of Technology, NL-5600 MB Eindhoven, The Netherlands.

° Present address: Bell Communications Research, 331 Newman, Springs Road, Box 7020, Red Bank, New Jersey 07701-7720.

[1] T. Ando, A. B. Fowler, and F. Stern, Rev. Mod. Phys. **54**, 437 (1982).
[2] *Polarons and Excitons* , edited by C. Kuper and G. Whitfield (Oliver and Boyd, Edinburgh, 1963).
[3] *Polarons in Ionic Crystals and Polar Semiconductors* , edited by J. T. Devreese (North-Holland, Amsterdam, 1972).
[4] *Polarons and Excitons in Polar Semiconductors and Ionic Crystals* , edited by J. T. Devreese and F. M. Peeters (Plenum, New York, 1984).
[5] W. Seidenbusch, G. Lindemann, R. Lassnig, J. Edlinger, and G. Gornik, Surf. Sci. **142**, 375 (1984).
[6] M. Horst, U. Merkt, W. Zawadzki, J. C. Maan, and K. Ploog, Solid State Commun. **53**, 403 (1985).
[7] H. Sigg, P. Wyder, and J. A. A. Perenboom, Phys. Rev. B **31**, 5253 (1985).
[8] M. A. Hopkins, R. J. Nicholas, M. A. Brummell, J. J. Harris, and C. T. Foxon, Superlattices and Microstructures **2**, 319 (1986); Phys. Rev. B **35** (1987).
[9] F. Thiele, U. Merkt, J. P. Kotthaus, G. Lommer, F. Malcher, U. Rössler, and G. Weimann, Solid State Commun. (1987).
[10] M. Horst, U. Merkt, and J. P. Kotthaus, Phys. Rev. Lett. **50**, 754 (1983).
[11] U. Merkt, M. Horst, and J. P. Kotthaus, Physica Scripta. T **13**, 272 (1986) and references there in.
[12] S. Das Sarma, Phys. Rev. B **27**, 2590 (1983); B **31**, 4034 (E) (1985).
[13] S. Das Sarma and A. Madhukar, Phys. Rev. B **22**, 2823 (1980).
[14] R. Lassnig and W. Zawadzki, Surf. Sci. **142**, 388, (1984).
[15] D. Larsen, Phys. Rev. B **30**, 4595 (1984).
[16] F. M. Peeters and J. T. Devreese, Phys. Rev. B **31**, 3689 (1985).
[17] F. M. Peeters, Xiaoguang Wu, and J. T. Devreese, Phys. Rev. B **33**, 4338 (1986).
[18] R. Lassnig, Surf. Sci. **170**, 549 (1986).
[19] Xiaoguang Wu, F. M. Peeters, and J. T. Devreese, Phys. Rev. B **34**, 8800 (1986).
[20] F. M. Peeters, Xiaoguang Wu, and J. T. Devreese, Physica Scripta. T **13**, 282 (1986).
[21] C. Kittel, *Introduction to Solid State Physics* , sixth edition, (John Wiley & Sons, New York, 1986).
[22] W. Götze and P. Wölfe, Phys. Rev. B **6**, 1226 (1972).
[23] C. S. Ting, S. C. Ying, and J. J. Quinn, Phys. Rev. B **16**, 5394 (1977).
[24] J. T. Devreese, J. De Sitter, and M. Goovaerts, Phys. Rev. B **5**, 2367 (1972).
[25] F. M. Peeters and J. T. Devreese, Phys. Rev. B **28**, 6051 (1983).
[26] S. Das Sarma, Phys. Rev. B **23**, 4592 (1981).
[27] W. Cai and C. S. Ting, Phys. Rev. B **33**, 3967 (1986).
[28] A. L. Fetter and J. D. Walecka, *Quantum Theory of Many-Particle Systems* , (McGraw-Hill, New York, 1971).
[29] N. J. Horing and M. M. Yildiz, Ann. Phys. (N.Y.) **97**, 216 (1976).
[30] M. L. Glasser, Phys. Rev. B **28**, 4387 (1983).
[31] A. H. MacDonald, J. Phys. C **18**, 1003 (1985).
[32] Xiaoguang Wu, F. M. Peeters, and J. T. Devreese, Phys. Status Solidi B (1987).
[33] Xiaoguang Wu, F. M. Peeters, and J. T. Devreese, (submitted for publication).
[34] Xiaoguang Wu, F. M. Peeters, and J. T. Devreese, Phys. Rev. B **34**, 2621 (1986).
[35] Xiaoguang Wu, F. M. Peeters, and J. T. Devreese, (submitted for publication).
[36] R. P. Feynman, R. W. Hellwarth, C. K. Iddings, and P. M. Platzman, Phys. Rev. **127**, 1004 (1962).
[37] F. M. Peeters and J. T. Devreese, Solid State Physics (eds. H. Ehrenreich and D. Turnbull), **38**, 81 (1984).
[38] M. A. Brummell, R. J. Nicholas, M. A. Hopkins, J. J. Harris, and C. T. Foxon, Phys. Rev. Lett. **58**, 77 (1987).
[39] D. Larsen, (to be published)

STRUCTURE AND ELECTRONIC PROPERTIES OF STRAINED Si/Ge SEMICONDUCTOR SUPERLATTICES

S. Ciraci

Department of Physics
Bilkent University
Ankara, Turkey

Inder P. Batra

IBM Research Division
Almaden Research Center
650 Harry Road
San Jose, California 95120-6099

ABSTRACT

The stability, growth, structural phase transitions, and the electronic properties of strained SiGe alloy and superlattices have been investigated by using self-consistent field pseudopotential method. The equilibrium structures of Si_n/Ge_n ($n \leq 6$) superlattices pseudomorphically restricted to the Si(001) surface are determined, and their formation enthalpies are calculated. A simple model for the formation enthalpy of superlattices is developed, whereby the activation barrier of the misfit dislocation is estimated. It is found that during the layer-by-layer growth, the energy of the topmost layer is lowered through the dimerization of atoms. The energy gap of all Si_n/Ge_n superlattices is found to be indirect. More significantly, the energy separation between the direct and indirect gap continues to decrease with increasing n, and is only 0.07 eV for n = 6. Extended conduction band states below the confined states point to a new feature of the band offset and quantum size effect. Localized states lying deep in the valence and conduction band continua are another novel result found in this study.

INTRODUCTION

Owing to its excellent etching and mechanical properties silicon has dominated the microelectronic technology. However, being an indirect gap semiconductor, silicon has been excluded from important laser applications. Improving electronic properties of this crystal has been a continuing interest of the material scientist. Recent developments in the fabrication of epitaxial heterostructures and semiconductor superlattices with 1-D and 2-D novel electronic properties have stimulated the idea of increasing carrier mobility in Si/Ge heterostructures by modulation doping. In an effort to compensate the deficiencies of silicon with more convenient materials, and to improve the existing technology the epitaxial growth of $Si/Si_{1-x}Ge_x$ has been achieved.[1,2] The lattice misfit of ≤4% is completely accommodated by the uniform lattice strain in the commensurate or pseudomorphic $Si/Si_{1-x}Ge_x$ layers. While the grown layers are in registry with the epilayer, the lattice constant in the perpendicular direction expands, leading to a tetragonal distortion. This way the planar compressive strain energy is partly relieved by the perpendicular expansive strain. Since the energy barrier associated with the reordering of atoms is too high, many defect free, strained layers can grow before the accumulated strain energy is relaxed by the generation of the misfit dislocation. Fiory *et al.*[3] were able to grow high quality, commensurate $Si_{1-x}Ge_x$ films of ~2500Å thickness when $x < 0.5$.

Recently, the growth of pure Ge (*i.e.*, x = 1) up to six layers pseudomorphically restricted to Si(001) substrate has been realized by Pearsall *et al.*[4] More importantly, they observed direct optical transitions in the Si_4/Ge_4 superlattice, which are found neither in constituent crystals, nor in $Si_{0.5}Ge_{0.5}$ alloy. Novel electronic structures, especially lowering of the difference of energy between the indirect and direct band gap in the Si/Ge superlattice have been predicted by self-consistent-field pseudopotential calculations of Ciraci and Batra[5] and Froyen *et al.*[6]

The stability of the Si/Ge heterostructures in spite of a large lattice mismatch is another interesting aspect. This became the focus of attention by a recent observation. Ourmazd and Bean[7] presented evidence for a neostructural order-disorder transition in the strained $Si/Si_{1-x}Ge_x$ superlattice. Apart from its fundamental and academic significance to the order-disorder transitions in alloys, this observation has important technological implications. The questions have arisen as to why and how the observed transition occurs, and how the long-range order affects the electronic properties of the superlattice. Theoretical studies[5,8] have started to investigate these issues. The band offset in the Si/Ge superlattices also has attracted much attention, and been treated in a number of recent publications.[9-15]

Clearly, the pseudomorphic Si/Ge superlattices present new conceptual ideas about the synthetic semiconductors and novel device applications. Considering a generalized formula for such a superlattice as $\{Si_{1-x}Ge_x\}_{n,a\|}/\{Si_{1-y}Ge_y\}_{m,a\|}$ we see that many degrees of freedom are available for controlling the properties of this system. The concentrations x and y of Ge in the sublattices, and the lateral lattice constant, $a_\|$, are variables, which set the band offset,[13] and provide excellent means to engineer the quantum well structure. The superlattice periodicity, n+m, (n as well as m itself) are also important parameters to control the character and the dimensionality of the confined states.[5,16,17] Furthermore, the modulation doping ranging from very low concentration to the formation of the impurity bands, brings about additional degrees of freedom in characterizing the devices based on these heterostructures. Effects related to the dimensionality of the confined carriers, novel electronic structure induced by the zone folding and lattice strain set an environment for impurity states, which is quite different from what we have seen so far.

This article deals with the stability, growth and the electronic structure of $Si/Si_{1-x}Ge_x$ and Si_n/Ge_n superlattices which have in-plane registry with the Si(001) surface. The primary objective is to understand them by using an *ab initio* method in order to develop a framework for the study of impurity states. Another objective is to demonstrate how and to what extent an *ab initio* method can be used to enhance our understanding.

METHOD AND PARAMETERS OF CALCULATIONS

Self-consistent-field (SCF) pseudopotential calculations[18] were performed within the framework of the local-density functional theory applied in momentum space.[19] Scalar relativistic effects were included via the use of nonlocal, norm-conserving ionic pseudopotentials given by Bachelet *et al.*[20] Ceperley-Alder exchange and correlation potential[21] has been used. Bloch states are expanded with the kinetic energy cut-off corresponding to $|\vec{k}+\vec{G}|^2 = 12$ Ry , which leads to a basis set consisting of $\sim$1200 plane waves in large unit cells. This energy cut-off is raised up to 18 Ry for calculating the electronic properties.

We have carried out total energy, electronic structure, interatomic force[22] and charge density calculations on Si, Ge, tetragonally distorted Ge, zincblende (zb) SiGe, and Si_n/Ge_n superlattices with $1 \leq n \leq 6$. To understand the origin of the observed neostructural phase transitions in the $Si_{1-x}Ge_x$ alloys we also investigated the "free floating" and strained SiGe in the rhombohedral structure. We have determined the equilibrium structures by total energy minimization. Then, we tested these equilibrium structures by calculating the interatomic forces.[22]

ENERGETICS AND STABILITY ANALYSIS FOR Si_n/Ge_n

All the structures we studied are grouped according to their tetragonal unit cells. For example, tetragonally strained Ge, zb-SiGe, Si_1/Ge_1, and Si_2/Ge_2 are studied in tetragonal unit cells including two atoms from each constituent. While Si_4/Ge_4 is studied in a tetragonal unit cell including four atoms from each constituent, because of different repeat periods Si_3/Ge_3 and Si_6/Ge_6 systems are treated in a unit cell consisting of 6 atoms each. The tetragonal unit cell and the corresponding Brillouin zone (BZ) are illustrated in Fig. 1. To assure commensurability, the lateral lattice constants of the pseudomorphic Si/Ge superlattices, $a_{\|}$, (in terms of cubic lattice constants) are taken to be equal to $a^0{}_{Si}$. This way a planar compressive strain, $\varepsilon = (a^0_{Ge} - a_{\|})/a^0_{Ge}$, is introduced. As seen in Fig. 1

the lattice constant perpendicular to epilayers, R_3 (or $a_\perp$ in terms of cubic lattice parameters) are determined by three types of interlayer spacings, *i.e.*, d_1(Si-Si), d_2(Si-Ge) and d_3(Ge-Ge). The change in the value of d_1 upon superlattice formation is, however, negligibly small. So fixing d_1 equal to $a^0_{Si}/4$, we concentrate on the tetragonally distorted Ge sublattice. The preferential accommodation of strain by Ge layers is consistent with higher values of force constants for Si relative to Ge. In the present fully relaxed calculations, d_2 is fixed by the minimization of the total energy of Si_1/Ge_1. The determination of the perpendicular lattice constant for Si_n/Ge_n with $n \geq 2$ required the optimization of the total energy with respect to two structural degrees of freedom (*i.e.*, optimization with respect to d_2 and d_3).

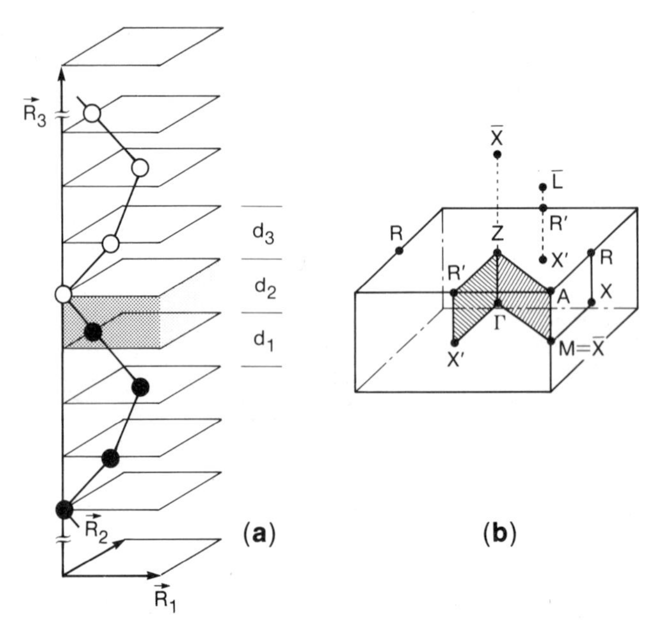

Fig. 1. (a) Tetragonal unit cell of the pseudomorphic Si_n/Ge_n superlattice with the filled and open circles denoting the position of Si and Ge atoms, respectively. $R_1 = R_2 = a^0_{Si}/\sqrt{2}$ and $d_1 = a^0_{Si}/4$. R_3, d_2 and d_3 are obtained by total energy minimization. (b) Superlattice Brillioun zone with the labelling of symmetry points. $\overline{X}$ and $\overline{L}$ correspond to X and L symmetry points of the BZ of Si with $\Gamma\overline{X} = 2\pi/a^0_{Si}$ and $X\overline{L} = \sqrt{3}\,\pi/a^0_{Si}$.

Calculated equilibrium values for the interlayer spacings are: $d_1 = 2.56$ a.u., $d_2 = 2.61$ a.u. and $d_3 = 2.70$ a.u. It should be noted that the calculated d_2 is close to the average value for Si and Ge, *i.e.* $(a^0_{Si} + a^0_{Ge})/8$; whereas the value of d_3 implies a tetragonal distortion, $\varepsilon_T = (a_\perp - a_\parallel)/a^0_{Ge}$ of ~5% in the Ge sublattice. For Ge epitaxially resricted to Si(001), *i.e.*, $a_{\|} = a^0_{Si}$ and denoted as Ge^{ep}, the interlayer spacing, $a_{\perp}/4$, is found to be nearly equal to d_3. The calculated equilibrium lattice constant of zb-SiGe, $a^0_{zb\text{-}SiGe}$, is 10.40 a.u.

In order to provide a consistent comparison of the energetics for the stability analysis, the total energy differences are calculated by using the same number of atoms, and similar unit cells. For example, the formation energy (or formation enthalpy at T = 0) of the strained Si_n/Ge_n superlattice is calculated

$$\Delta E^{f,s}(Si_n/Ge_n) = E^s_T(Si_n/Ge_n) - [E^0_T(Si_{2n}) + E^0_T(Ge_{2n})]/2$$

where the total energies of the constituent (strain-free) crystals $E^0_T(Si_{2n})$ and $E^0_T(Ge_{2n})$ are calculated in a unit cell corresponding to that of Si_n/Ge_n, but with the equilibrium lattice constants, (a^0_{Si} and a^0_{Ge}), fixed for bulk crystals. The formation energy of the (strain free) zb-SiGe is defined as

$$\Delta E^{f,0}(zb - SiGe) = E^0_T(zb - SiGe) - [E^0_T(Si) + E^0_T(Ge)]/2,$$

and calculated to be ~1.05 mRy per atom-pair. This is 0.27 mRy smaller than the value reported by Martins and Zunger.[8] The calculated formation energies of the strained superlattices are given in Table 1. Since the formation energy $\Delta E^f \geq 0$ indicates instability, the decomposition into constituent crystals (*i.e.*, segregation) is favored, as long as the kinetic of the reaction permits. Or else the strain energy accumulated in the Ge sublattice can be relieved by the creation of a misfit dislocation.

Table 1. Calculated formation energies of the pseudomorphic Si_n/Ge_n superlattices. The unit of energy is mRy per formula Si_n/Ge_n unit.

n = 1	n = 2	n = 3	n = 4	n = 6
1.46	1.73	5.04	7.03	9.29

From these *ab initio* results we develop a simplified picture of the superlattice formation energy, which has two major components. These are the strain and interfacial energies. The strain energy per atom in the Ge sublattice is defined as

$$\xi = [E^s_T(Ge_n) - E^0_T(Ge_n)]/n,$$

in terms of the energy, $E_T^s(Ge_n)$ of the epitaxial Ge^{ep}. It is found to be $\xi = 1.46$ mRy per Ge^{ep} atom. Using the calculated $\Delta E^{f,0}(zb - SiGe)$ we estimate the interfacial energy, $\chi = \Delta E^{f,0}(zb - SiGe)/2$, to be ~0.5 mRy per atom. Then, the formation energy of the superlattice is defined as

$$\Delta E^{f,s}(Si_n/Ge_n) \simeq 2\chi + n\xi .$$

The formation energies estimated from this simple model are compared with our *ab initio* results, displaying almost perfect match for $n \geq 3$. The formation energy of Si_7/Ge_7 is predicted to be ~11.3 mRy. As revealed from experimental works,[4] only six epitaxial Ge layers can grow on Si(001), and beyond that thickness misfit dislocations start to form. The above model yields an estimate for the energy barrier to form a misfit dislocation to be $9.8 < \Delta Q < 11.3$ mRy/cell.

PHASE TRANSITIONS IN SiGe ALLOYS

Having discussed the stability of superlattices, we next consider the stability of the SiGe alloy, which has also a positive energy of formation ($\Delta E^{f,s} > 0$). Therefore, a phase transition to a different structure, or decomposition into constituent crystals (*i.e.*, segregation) is expected, as long as the kinetics of the reaction permits. Such a segregation has not been observed so far due perhaps to relatively lower kinetic energy of atoms to overcome the activation barrier at low temperature. This situation at T = 0, however, may be different beyond some critical temperature, T_c, where the system at hand may be stabilized by the entropy of disorder leading to a negative enthalpy of formation. The strain energy build in the alloy upon the epitaxial growth appears to induce the observed phase transition. Ourmazd and Bean[7] proposed two variants for the ordered phase, both having the rhombohedral structure. These structures can be viewed as the bilayer segregation along the offset [111] direction in the strained SiGe alloy. In the first structure, denoted RH_1, Si(Ge) has three heteropolar (*i.e.*, Si-Ge) and one homopolar bond (*i.e.*, Si-Si, or Ge-Ge). The widely spaced planes along the [111] direction[5,7] are formed by atoms of the same kind. In the second type of structure, RH_2, Si(Ge) has one heteropolar, and three homopolar bonds, and pairs of widely spaced planes are formed by atoms of the opposite kind.

Our total energy calculations yield that the strained RH_1 structure has lower energy (*i.e.*, by 1 mRy/cell more favorable) relative to the disordered structure representing the $Si_{0.5}Ge_{0.5}$ alloy. This explains why the order- disorder transition can occur. The equilibrium structure (free from the strain imposed by the substrate) of RH_1 is optimized in the primitive (four atom) unit cell. The calculated total energy is found to be 0.75 mRy/atom-pair higher than the average equilibrium of bulk Si and Ge, however. According to this result, the ordered RH_1 structure is stable relative to the strained alloy, but is unstable relative to the decomposition into Si and Ge. It becomes even more unstable under the strain introduced by lateral epitaxy. Since we find the total energy of the ordered phase RH_2 to be higher than that of RH_1, both being under the strain of the lattice, we conclude that RH_2 is unfavorable as far as the order-disorder transition is concerned. An important result we find is that forming a strained Si_4/Ge_4 superlattice as described in Fig. 1 is even more favorable than the strained RH_1 phase with an energy benefit of ~0.5 mRy/atom-pair. The degree of instability of these phases in terms of their formation energies relative to the average energy of Si and Ge constituents is ordered as

$$\Delta E^{f,s}(Si_{0.5}Ge_{0.5}) > \Delta E^{f,s}(RH_2) > \Delta E^{f,s}(zb - SiGe) > \Delta E^{f,s}(RH_1) > \Delta E^{f,s}(Si_4/Ge_4) >$$

$$\Delta E^{f,0}(RH_1) > 0$$

Martins and Zunger[8] have studied the stability of the epitaxial RH_1 phase relative to the epitaxial zb-SiGe and the average of Si and Ge^{ep}. The formation energies they calculated for the strain-free zb-SiGe and RH_1 are in agreement with the present results. They found the epitaxial RH_1 has lower energy relative to the average energy of Si and Ge^{ep}. Our results $\Delta E^{f,s}(RH_1) > \Delta E^{f,s}(Si_4/Ge_4)$, however, indicate that the epitaxial RH_1 phase is under an additional strain of the lattice, as would be expected.

The ordering in the SiGe alloy induces changes in the electronic structure. While the higher and lower-lying valence and conduction band states are not much affected, the relative and absolute positions of the states split off from the band edges undergo significant changes. The ordering via segregation leads to even more significant changes, which will be discussed in Sec. VI.

GROWTH OF Ge ON Si(001) SURFACE

Following the stability analysis we would like to address the microscopic aspects of the growth process on the Si(001) substrate. Assuming a layer-by-layer growth of the adatoms (Si or Ge) in MBE, the important question we explore is the atomic structure of the topmost grown layer. This has relevance for the quality of the heterostructures, and the phase transformations, as well. Several studies in the past decade have established that the free Si(001) surface undergoes a (2×1) reconstruction with a dimer bond formation between adjacent surface atoms. The energy gain through this reconstruction is ~0.8 eV per surface atom.[23] Since the grown Si layers are strain-free, the same atomic configuration is expected to occur in the topmost grown Si layer. Accordingly, the last Si layer has to grow subsequent to the breaking of dimer bonds in the subsurface layer, but it itself forms dimer bonds eventually.

The dimer bond is also the essential feature of the Ge(001) surface, which shows a similar (2×1) reconstruction. However, owing to the lattice strain, the situation, whether the dimer bond does occur in the epitaxial Ge layers grown on the Si(001) surface, is not so obvious. The structure of the topmost grown Ge layer is explored by the total energy calculations. To this end we optimized the position of Ge atoms adsorbed on Si(001) surface as if they continue the bulk structure. The geometry optimization has fixed the Ge-Si interlayer distance close to that of the average of Si and unstrained Ge. In the second step, a possible (2×1) reconstruction geometry (with the dimer bond) for the strained Ge layer is constructed. The total energy of this reconstructed surface is found to be $\sim$0.5 eV (per surface Ge) lower than that of the ideal structure. This clearly indicates that the ideal structure undergoes a reconstruction, and may form dimer bonds (if there is no other structure with even lower energy). Unfortunately, the extended defects can not be considered within the present method. Following the above arguments, it is anticipated that Ge layers grow by breaking the existing dimer bonds, and form new ones in the last grown layer.

In view of the above energetics arguments, the reconstruction becomes important in the growth of the $Si_{1-x}Ge_x$ layer. Since Si-Si and Ge-Ge dimer bonds form, the Si-Ge dimer bond can also form. However, as discussed in the previous section, the Si-Ge heteropolar bond is energetically unfavorable as compared to the homopolar average, implying a preferential adsorption and a selective dimerization. Also note that the Si-Si dimerization releases $\sim$0.5 eV (per bond) more energy as compared to the Ge-Ge dimerization. The selective dimerization, on the other hand, may induce short range order, or domain structure in each grown layer. Taking the high mobility (and thus sizeable kinetic energy) of the adatoms during the growth, this seems to be likely. Note that the ordered RH_1 structure can be formed by short range rearrangements of Si and Ge in the grown layer.[5]

ELECTRONIC STRUCTURE

SCF-pseudopotential method within the local density functional approach underestimates the conduction band energies. The error is $\sim$0.5 eV for both Si and Ge. Therefore, by applying a constant shift to the conduction band energies the results of the present calculations can be used to explore the electronic structure of Si_n/Ge_n.

Silicon substrate has six minima along the $\bar{\Delta}$ directions giving rise to an indirect band gap of $\sim$1.1 eV (the symmetry points, and directions of the fcc BZ denoted by bar). The direct band gap is large, and the energy difference between the direct and indirect band gap, $\Delta E_g = E_{g,d} - E_{g,i}$, is 2.3 eV. Also Ge is an indirect band gap semiconductor, except that the conduction band minima occur at the eight $\bar{L}$ -points. The superlattice formation has dramatic effects on the electronic structure through zone folding,[24] lattice strain, and band lineup. The O_h symmetry of the bulk Si is broken by a superlattice formation (or by uniaxial strain), and the states with $\vec{k}$ parallel to [001] are folded along the ΓZ-direction of the superlattice BZ (see Fig. 1). This gives rise to the lowering of ΔE_g. In Fig. 2 the energy bands of Si_4 and Ge_4 calculated in the tetragonal BZ clearly show this effect. For example, in addition to four equivalent conduction band minima along ΓM direction

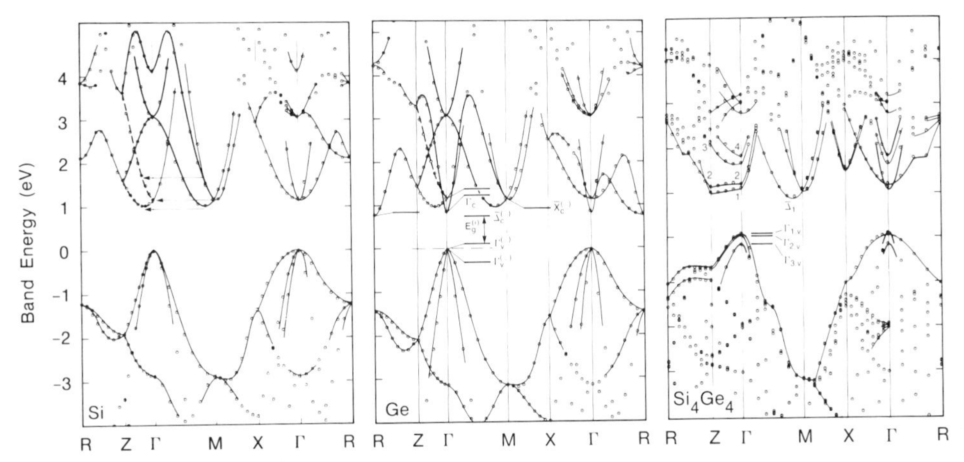

Fig. 2. Energy band structure of Si_4, Ge_4, and Si_4/Ge_4. Strain induced splittings, and shifts of the band states of the epitaxial Ge are shown by bars. The arrows indicate the zone folding states. The zero of energy is taken at the average of the three highest valence bands

(which correspond to $\overline{\Delta}_{min}$), remaining two minima along the superlattice direction are folded near the Γ point. The tetragonal strain in the Ge sublattice splits the bands at the top of the valence band by 0.41 eV, and raises the conduction band minima at the R-points (which are equivalent to the $\overline{L}$ -points of the fcc BZ), and lowers the minima along the ΓM-direction below the lowest conduction band state at R. The net effect is then the lowering of $E_{g,i}$ (which occurs now between Γ and $\overline{\Delta}_{min}$) and the increasing of ΔE_g in the strained Ge sublattice. In addition, owing to the superlattice formation Ge bands are raised by the band offset. Since the bands of sublattices are shifted relative to each other, some states in one sublattice can not find the matching partner in the adjacent sublattice. As a results, some states with $\vec{k}$ along superlattice direction are confined and lower their dimensionality. Some states have extended nature, but split due to different potentials in Si and Ge sublattices. All these effects are taken into account self-consistently in our calculations.

The confined character of the conduction band states begin to appear for the superlattice n = 3, and increases with increasing n. In Fig. 2 the flat bands along ΓZ of Si_4/Ge_4 evolve from the lowest conduction band states of Si, $\overline{\Delta}_{c,1}$, but confined in the Si sublattice. Their 2D-charecter is apparent along the XΓ direction, where they form subbands. While the states of the highest band along RZ deep in the valence band are localized in the the Ge sublattice, those of the lowest conduction band are confined in Si.

The charge density contour plots illustrated in Fig. 3 reveal the degree of confinement of the superlattice states. The topmost three states of the valence band have

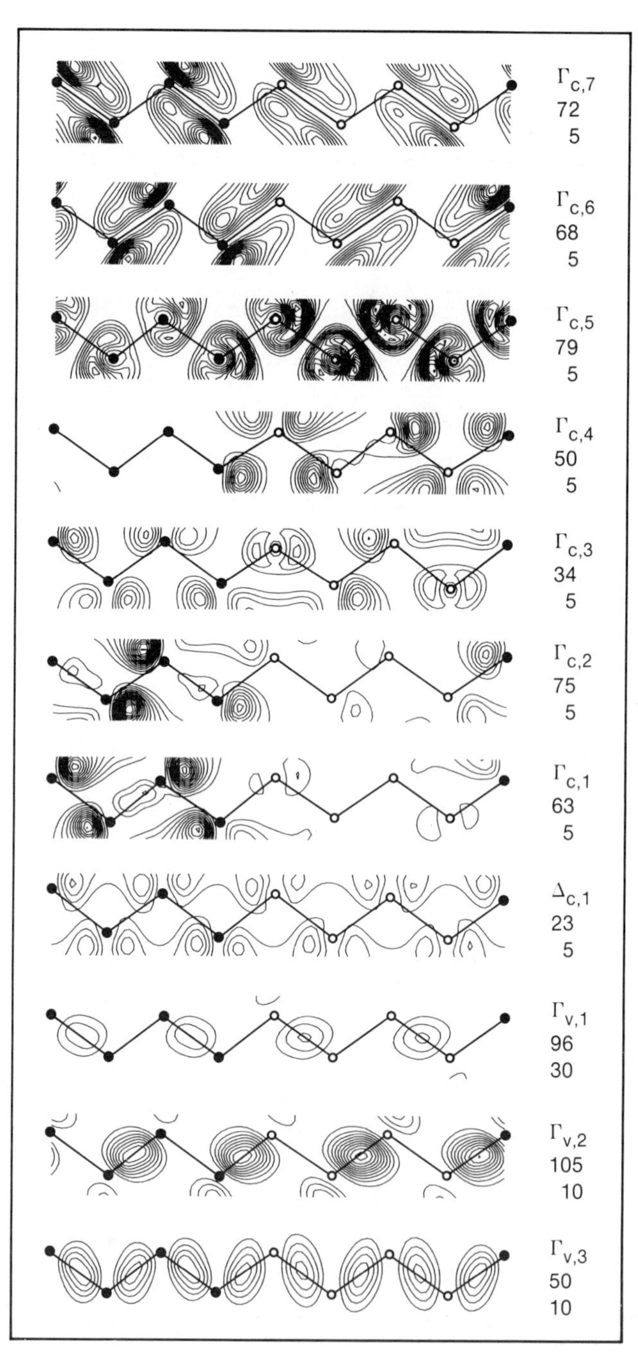

Fig. 3. Contour plots of the states for Si_4/Ge_4. $\Gamma_{v,n}$ and $\Gamma_{c,n}$ indicate valence and conduction band states at Γ, respectively. $\Delta_{c,1}$ is the lowest conduction band state along ΓM direction. Maximum value of charge density and contour spacings are given for each panel in units of 10^{-4} a.u.

higher weight in Ge, but are not fully localized, because of the quantum size effect (*i.e.*, due to the small width of the potential well and also small hole effective mass). Similarly, the lowest conduction band states near $\Delta_{c,1}$ have extended nature, because of the small transverse electron effective mass, $m^*_t = 0.19m_0$ leading to a significant quantum size effect. However, since the longitudinal effective mass, $m^*_l \cong m_0$, the first two conduction bands along ΓZ are confined in Si.

In Fig 4. we present the states of Si_n/Ge_n, which are relevant to our discussion. As seen the band gaps are indirect for all n. However, the energy difference between the

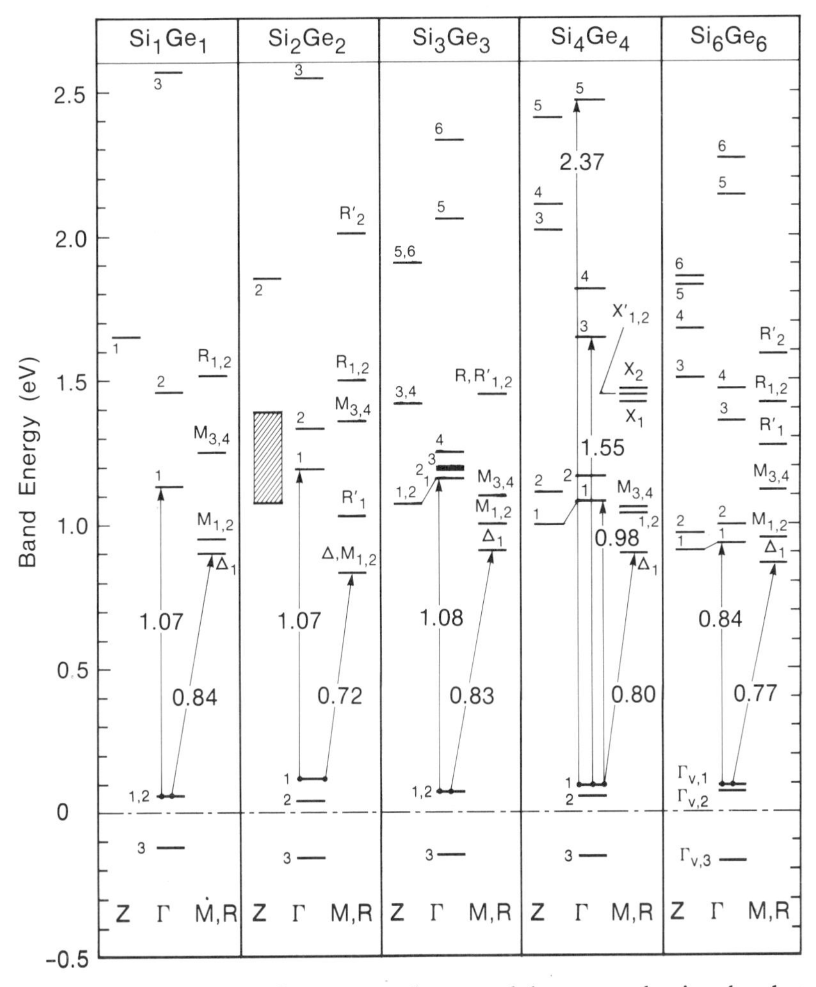

Fig. 4. Energy level diagram for upper valence and lower conduction band states of Si_n/Ge_n. Direct and indirect transitions are shown by arrows. The average energy of the three highest valence bands is taken to be zero of energy.

direct and indirect gap, ΔE_g, which is 2.3 eV in bulk Si, has reduced to 0.07 eV in Si_6/Ge_6. It is also seen that $E_{g,d}$ has an inverse proportioanality with n (*i.e.*, with the width of the quantum well) the exponent being between 1 and 2. Interestingly, by going from n = 3 to n = 6, the indirect band, itself is reduced, but the minimum gap is obtained at n = 2. Three direct transitions of Si_4/Ge_4 shown in Fig. 4 (0.98, 1.55 and 2.37 eV) are in agreement with those observed in electroreflectance spectroscopy[4] (0.76 $\mp$ 0.13, 1.25 $\mp$ 0.13 and 2.31 $\mp$ 0.12 eV)

CONCLUSIONS

The SCF-total energy calculations indicate that the strained Si_n/Ge_ns uperlattices are unstable relative to the decomposition into constituents. The energy barrier to destroy the epitaxy by misfit dislocation is predicted to be ~10 mRy/cell. The SiGe alloy, which has also a positive enthalpy of formation at T = 0, lowers energy by transforming into an ordered, and strained rhombohedral structure, which is also only a metastable structure. The order-disorder transformation induces changes in the electronic structure.

All strained Si_n/Ge_n ($1 < n \leq 6$) superlattices are found to be indirect band gap semiconductors. However, the minimum gap, as well as the direct-indirect gap difference are reduced with increasing n. Our results indicate the possibility that a SiGe superlattice can be made a direct band gap semiconductor by varying the sublattice periodicities (n and m), and also the lateral lattice constant.[13] It is suggested that the 2-D impurity bands to be produced by δ-doping in sublattices is another possibility to make the SiGe system a direct band gap material.

Conduction band states confined in Si, and the topmost valence band states localized in Ge suggest a type-II staggered band lineup. However, the finding that the lowest conduction band states have extended character, is a situation never considered in the band offset problem. This alone suggests that simple theories based on the effective mass approximation can not describe the electronic structure fully, when the superlattice periodicity becomes small. Moreover, because of the lowest conduction band states of extended character, caution has to be exercised in calculating the shallow impurity states within the effective mass approximation applied in quantum well structure.

REFERENCES

1. E. Kasper, H. J. Herzog, and H. Kimble, *App. Phys.* **8**, 199 (1975).
2. J. C. Bean, *J. Vac. Sci. Technol.* **A1**, 540 (1983).
3. A. T. Fiory, J. C. Bean, L. C. Feldman, and I. K. Robinson, *J. App. Phys.* **56**, 1227 (1984).
4. T. P. Pearsall, J. Bevk, L. C. Feldman, J. M. Bonar, J. P. Mannaerts, and A. Ourmazd, *Phys. Rev. Lett.* **58**, 729 (1987).

5. S. Ciraci and I. P. Batra, *Phys. Rev. Lett.* **58**, 2114 (1987).
6. S. Froyen, D. M. Wood, and A. Zunger, *Bull. of Am. Phys. Society* **32**, 906 (1987).
7. A. Ourmazd and J. C. Bean, *Phys. Rev. Lett.* **55**, 765 (1985).
8. J. L. Martins and A. Zunger, *Phys. Rev. Lett.* **56**, 1400 (1986).
9. W. A. Harrison, *J. Vac. Sci. Technol.* **14**, 1016 (1987); *ibid* **B3**, 1231 (1985)
10. J. Tersoff, *Phys. Rev. B.* **30**, 4874 (1984).
11. G. Margaritando, A. D. Katnani, N. G. Stoffel, R. R. Daniels, and Te-Xiu Zhao, *Solid State Commun.* **43**, 163 (1982).
12. C. H. Van de Walle and R. M. Martin, *J. Vac. Sci. Technol.* **B3**, 1256 (1985); *ibid: Phys. Rev.* **B34** 5621 (1986).
13. G. Abstreiter, H. Brugger, T. Wolf, H. Jorke, and H. Herzog, *Phys. Rev. Lett.* **54**, 2441 (1985).
14. Su-Huai Wei and A. Zunger, *Phys. Rev. Lett.* **59**, 144 (1987).
15. I. Morrison, M. Jaros, and K. B. Wong, *Phys. Rev.* **B35**, 9693 (1987).
16. S. Ciraci and I. P. Batra, *Phys. Rev.* **B35**, 1225 (1987).
17. I. P. Batra, S. Ciraci, and J. Nelson, *J. Vac. Sci Technol* **B5**, 1300 (1987).
18. M. Schluter, J. R. Chelikowsky, S. G. Louie and M. L. Cohen, *Phys. Rev.* **B12**, 4200 (1975).
19 J. Ihm, A. Zunger and M. L. Cohen, *J. Phys.* **C12**, 4409 (1979); M. T. Yin and M. L. Cohen, *Phys. Rev. Lett.* **45**, 1004 (1980).
20. C. B. Bachelet, D. R. Hamann and M. Schluter, *Phys. Rev.* **B26**, 419 (1982).
21. D. M. Ceperley and B. J. Alder, *Phys. Rev. Lett.* **45**, 566 (1980).
22. See for the method and related references: I. P. Batra, S. Ciraci, G. P. Sirivastava, J. S. Nelson, and C. Y. Fong, *Phys. Rev.* **B34**, 8246 (1986).
23. F. F. Abraham and I. P. Batra, *Surf. Sci.* **163**, L572 (1985).
24. L. Esaki and R. Tsu, *IBM J. Res.Develop.* **14**, 61 (1970).

THEORY OF RAMAN SCATTERING FROM PLASMONS POLARITONS IN GaAs/$Al_xGa_{1-x}As$ SUPERLATTICES

M. Babiker, N.C. Constantinou and M.G. Cottam*

Physics Department
University of Essex,
Colchester CO4 3SQ, UK

ABSTRACT

When layers of the alloy in epitaxially-grown GaAs/$Al_xGa_{1-x}As$ superlattices are modulation-doped with Si during the growth process, the donor electrons lead to the assembly of a periodic array of electron layers separated by dielectric layers. Such a system of electrons is capable of exhibiting plasma oscillations, the existence of which has been substantially confirmed by Raman measurements. The theory pertaining to this experimental situation is examined here using linear-response methods which we apply to two theoretical models of the superlattice. The first model assumes a strictly two-dimensional picture of the electron layers and this turns out to be a good approximation for a fairly comprehensive analysis of the various aspects exhibited by the first results due to Olego et al. The second model incorporates the effects of a finite thickness for the electron layers. This is shown to lead to better agreement with experiment. The theory yields predictions on the energy, integrated intensity and shape of the observed Raman peak. Furthermore, it leads to the prediction of a second plasmon branch which we suggest may have already been exhibited by experimental results due to Olego et al. and Sooryakumar et al.

1. INTRODUCTION

Impurity donors or acceptors are the main source of free carriers (electrons or holes) in semiconductor systems. In the most widely discussed superlattice systems, namely GaAs/$Al_xGa_{1-x}As$ superlattices, the carriers are confined primarily in the GaAs layers, and are spatially separated from the impurity (Si) donors. This is achieved by modulation doping[1] which ensures that the impurities are distributed only in the $Al_xGa_{1-x}As$ layers. The conduction band profile of the superlattice is effectively a series of GaAs quantum wells to which the donor electrons become transferred. The electrons thus form an array of quasi-two-dimensional layers separated by essentially dielectric layers of the alloy $Al_xGa_{1-x}As$. Areal electron densities in excess of 5×10^{15} m^{-2} have been achieved in this manner using molecular beam epitaxy (MBE).

* Present address: Physics Department, University of Western Ontario, London, Ontario NA6 3K7, Canada.

Plasma oscillations are only one feature among a host of new properties that such a system of electrons can exhibit. Experimental work on the exposition of the elementary excitations of quasi-two-dimensional systems of electrons began after a suggestion by Burstein et al.[2] that resonant inelastic light scattering may be a useful tool of investigation in this context. The first experimental results on light scattering by the plasma oscillations in superlattices were reported by Olego et al.[3] These researchers were able to identify clearly a low-energy plasmon peak as a Stokes' shift in a low-temperature polarised Raman scattering experiment performed on two separate $GaAs/Al_xGa_{1-x}As$ superlattices. Their work also gave rise to the first experimental dispersion curve showing the plasmon energy as a function of the 'in-plane' wavevector along the superlattice interfaces. Since then, further light scattering measurements were made on similar superlattices and on finite multilayers by Fasol et al.,[4] Pinczuk et al.[5] and by Sooryakumar et al.[6]

Theoretical work on the plasma oscillations for such an array in fact preceded the experimental work. It began with Fetter's[7] treatment of strictly two-dimensional electronic sheets. Fetter's dispersion relation was later derived by Das Sarma and Quinn[8] and by Bloss and Brody,[9] among others, also assuming a sheets model for the superlattice electronic system.

It is quite significant that the experimental curve obtained by Olego et al.[3] substantially confirms the theoretical dispersion relation derived kinematically on the basis of the sheets model, especially at long wavelengths. It appears that each of those superlattice systems provides a good practical approximation to the idealised electron sheets model, at least as regards the low-energy plasmon branch observed by Olego et al.[3] It therefore became clear that it is useful to investigate further the consequences of this strictly two-dimensional picture as a model in its own right as well as examining more realistic models.

An important aspect of the problem is the development of dynamical theories to incorporate other characteristics exhibited by Raman measurements of the type performed by Olego et al.[3] These include the relative intensities, polarisation selection rules and the shape of the Raman peak, in addition to the dispersion relation which can be derived on kinematical grounds alone. Some of these light scattering characteristics have received attention in the theoretical treatments by Hawrylak et al.,[10] Jain and Allen[11,12] and by Katayama and Ando.[13] The dynamical aspects can in fact be handled efficiently by linear-response techniques, which provide a well-defined theoretical apparatus easily adaptable to deal with the situation in the context of superlattices.[14-18] Here we show how linear response theory can be used to derive the light scattering characteristics for the type of experiment performed by Olego et al.[3] assuming the sheets model. We show that in addition to recovering the correct form of the dispersion relation for this model, our theory provides insight into the scattering mechanism, and rigorously describes the way in which the plasmon spectral lineshape and integrated intensity depend on the polarisation directions and the scattering geometry.

Olego et al.[3] have also noted that there are small differences between the theoretical dispersion curve and the experimental one. Camley and Mills[19] and Wasserman and Lee[20] attributed the discrepancy to the assumption of strictly two-dimensional electron layers. They showed that when the derivation incorporates finite electron layers a better agreement with the experimental curve is achieved. A linear response theory corresponding to the finite electron layer model gives rise to corresponding changes in the light scattering characteristics, as we have shown in our recent work.[16-18] Furthermore the finite thickness theory predicts a second branch in the low energy region which, we suggest, may have already been observed by Olego et al.[3] and by Sooryakumar et al.[6]

In §2 we systematise the procedure for linear-response theory applicable to light scattering, emphasising the scattering mechanism and the way in which response or Green functions enter the light scattering cross section. This also leads to derivations of the dispersion relation and the light scattering intensity. In §3 we derive Greens functions appropriate for the sheets model, investigate the theoretical predictions for light scattering and compare with the results by Olego et al.[3] In §4 we apply the formalism to the case of the finite electron layer model and compare the various results with the experimental data by Olego et al.[3] and Sooryakumar et al.[6] Section 5 contains our conclusions and further comments.

2. LIGHT SCATTERING AND LINEAR-RESPONSE THEORY

A typical experimental arrangement involving polarised Raman scattering by superlattice excitations is shown in fig. 1. This shows the incident and scattered light beams in the same vertical plane (xz plane) at angles θ and ϕ respectively to the z axis. The shaded regions represent the GaAs layers containing the electrons. In general the layer thicknesses are assumed to be d_1 for AlGaAs layers and d_2 for the GaAs layers, thus creating a periodicity of length $D = d_1 + d_2$. In the sheets model $d_2 = 0$ and in the finite thickness model both d_1 and d_2 are finite.

The general expression for the scattering cross section is written in the standard form (e.g. see refs. 21-22)

$$\sigma = LA_o \cos\phi[n(\omega)+1)]\,\mathrm{Im}\sum_{\mu,\nu,\nu',\delta,\delta'}\int dq\int dq' \frac{\langle\langle \chi^*_{\delta'\nu'}(q');\chi_{\delta\nu}(q)\rangle\rangle_{K_{\parallel},\omega}}{(q'+K^*_z)(q+K_z)}$$
$$\times\ (f^{\nu}g^{\mu\delta})(e^{\mu}_S)^2 e^{\nu}_I e^{\nu'}_I (f^{\nu'}g^{\mu\ \delta'})^* \tag{1}$$

where A_o is the constant cross-sectional area of the scattered beam, $\underline{e}_I$ and $\underline{e}_S$ are unit polarisation vectors of the incident and scattered

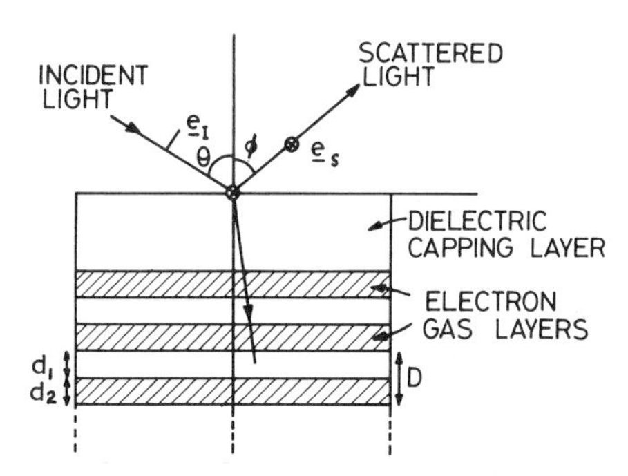

Figure 1. The assumed light scattering geometry including a schematic illustration of the superlattice

light at frequencies ω_I and ω_S, respectively and the summation is over Cartesian components. The L is a proportionality factor which is effectively a constant across a given spectrum and the f and g coefficients are standard factors which describe the transmission of the incident and scattered beams, respectively, through the sample surface. $n(\omega)$ is the Bose-Einstein's factor:

$$n(\omega) = [\exp(h\omega/k_B T) - 1]^{-1} \tag{2}$$

where T is the temperature. The integrations over q and q' refer to real wavevector components perpendicular to the surface. K_z is the light scattering wavevector in the z-direction; it is complex in general (i.e. $\mathrm{Im}\, K_z \neq 0$) corresponding to absorption of light as it penetrates the sample. $\underline{K}_\parallel$ is the in-plane wavevector taken along the x axis. Assuming $\omega_I \simeq \omega_S$ we can write

$$K_z = (\frac{2\pi}{\lambda})\ (\varepsilon-\sin^2\theta)^{\frac{1}{2}} + (\varepsilon-\sin^2\phi)^{\frac{1}{2}} \quad ; \; K_\parallel = (\frac{2\pi}{\lambda})(\sin\theta-\sin\phi) \tag{3}$$

where λ is the wavelength of the incident light in vacuo. Note that $K_\parallel$ is real in conformity with the translational invariance symmetry of the system along the layers. ε is the complex dielectric function at wavelength λ.

The cross section in eq. (1) contains Green functions involving components of the dynamic susceptibility χ where (e.g. see refs. 21 and 23)

$$\chi_{\delta\nu} = \sum_\gamma b_{\delta\nu\gamma} E^*_\gamma(q) \tag{4}$$

where $\underline{E}(q)$ is the electric field associated with the excitation at wavevector component q. Here we restrict attention to the case of a medium with cubic symmetry (as in GaAs-based systems). In this case the electro-optic coefficients $b_{\delta\nu\gamma}$ have a fixed value b if δ, ν and γ are all different, and are zero otherwise.[21] Thus the evaluation of the cross section requires the electric field Green functions $\langle\langle E_\mu(q');E^*_\nu(q)\rangle\rangle_{K_\parallel,\omega}$. It is possible in the present application to express the result in terms of the related functions $G_{\mu\nu}(K_\parallel,q)$ defined by

$$\langle\langle E_\mu(q');E^*_\nu(q)\rangle\rangle_{K_\parallel,\omega} = \frac{(2\pi)^3}{\varepsilon_o \overline{V}}\,\delta(q-q')G_{\mu\nu}(K_\parallel,q) \tag{5}$$

where $\overline{V}$ is the sample volume and ε_o is the premittivity of free space.

For simplicity we examine the case where the incident and scattered beams are in the x-z plane (i.e. perpendicular to one of the axes of cubic symmetry of the crystal). For comparison with some of the experimental data, this needs to be generalised as we discuss later. We also assume that the incident and scattered beams are transverse magnetic and transverse electric respectively, implying $\underline{e}_I = (\cos\theta,0,\sin\theta)$ and $\underline{e}_S = (0,1,0)$, although other polarisations are possible (ref. 18). We find

$$\sigma \propto (\frac{|b|^2}{K''_z})A_o\cos\phi[n(\omega)+1]|g^{yy}|^2\mathrm{Im}[R_1(\theta,\phi)] \tag{6}$$

$$R_1(\theta,\phi) = |f^x|^2\cos^2\theta G_{zz}(K_\parallel,-K^*_z) + |f^z|^2\sin^2\theta G_{xx}(K_\parallel,-K^*_z)$$

$$+ \sin\theta\cos\theta[(f^x)^* f^z G_{xz}(K_\parallel,-K^*_z) + (f^z)^* f^x G_{zx}(K_\parallel,-K^*_z)]. \tag{7}$$

K''_z denotes $\mathrm{Im}(K_z)$ and represents the reciprocal penetration depth of the light. Also the f and g factors appearing in eqs. (6) and (7) are given by

$$f^x = \frac{2(\varepsilon-\sin^2\theta)^{\frac{1}{2}}}{[\varepsilon\cos\theta + (\varepsilon-\sin^2\theta)^{\frac{1}{2}}]} \quad ; \quad f^z = \frac{2\cos\theta}{[\varepsilon\cos\theta + (\varepsilon-\sin^2\theta)^{\frac{1}{2}}]} \quad ; \tag{8}$$

$$g^{yy} = [\cos\phi + (\varepsilon-\sin^2\phi)^{\frac{1}{2}}]^{-1} \ . \tag{9}$$

As we see later the Green functions G_{xx}, G_{xz}, G_{zx} and G_{zz} have poles from which we obtain directly the dispersion relations. In general we can write

$$G_{\mu\nu} = (U_{\mu\nu} + \frac{V_{\mu\nu}}{\Delta}), \tag{10}$$

so that $\Delta = 0$ yields the dispersion relation.

The integrated light scattering intensity for a Raman peak at frequency ω_o can be deduced from equation (7). This is best done by considering the limit in which the optical absorption becomes negligible (corresponding to $\mathrm{Im}\ \varepsilon \to 0$). The resonance peak at ω_o is then represented by a delta function and its strength is just the integrated intensity. The result is

$$I(\omega_o) = |b|^2 A_o \cos\phi[n(\omega_o)+1]|g^{yy}|^2 \Big\{ (\frac{d\Delta}{d\omega})^{-1} \Big[|f^x|^2 \cos^2\theta V_{zz} + |f^z|^2 \sin^2\theta V_{xx} + \sin\theta\cos\theta(f^{x*}f^z + f^{z*}f^x)V_{xz} \Big] \Big\}_{\substack{\omega=\omega_o \\ q=K_z^o}} \ . \tag{11}$$

The evaluation of the Green functions $G_{\mu\nu}$ (and the related functions $V_{\mu\nu}$) is the only additional requirement for the complete specification of the cross section and the integrated intensity. The derivations are considered in the context of the superlattice model as we now discuss.

3. THE SHEETS MODEL

In this model the superlattice is taken to consist of sheets of electrons occupying infinitesimally thin GaAs layers (i.e. $d_2 = 0$ in figure 1), separated by layers of an isotropic dielectric of thickness d_1. The derivation of the Green functions are based on the equation

$$\nabla^2 \underline{E} + (\varepsilon_b\omega^2/c^2)\underline{E} = -(\omega^2/\varepsilon_o c^2)\underline{P} - (1/\varepsilon_o\varepsilon_b)\underline{\nabla}(\underline{\nabla}.\underline{P}) \tag{12}$$

where $\underline{E}$ is the electric field vector and ε_b is the background dielectric constant. $\underline{P}$ is an externally applied polarisation field in the form

$$\underline{P} = \underline{P}^0 \exp[i(qz + K_{\parallel}x - \omega t)], \tag{13}$$

with ω the frequency. A complete specification of the electric field vector in the superlattice is obtained by writing the solutions of eq. (12) as a sum of a homogeneous part and a particular integral, making use of the transversality condition on the homogeneous part, imposing the electromagnetic boundary conditions at one of the interfaces and applying Bloch's theorem on the macroscopic periodicity. The boundary conditions at any interface are the continuity of tangential $\underline{E}$ and the jump condition on the magnetic field due to the current in the electron sheets. Thus we require

$$\begin{aligned} &E_x^{(1)} = E_x^{(2)} , \\ &(\varepsilon_o c^2/i\omega)\hat{z}x(\underline{\nabla} \times \underline{E}^{(2)} - \underline{\nabla} \times \underline{E}^{(1)}) = s(E_x^{(2)}\hat{x} + E_y^{(2)}\hat{y}) \end{aligned} \tag{14}$$

where superscripts (1) and (2) label the $Al_xGa_{1-x}As$ layers adjacent to any interface; $\hat{x}$, $\hat{y}$ and $\hat{z}$ are unit vectors and s is defined by

$$s = in_o e^2/m^*\omega \tag{15}$$

where n_o is the areal electron density and e and m* are the electronic charge and effective mass. Bloch's theorem is such that

$$\underline{E}(z+D) = e^{iqD}\underline{E}(z) \tag{16}$$

where D is the superlattice periodicity length (in this model $D = d_1$). The appropriate Green functions are obtained in the form

$$G_{xx}(K_\parallel, q) = \frac{k_z^2}{\varepsilon_b(q^2-k_z^2)}\left\{1 - \frac{\Omega^2 k_z^2}{\omega^2(q^2-k_z^2)\Delta}\right\}; \tag{17}$$

$$G_{xz}(K_\parallel, q) = G_{zx}(K_\parallel, q) = -\frac{K_\parallel q}{k_z^2}\, G_{xx}(K_\parallel, q); \tag{18}$$

$$G_{zz}(K_\parallel, q) = \frac{1}{\varepsilon_b(q^2-k_z^2)}\left\{\frac{\varepsilon_b\omega^2}{c^2} - q^2 - \frac{\Omega^2 K_\parallel^2 q^2}{\omega^2(q^2-k_z^2)\Delta}\right\} \tag{19}$$

where

$$k_z = (\varepsilon_b\omega^2/c^2 - K_\parallel^2)^{\frac{1}{2}} \tag{20}$$

and Ω is a characteristic frequency defined by

$$\Omega = (n_o e^2/\varepsilon_o\varepsilon_b m^* D)^{\frac{1}{2}}. \tag{21}$$

The function Δ appearing above is explicitly given by

$$\Delta = 1 - \left(\frac{\Omega^2 k_z D \sin k_z D}{2\omega^2[\cos qD - \cos k_z D]}\right) \tag{22}$$

For further details of the procedure leading to these results see refs. 14 and 15. The decomposition put forward in eq. (10) follows immediately; the poles of the Green functions are those satisfying $\Delta = 0$. This corresponds to the dispersion relation which can be obtained kinematically for p-polarised (i.e. transverse magnetic) modes. At low frequencies such that $\omega < \Omega$ we have from eq. (20) $k_z \simeq iK_\parallel$ and the condition $\Delta = 0$ becomes equivalent to

$$\omega^2 = \frac{\Omega^2 K_\parallel D \sinh K_\parallel D}{2[\cosh K_\parallel D - \cos qD]}, \tag{23}$$

which is precisely the well known dispersion relation (e.g. ref. 8) corresponding to the non-retarded limit. For long wavelengths such that $K_\parallel D \ll 1$ we have the linear-dependence form

$$\omega = K_\parallel\left(\frac{\Omega^2 D^2}{1 - \cos qD}\right)^{\frac{1}{2}}. \tag{24}$$

The cross section based on this sheets model is obtained by substituting from eqs. (17) - (19) into eq. (7) and the corresponding light scattering intensity similarity by substitution in eq. (11) of the appropriate $V_{\mu\nu}$.

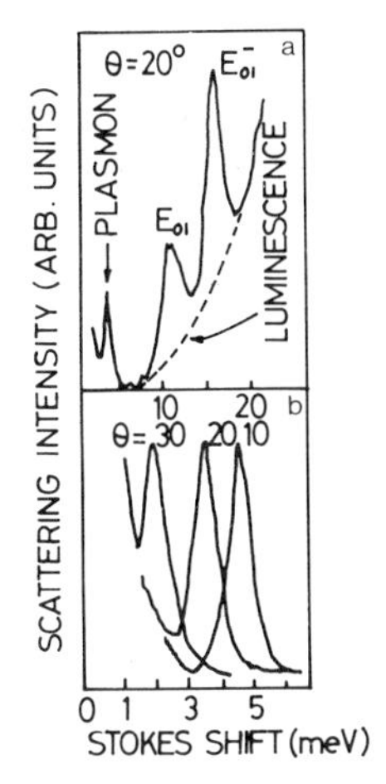

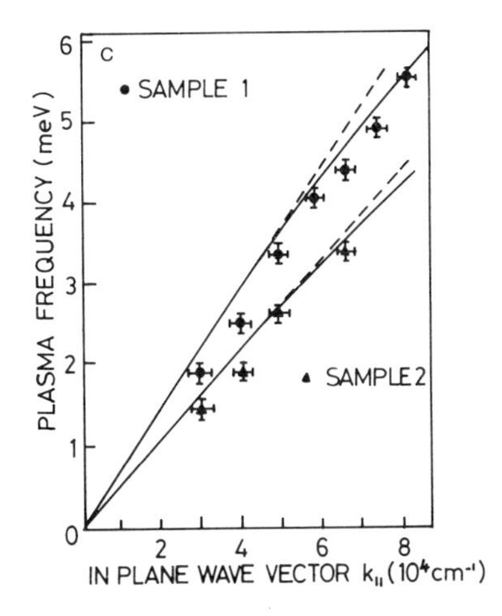

Figure 2 (a) Inelastic light scattering spectrum (b) Shifting of the low energy plasmon peak to lower energies with increasing θ and (c) dispersion curves for the two samples; the solid lines were calculated using eq. (23) and the dashed lines represent the evaluations of eq. (24). These results are those reported by Olego et al. (ref. 3) courtesy of Physical Review.

In their experiment, Olego et al.[3] used a 90° geometry which with reference to Fig. 1, is such that $\theta+\phi = 90^{\circ}$. They also used a dye laser source with $\lambda \simeq 780$ nm. Their results for the two samples labelled 1 and 2 are shown in figure 2.

In order to compare the predictions of our theory with those results we consider for definiteness the case of sample 1. The appropriate parameters are

$$D = 89 \text{ nm};\ n_o = 7.3 \times 10^{15}\ \text{m}^{-2};\ m^* = 6.37 \times 10^{-32}\ \text{kg};$$

$$\varepsilon_b \simeq 13.1;\ T = 10\text{K}. \qquad (25)$$

These yield for the parameter Ω (eq. 21)

$$\hbar\Omega \simeq 11.1 \text{ meV}. \qquad (26)$$

At the wavelength $\lambda = 780$ nm the complex dielectric constant entering K_z (eq. 3) is approximately $12.96 + 0.5i$ as deduced from ref. 24.

The predicted lineshapes for the three different values of θ are shown in figure 3 where each curve is plotted in the same arbitrary units. For $\theta = 20^{\circ}$ the resonance occurs at $h\omega \simeq 3.8$ meV and its width (FWHM) is of the order 0.5 meV. The finite width arises from the opacity broadening, i.e. optical absorption allows a spread of values of the z-component of the plasmon wavevector and correspondingly there is a spread of frequencies due to the dispersion of the mode. When compared with the results of Olego et al.[3] shown above, the agreement is indeed quite good.

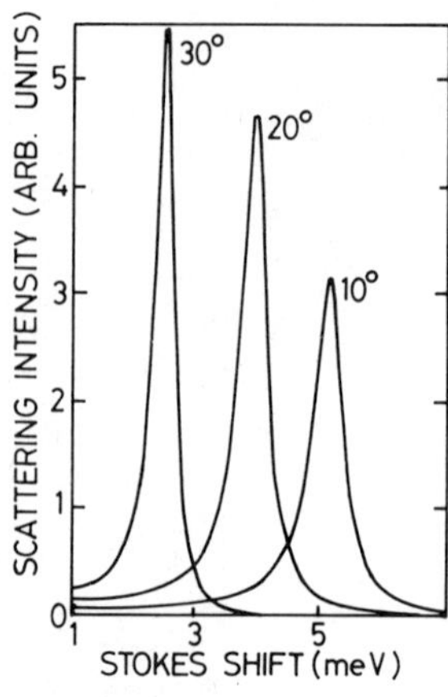

Figure 3. Theoretical spectra for light scattering from the superlattice bulk plasmons for a right-angle scattering geometry with λ = 780 nm. See the text for the assumed values of other parameters. The values on the curves are of θ.

It was not possible to compare the integrated light scattering intensities with those deduced from figure 2 since the three curves shown were plotted by Olego et al. in uncorrelated units. Nevertheless theoretical predictions on the relative intensities can be made on the basis of eq. (11). For brevity we consider discussions on relative intensities only in the context of the finite thickness model in the next section.

4. FINITE ELECTRON LAYER MODEL

In this model the superlattice is taken to consist of alternate layers of material 1 ($Al_xGa_{1-x}As$) and material 2 (GaAs), both with finite layer thicknesses d_1 and d_2 are in figure 1. We denote the dielectric functions by $\varepsilon_1(\omega)$ and $\varepsilon_2(\omega)$ such that

$$\varepsilon_1(\omega) = \varepsilon_b \quad ; \quad \varepsilon_2(\omega) = \varepsilon_b\left[1 - \frac{\omega_p^2}{\omega(\omega+i\gamma)}\right] \tag{27}$$

where as before ε_b is the background dielectric constant, and

$$\omega_p = (ne^2/\varepsilon_o\varepsilon_b m^*)^{\frac{1}{2}} \tag{28}$$

is the effective bulk plasma frequency of the electrons confined within the GaAs layers. Here n is an effective <u>volume</u> density of the electrons and γ is a damping parameter introduced in the familiar manner.[25]

The derivation of the Green functions for this model follows similar steps to those in the previous section. The solutions of the basic electric field equations must, however, be found in both types of layer and the boundary conditions now require the continuity of tangential $\underline{E}$ and tangential $\underline{H}$ at two successive superlattice interfaces. We obtain the following Green functions

$$G_{xx}(K_\parallel, q) = \frac{1}{D(q^2-k_1^2)}\left[\frac{k_1^2 d_1}{\varepsilon_1} + \frac{\varepsilon_o \Gamma_{xx}(1,2)}{\Delta}\right] + (1 \leftrightarrow 2) \quad ; \tag{29}$$

$$G_{xz}(K_\parallel, q) = G_{zx}(K_\parallel, q) = \frac{1}{D(q^2-k_1^2)}\left[-\frac{K_\parallel q d_1}{\varepsilon_1} + \frac{\varepsilon_o \Gamma_{xz}(1,2)}{\Delta}\right] + (1 \leftrightarrow 2); \tag{30}$$

$$G_{zz}(K_\parallel, q) = \frac{1}{D(q^2-k_1^2)}\left[\left(\frac{\varepsilon_1 \omega^2}{c^2} - q^2\right)\frac{d_1}{\varepsilon_1} + \frac{\varepsilon_o \Gamma_{zz}(1,2)}{\Delta}\right] + (1 \leftrightarrow 2) \tag{31}$$

where $(1 \leftrightarrow 2)$ denotes corresponding terms with the labels 1 and 2 interchanged. k_1 and k_2 are defined by

$$k_j = \left(\frac{\omega^2 \varepsilon_j}{c^2} - K_\parallel^2\right)^{\frac{1}{2}}. \tag{32}$$

Δ is the analogue of that in the previous section. Here we have

$$\Delta = \cos qD - \cos k_1 d_1 \cos k_2 d_2 + \tfrac{1}{2}\left(Z + \frac{1}{Z}\right)\sin k_1 d_1 \sin k_2 d_2 \tag{33}$$

with

$$Z = (\varepsilon_1 k_2 / \varepsilon_2 k_1). \tag{34}$$

The functions Γ_{xx}, Γ_{xz} and Γ_{zz} are complicated functions of $K_\parallel$ and q and for brevity we do not quote them here. The reader is referred to refs. 17 and 18 for these and for further details of the derivation.

When the Green functions (29) - (31) are substituted in eqs. (6) and (7) we obtain the light scattering cross section for this model. Thus we can in principle obtain all the light scattering characteristics of the Olego et al. experiment using the parameters (eq. 25) for their sample 1. However the above formalism simplifies for the case of small ω or $K_\parallel$ large (i.e. the non-retarded limit). For $\omega^2\varepsilon_j/c^2 \ll K_\parallel^2$ (j = 1,2) both k_1 and k_2 become equal to iK , then the dispersion relation $\Delta = 0$ becomes

$$\cos qD = \cosh K_\parallel d_1 \cosh K_\parallel d_2 - \tfrac{1}{2}\left(\frac{\varepsilon_1}{\varepsilon_2} + \frac{\varepsilon_2}{\varepsilon_1}\right)\sinh K_\parallel d_1 \sinh K_\parallel d_2. \tag{35}$$

An alternative form of this dispersion relation is (e.g. ref. 19)

$$\frac{\varepsilon_1}{\varepsilon_2} = -\beta \pm (\beta^2 - 1)^{\frac{1}{2}} \tag{36}$$

where

$$\beta = \frac{\cosh K_\parallel d_1 \cosh K_\parallel d_2 - \cos qD}{\sinh K_\parallel d_1 \sinh K_\parallel d_2}. \tag{37}$$

These lead to the solutions

$$\omega^{\pm}(K_{\parallel},q) = \omega_p[1 + \beta \mp (\beta^2-1)^{\frac{1}{2}}]^{-\frac{1}{2}}, \qquad (38)$$

i.e. there are two branches. The lower or acoustic branch corresponding to ω^- is just the analogue of that obtained in the sheets model. This can be seen by taking the limit $d_2 \to 0$ assuming a constant areal electron charge density n_o (so that $n_o = nd_2$). The correct limit follows on noting that

$$\lim_{d_2 \to 0}(\varepsilon_2 d_2) = (-n_o e^2/\varepsilon_o m^* \omega^2) \qquad (39)$$

where we have assumed $\gamma = 0$. This leads to the right hand side of eq. (23) from $\omega^-(K_{\parallel},q)$. The Green functions (29) - (31) reduce to (17) - (19) in this limit.

The occurrence of the upper branch or optic branch is associated with finite d_2. The dependence of $\omega^{\pm}$ on $K_{\parallel}$ is illustrated in figure 4 for fixed D and for various electron layer thicknesses. We have introduced the parameter δ by

$$\delta = d_2/D. \qquad (40)$$

All the other parameters are assigned in accordance with the Raman results by Olego et al.[3] on their sample 1, as in eq. (25).

We note from figure 4 that the acoustic plasmon curve for $\delta = 0.29$ (corresponding to sample 1) gives a much better agreement than does $\delta = 0.0$

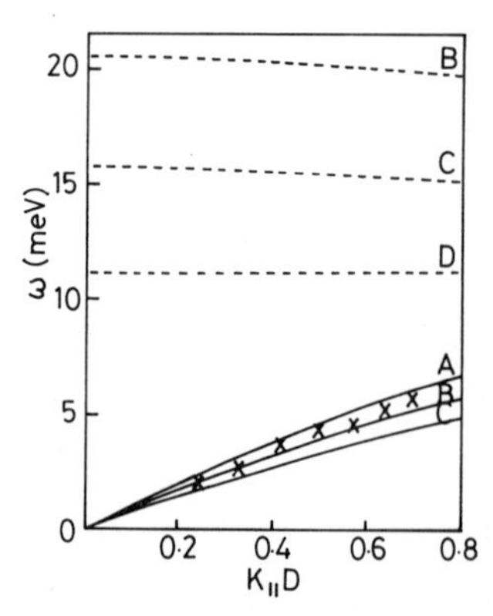

Figure 4. Theoretical dispersion curves (ω versus $K_{\parallel}D$ for fixed qD) for the acoustic branch and optic branch. The corresponding experimental results for $\delta = 0.29$ are shown by the crosses. The theory curves are given for the following values of δ: line A, 0.0; lines B, 0.29; lines C, 0.6 and line D, 1.0.

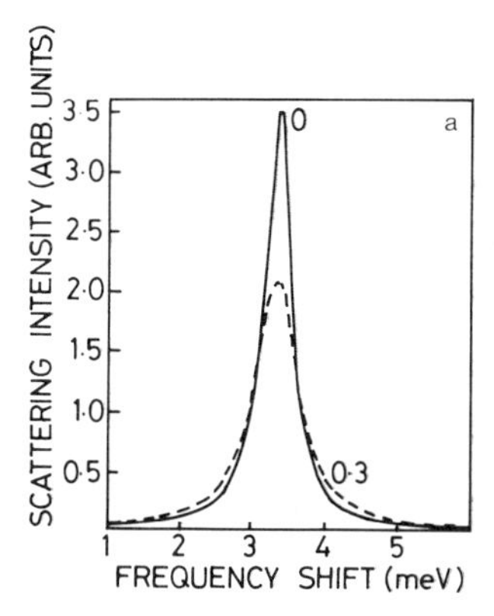

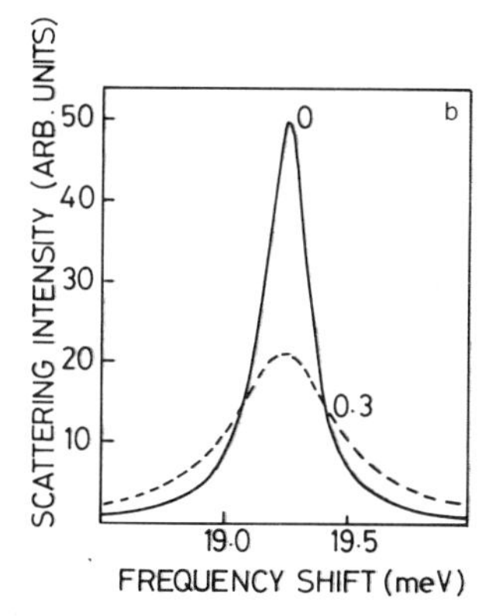

Figure 5. (a) Comparison between the theoretical spectra for the acoustic plasmon obtained with $\gamma = 0$ and $\gamma = 0.3$ meV for the case $\theta = 20^o$ in the 90° geometry appropriate for the Olego et al. experiment on their sample 1. (b) Comparison of theoretical spectra due to the optic plasmon with $\gamma = 0$ and $\gamma = 0.3$ for the case $\theta = 45^o$ in a 180° geometry. The parameters are as in (a).

corresponding to the sheets model. The predicted optic plasmon frequency is close to that of a peak at 22 meV. We argue later that this peak may have already been exhibited by the data due to Olego et al.[3] and Sooryakumar et al.[6]

Figure 5(a) shows the predicted spectral lineshapes for scattering from the acoustic plasmon for $\delta = 0.29$ and $\theta = 20^o$. The figure shows the cases with $\gamma = 0$ and $\gamma = 0.3$. The effects of including an intrinsic damping is to lower the peak height and broaden the spectrum; the width has been increased to about 0.8 meV for $\gamma = 0.3$ in close agreement with experiment. A value of γ between 0.1 and 0.3 meV due to collisonal damping has been quoted in earlier work as realistic for GaAs (e.g. refs. 11 and 12).

In the case of light scattering from the optic plasmon we have considered a 180° geometry (with $\theta = -\phi = 45^o$) on the same sample. The optic plasmon resonance occurs at $\hbar\omega \simeq 19.25$ meV as shown in figure 5(b). The effect of finite γ is seen to be more pronounced here.

The integrated light scattering intensity is in principle given by eq. (11) with appropriate substitutions for $V_{\mu\nu}$ and Δ. The result fortunately also simplifies in the non-retarded limit. The integrated intensities $I^{\pm}$ for Stokes scattering from the plasmon peaks at $\omega^{\pm}(K_{\|},K_z)$ are found to be

$$I^{\pm} \propto \frac{W[n(\omega^{\pm})+1]\,|S^{\pm}|}{\omega^{\pm}(\beta^2-1)^{\frac{1}{2}}\sinh K_{\|}d_1 \ \sinh K_{\|}d_2} \tag{41}$$

where

$$S^{\pm} = [\xi_1\sinh(K_{\|}d_2) - (\beta \pm (\beta^2-1)^{\frac{1}{2}})\xi_2\sinh(K_{\|}d_2)]/K_{\|}d_2 \tag{42}$$

with

$$\xi_j = \cosh(K_{\|}d_j) - \cos(K_z d_j). \tag{43}$$

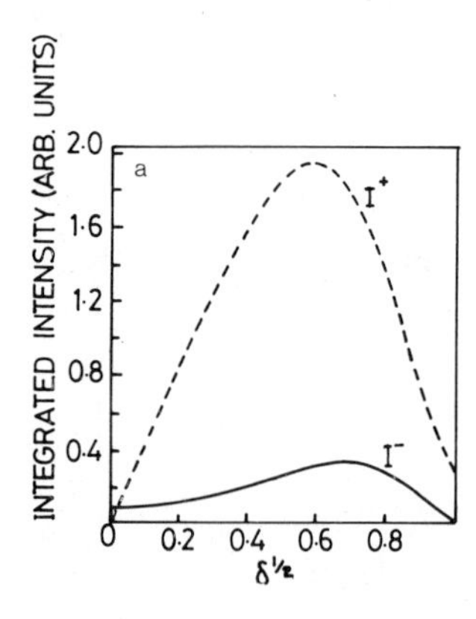

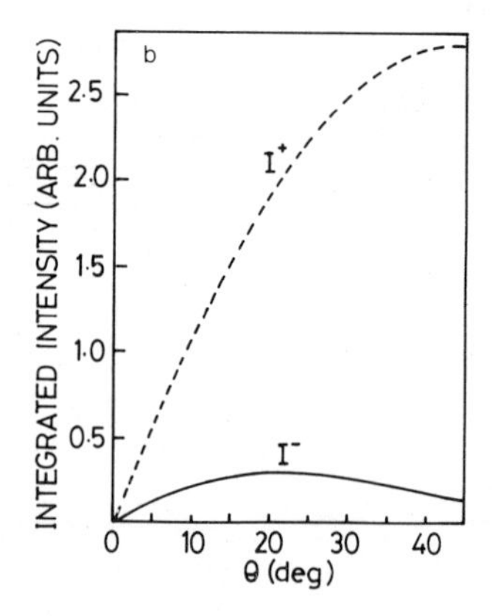

Figure 6. (a) Variations with $\delta^{\frac{1}{2}}$ in the integrated intensities of the acoustic and optic plasmon peak for $\theta = 20^{\circ}$ in the 90° geometry (b) Variation of the integrated intensities with θ for $\delta = 0.29$ and in the 90° geometry.

W is given by

$$W = \frac{K_{\parallel}^2[f^x K_z \cos\theta - f^z K_{\parallel} \sin\theta]^2 |g^{yy}|^2 \cos\phi}{(K_{\parallel}^2 + K_z^2)^2} . \tag{44}$$

For small $\delta \ll 1$ we find for the ratio I^+/I^-

$$\frac{I^+}{I^-} \simeq \left|\frac{n(\omega^+)+1}{n(\omega^-)+1}\right| \left|\frac{K_{\parallel} D \sinh K_{\parallel} D}{2\xi_1}\right|^{\frac{1}{2}} \left(\frac{K_z}{K_{\parallel}}\right)^2 \delta^{\frac{1}{2}} . \tag{45}$$

This shows that the integrated intensity I^+ of the optic plasmon is predicted to vanish as $\delta \to 0$. For non-zero δ I^+ is expected to become comparable with or even exceeds I^- because the ratio $(K_z/K_{\parallel})$ may be large. Figure 6(a) shows the variation of I^+ and I^- with $\delta^{\frac{1}{2}}$. We have assumed a 90° geometry with $\theta = 20^{\circ}$ and the same polarisations. Here the ratio $(K_z/K_{\parallel}) \simeq 11.0$ and this enhances the intensity ratio. Figure 6(b) shows the variation of I^+ and I^- with θ for $\delta = 0.29$.

In Table 1 we summarise the comparison between the predictions of the theory and experimental results from Olego et al.[3] on their sample 1 and from Sooryakumar et al.[6] The table contains data for the optic as well as the acoustic branch. In carrying out the theory for this table we have generalised the assumptions concerning the scattering geometry (and in particular, the polarisation directions relative to the crystal axis) to compare with the experimental situation. For further details see reference 31.

It can be seen from the table that the agreement between the theory and experiments is quite good for the acoustic branch as regards position of the plasmon peak. The agreement is not as good however as regards the optic plasmon energies. However the predictions of the intensity ratio is in close correspondence with both sets of measurements. This lends support to our interpretation of the observed high intensity peak in the observed

Table 1. Comparison of experiment and theory for the plasmon energies and relative Raman intensities

$K_{\parallel}D$	Source	Plasmon energy (meV) acoustic (ω^-)	optic (ω^+)	Intensity ratio I^+/I^-
0.3	Experiment (ref. 3)	3.4	16.0	7.3
	Theory	3.4	20.3	7.1
0.32	Experiment (ref. 6)	2.3	22.3	9.4
	Theory	2.6	20.0	9.3
0.47	Experiment (ref. 6)	3.6	22.3	6.8
	Theory	3.8	19.9	6.6
2.37	Experiment (ref. 6)	10.4	23.2	1.8
	Theory	10.2	17.5	2.2

data as the optic plasmon. It also confirms the usefulness of the linear response theory in exhibiting the characteristics of Raman measurements in a fairly straightforward manner.

5. CONCLUSIONS AND FURTHER COMMENTS

We have considered the linear-response theory of light scattering from plasma oscillations in GaAs/$Al_xGa_{1-x}As$ superlattices with particular reference to the experiments by Olego et al.[3] and Sooryakumar et al.[6] The theory was developed first for an idealised sheets model. This was shown to describe fairly well the relevant light scattering characteristics corresponding to the lowest plasmon branch observed experimentally. The corresponding theory for a model with finite electron layers was discussed in §4 and was shown to lead to better agreement with experiment as regards this branch. It also led to the prediction of a second (optic) branch which, we suggest, may have already been observed. Direct comparison with experimental results gives consistent agreement for the intensity ratio I^+/I^-. There are discrepancies only as regards the mode energy of the optic branch. This could be attributed to uncertainties in the parameters, and the possibility that the quantisation of the electron states in the GaAs quantum wells may lead to an effective bulk plasma frequency different from the one we have employed here, as discussed for example by Bloss.[26] However we maintain that such considerations should not produce much change in our prediction of the intensity ratio. An alternative explanation for the upper branch (put forward in ref. 6 on a qualitative basis) is that it may be due to coupling of LO phonons and plasmons.

The formalism can in principle be extended to the case of semi-infinite superlattices or to structures with a finite number of multilayers. The possibility of a surface mode in a truncated superlattice was considered by Giuliani and Quinn.[27] Constantinou and Cottam[28] considered a more general boundary condition on the superlattice uppermost layer. This was followed by essentially similar work due to Jain and Das Sarma.[29] As far as we know such surface modes have not yet been observed.

Finally it is worth mentioning that, due to improvement of resolution in both energy and wavevector measurements, plasma properties excitable by fast electrons may be investigated in the case of superlattices by electron loss spectroscopy. The method is readily applicable to metallic superlattices as discussed recently by Babiker.[30] It remains to be seen whether such a method is also suitable for probing plasmons in semiconductor superlattices.

REFERENCES

1. R. Dingle, L. Stormer, A.C. Gossard and W. Wiegmann, Appl. Phys. Lett. 33, 665 (1978).
2. E. Burstein, A. Pinczuk and S. Buchner, Physics of Semiconductors, ed. B.L.M. Wilson (IOP: London 1979) p. 1231.
3. D. Olego, A. Pinczuk, A.C. Gossard and W. Wiegmann, Phys. Rev. B25, 7867 (1982).
4. G. Fasol, N. Mestres, H.P. Hughes, A. Fischer and K. Ploog, Phys. Rev. Lett. 56, 2517 (1986).
5. A. Pinczuk, M.G. Lamont and A.C. Gossard, Phys. Rev. Lett. 56, 2092 (1986).
6. R. Sooryakumar, A. Pinczuk, A.C. Gossard and W. Wiegmann, Phys. Rev. B31, 2758 (1985).
7. A.L. Fetter, Ann. Phys. NY 88, 1 (1974).
8. S. Das Sarma and J.J. Quinn, Phys. Rev. B25, 7603 (1982).
9. W.L. Bloss and E.M. Brody, Solid State Commun. 43, 523 (1982).
10. P. Hawrylak, J.W. Wu and J.J. Quinn, Phys. Rev. B32, 5169 (1985).
11. J.K. Jain and P.B. Allen, Phys. Rev. Lett. 54, 947 and 2437 (1985).
12. J.K. Jain and P.B. Allen, Phys. Rev. B32, 997 (1985).
13. S. Katayama and T. Ando, J. Phys. Soc. Japan 54, 1615 (1985).
14. M. Babiker, N.C. Constantinou and M.G. Cottam, Solid State Commun. 57, 887 (1986).
15. M. Babiker, N.C. Constantinou and M.G. Cottam, J. Phys. C19, 5849 (1986).
16. M. Babiker, N.C. Constantinou and M.G. Cottam, Solid State Commun. 59, 751 (1986).
17. M. Babiker, N.C. Constantinou and M.G. Cottam, J. Phys. C20, 4581 (1987).
18. M. Babiker, N.C. Constantinou and M.G. Cottam, J. Phys. C20, 4597 (1987).
19. R.E. Camley and D.L. Mills, Phys. Rev. B29, 1695 (1984).
20. A.L. Wasserman and Y.I. Lee, Solid State Commun. 54, 855 (1985).
21. D.L. Mills, Y.C. Chen and E. Burstein, Phys. Rev. B13, 4419 (1976).
22. M.G. Cottam and A.A. Maradudin, in Surface Excitations, eds. V.M. Agranovich and R. Loudon (North Holland: Amsterdam 1984) Ch. 1.
23. W. Hayes and R. Loudon, Scattering of Light by Crystals (Wiley: NY 1978).
24. J.S. Blakemore, Appl. Phys. 53, R123 (1982).
25. B. Fischer, N. Marchall and H.J. Queisser, Surf. Sci. 34, 50 (1973).
26. W.L. Bloss, Solid State Commun. 48, 927 (1983).
27. G.F. Giuliani and J.J. Quinn, Phys. Rev. Lett. 51, 919 (1983) and Surf. Sci. 142, 433 (1984).
28. N.C. Constantinou and M.G. Cottam, J. Phys. C19, 739 (1986).
29. J.K. Jain and S. Das Sarma, Phys. Rev. B35, 918 (1987).
30. M. Babiker, J. Phys. C19, L773 (1986).
31. N.C. Constantinou, University of Essex Thesis, 1987.

INDEX

Acceptor(s), 1, 11, 12, 14, 15, 16, 17, 18, 44, 138, 141 144, 159, 162, 163, 164, 165, 167, 172, 199, 207, 208, 210, 211, 219, 226, 227, 231, 241, 245, 271, 272, 273, 277, 279, 280, 281, 282, 333
  atoms, 157
  band(s), 166, 167
  binding energies (energy), 14, 161, 162, 211, 272, 279, 280, 281, 282
  concentration(s), 13, 15, 16, 161, 162, 163, 223
  energies, 15, 162
  impurity, 207
  impurity band, 162, 216
  ionization energies, 161, 162
  levels, 195, 215
  state(s), 15, 157, 162, 164, 198, 207, 222, 272, 279, 281
  wave functions, 162, 163, 281
AlGaAs/GaAs(Al_xGa_{1-x}/GaAs), 23, 135, 139, 140, 141, 143, 144, 168, 220, 222, 223, 226, 275, 277, 281
  n-, 136
Auger, 47, 48

Back-Gate voltage, 297, 298, 299, 300, 301, 302, 303, 304
Band
  lineup(s), 119, 190, 326, 330
  offset, 36, 80, 83, 108, 113, 114, 115, 118, 119, 171, 180, 223, 319, 320, 327, 330
Beryllium (Be), 13, 19, 20, 21, 22, 23, 25, 26
Bound state, 13

Capture, 15, 83, 110, 113, 122, 126, 133, 163, 229, 235, 236, 237
  cross section, 83, 110, 113, 114, 116, 122, 132
  efficiency, 11
Chemical characterization, 63, 66
Confined states, 319
  QW state, 12
Coulomb interaction(s), 12, 85, 86, 96, 105, 128, 157, 245, 246, 247, 250,

Cryogenic temperature, 135
Cyclotron
  resonance, 58, 59, 224, 304, 307, 308, 312
  (resonance) (effective) mass, 224, 303, 304, 307, 308, 309, 310, 311, 312, 313, 315, 316, 317
  (resonance) frequency, 123, 308, 309, 310, 315

Deep donor level, 122, 126, 127
Deep impurities (impurity), 135, 175, 176, 178, 179
Deep impurity levels, 175
Deep level(s), 77, 78, 80, 82, 87, 107, 110, 114, 127, 129, 131, 137, 138, 139, 140, 143, 147, 175, 176, 178, 179, 180, 182, 185
Deep level transient spectroscopy (DLTS), 82, 84, 107, 108, 110, 111, 112, 113, 115, 116, 117, 118, 122, 128, 132, 139, 140, 143
Deep trap(s), 110, 180
Defect levels, 3
δ-doped, 13, 144, 163, 168
  layers, 160
δ-doping, 3, 144, 168, 330
  layer, 1, 2, 3
δ-layer, 1, 2, 3, 4, 6, 7, 9
δ(z)-confinement, 3
Density of states, 52, 97, 98, 111, 159, 161, 162, 167, 169, 170, 172, 223
Diamagnetic shift, 285, 287, 288, 289, 290, 291, 293, 294
Dipole(s), 13, 147, 148, 152
  matrix elements, 162
Donor(s), 1, 2, 3, 12, 13, 58, 85, 87, 92, 94, 96, 99, 109, 121, 122, 126, 129, 132, 138, 139, 141, 142, 144, 152, 159, 163, 164, 165, 166, 167, 168, 170, 171, 172, 179, 198, 202, 203, 204, 210, 219, 226, 233, 245, 256, 333
  atoms, 157
  band(s), 166, 170, 171
  binding energies (energy), 13, 162, 201

Donor(s) (continued)
concentration, 163, 164
doping level, 129
electron, 150
impurities (impurity), 122, 152, 199, 204, 257
impurity atom, 131
impurity band(s), 172, 214
impurity level, 109
impurity potentials, 162
layers, 143, 165
level(s), 109, 126, 127, 128, 131, 195, 201
potentials, 162
state(s), 3, 126, 147, 148, 150, 153, 154, 155, 157, 168, 169, 195, 196, 199, 207, 222
wave function, 163, 202, 204
Doping modulation, 9, 320, 333
Doping superlattice(s), 159, 160, 161, 166, 212, 214, 215, 216, 217, 240, 241
D°X, 45
DX, 128, 132
center(s), 7, 8, 32, 77, 135, 138, 142, 143, 144, 178
centre, 121, 122, 123, 126, 127, 131, 132
donor levels, 132
donor states, 126
level(s), 8, 121, 123, 127, 128, 130, 131, 132
state(s), 121, 132, 133
traps, 7

Effective mass approximation(s) (EMA), 147, 152, 157, 195, 196, 214, 217, 330
Effective mass theory, 33, 175, 195, 196, 198, 212, 219, 220, 221, 226, 227
Electrical characterization, 107, 118, 230
Electron-hole recombination, 32, 159, 216
Electronic band structure, 197
Electronic structure(s), 33, 152, 159, 160, 246, 249, 255, 321, 325, 326, 330
Electronic subband(s), 212, 213, 217
Electronic transport, 118
Envelope (wave) function(s), 11, 12, 15, 161, 172, 176, 196, 199, 207, 214, 220, 227, 278, 279
Exciton(s), 32, 34, 43, 44, 45, 165, 219, 227, 231, 233, 241, 242, 271, 272, 273, 275, 277, 278, 279, 281, 285, 286, 287, 288, 290, 293, 294
binding energy (energies), 14, 222, 286, 288, 289, 294
Exciton (excitonic) recombination, 11, 32, 242

Fibonacci superlattices, 29, 30, 31, 37, 38, 39
Field-effect transistor (FET), 135, 139, 246, 271
heterostructure (HFET), 135, 136, 137, 138, 139, 140, 141, 142, 143, 144
Frohlich-coupling, 307

Gallium Arsenide (GaAs), 1, 2, 3, 4, 14, 16, 17, 18, 19, 20, 21, 22, 25, 26, 30, 31, 32, 33, 35, 36, 37, 38, 69, 77, 80, 82, 99, 108, 109, 110, 113, 115, 116, 121, 122, 128, 131, 132, 135, 136, 137, 138, 140, 141, 142, 144, 150, 151, 152, 161, 164, 166, 169, 179, 180, 181, 182, 183, 184, 202, 203, 204, 205, 206, 207, 208, 210, 211, 215, 216, 217, 219, 220, 222, 223, 230, 231, 236, 243, 245, 252, 256, 257, 258, 259, 260, 261, 262, 263, 264, 265, 271, 273, 275, 278, 280, 281, 285, 286, 289, 290, 308, 333, 335, 337, 340, 345
GaAs/AlGaAs (GaAlAs), 7, 229, 245, 249, 255, 265
GaAs/$Al_xGa_{1-x}As$($Ga_{1-x}Al_xAs$/GaAs) (GaAs-GaAlAs)(Al_xGa_{1-x}/GaAs)
heterojunctions, 85, 86, 102, 118, 297, 298, 299
heterostructures, 33, 223, 224, 226, 297, 299, 303, 304, 307, 308, 310, 312, 313, 315
interfaces, 108, 118, 119, 184
superlattice(s) (SL), 25, 77, 84, 107, 112, 117, 119, 142, 180, 181, 182, 183, 184, 317, 333, 334, 345
GaAs-(GaAl)As, 11, 31, 32, 33, 54, 82, 84, 108, 229, 245, 255, 285
$Ga_{1-x}In_xAs$-InP superlattices, 43, 45, 46, 47, 48, 50, 51, 53, 54, 55, 57

Hall
coefficient, R_H 88
effect measurements, 96, 249, 250, 253
experiments, 21
measurements, 20, 60, 131, 142, 252, 253, 273, 301
mobility, 20, 223, 224
plateau(s), 58, 297, 299, 300
resistance, 7, 297, 298, 300, 301, 303
resistivity, 96, 99, 298
Heterojunction bipolar transistor (HBT), 19, 23, 24, 25, 26
Heterostructures, 14, 29, 30, 32, 33, 37, 45, 77, 78, 80, 81, 82, 89, 139, 147, 219, 220, 222, 225, 227, 256, 260, 297, 320, 321, 325
Hole
heavy (HH), 13, 15, 52, 60, 162, 170, 171, 198, 207, 210, 211, 220, 271, 279, 280, 281, 285, 286, 287, 290

Hole (continued)
291, 292, 293, 294
heavy exciton, 287, 288, 289, 290, 291, 293
light, 13, 52, 162, 198, 207, 210, 220, 279, 280, 281, 285, 286, 287, 290, 292, 293
light exciton, 287, 288, 290
state(s), 154, 157, 292
subband(s), 216, 222, 291, 292, 293, 294
Hot electron(s), 229, 230, 235, 236, 237, 239, 240, 242, 243, 245
Hubbard (repulsive) correlation energy, U_H 169, 170
Hydrostatic pressure, 7, 85, 86, 87, 89, 102, 121, 122, 127, 128

Impurity (impurities), 1, 9, 11, 12, 13, 15, 16, 17, 19, 21, 26, 34, 36, 46, 58, 77, 82, 94, 98, 118, 126, 142, 147, 149, 150, 157, 159, 160, 163, 164, 167, 168, 170, 172, 175, 176, 178, 179, 180, 184, 185, 196, 200, 202, 204, 206, 213, 219, 220, 222, 226, 227, 231, 232, 236, 246, 255, 257, 259, 271, 277, 279, 297, 308, 333
-assisted scattering, 257
-assisted tunneling, 259
band(s), 149, 152, 159, 162, 165, 166, 167, 169, 170, 171, 172, 330
concentration(s), 14, 20, 21, 22, 164, 166, 172, 212, 277
level(s), 21, 119, 160, 175, 189, 190, 204
pair(s), 165, 172
state(s), 11, 13, 147, 153, 159, 160, 166, 167, 200, 217, 219, 226, 227, 228, 321
wave functions, 12, 197, 199, 200, 205
$In_xGa_{1-x}As/GaAs$, 271, 272, 273, 278, 282
Interface, 11, 12, 14, 15, 16, 17, 18, 21, 22, 32, 33, 43, 45, 46, 47, 48, 59, 60, 63, 66, 67, 68, 69, 71, 72, 73, 82, 84, 92, 107, 108, 110, 117, 118, 119, 147, 148, 149, 152, 154, 157, 180, 182, 220, 222, 225, 229, 249, 251, 271, 275, 277, 285, 287, 288, 290, 292, 294, 301, 302, 304, 334, 337, 338, 341
state(s), 98, 118, 154, 159, 195
Irradiation-induced defects, 108, 113, 119

k·p
coupling, 222, 224, 226
interaction, 219
perturbation method, 198
theory, 315

Landau level(s), 86, 99, 102, 219, 222, 286, 298, 304, 308, 309, 312, 315, 316

Lineshape(s), 11, 13, 14, 15, 16, 18, 297, 298, 334, 339, 343
Linhard
function (formula), 245, 246, 252
response function, 250, 253
Localization, 12, 39, 78, 85, 86, 88, 92, 97, 100, 101, 105, 152, 170, 268

Magnetic freeze-out, 85, 90, 92, 104
Magneto
-donor(s), 86, 88, 92, 94, 95, 96, 102, 104
-optics, 285
resistance(s), 4, 5, 6, 7, 8, 88, 137, 299, 300, 301, 303
resistivity, 298
transport, 4, 7
Many-particle
character, 308
effects, 308, 313
Metal non-metal transition (MNMT), 85, 89, 90, 91, 92, 93, 94, 96
Metal-insulator transition, 7, 109
Metalorganic chemical vapor deposition (MOCVD), 1, 29, 44, 48, 51, 52, 85, 195
low pressure (LPMOCVD), 43, 44, 45, 46, 47, 48, 49, 53, 54, 55, 56, 57, 58, 59, 60
Microfabrication, 255, 256, 268
Microwave, 54, 135
Modulation
doped, 11, 223, 229, 231, 233, 249, 333
(doped) heterojunctions (MDH), 11, 45, 58, 245
(doped) heterostructures 43, 219, 221, 222
(doped) (multiple) quantum well(s), 245, 246, 248, 249, 251, 252, 253
Molecular beam epitaxy (MBE), 1, 2, 11, 14, 19, 20, 21, 22, 26, 29, 30, 33, 39, 67, 68, 85, 107, 108, 117, 119, 121, 122, 135, 195, 223, 230, 255, 256, 257, 265, 271, 286, 299, 325, 333
Monolayer superlattice(s), 29, 32, 33
Mott transition, 94
Mott-Hubbard transition (Hubbard-Mott transition), 159, 160, 167, 169, 170, 172
Multiple (multi) quantum well(s) (MQW), 45, 52, 53, 54, 108, 109, 200, 204, 214, 243, 245, 246, 249, 250, 274, 286

Negative differential resistance (NDR), 43, 54, 57, 60, 229, 235, 236, 237, 256, 257, 259, 266
n-i-p-i (nipi), 152, 157, 165, 166, 167, 168, 169, 171, 172
crystal(s), 160, 162, 166, 195, 212, 214

Negative differential resistance (NDR) (continued)
doping superlattice(s), 159, 162, 163, 172
structure(s), 160, 163, 164, 166, 167, 172

Persistent photoconductivity (PPC), 121, 122, 123, 126, 132, 297, 298, 300
Photoemission, 141, 230
Photoluminescence (PL), 11, 13, 14, 16, 17, 34, 35, 36, 43, 44, 52, 53, 54, 55, 114, 115, 163, 167, 217, 229, 230, 231, 232, 233, 234, 236, 237, 238, 239, 240, 241, 242, 271 272, 273, 274, 275, 276, 277, 278 279, 281, 282, 286, 333, 340, 342 343, 345, 346
Plasmon(s), 245, 246, 247, 248, 249, 250, 251, 253, 334, 339, 340, 344
(resonance) frequency, 246, 343, 345
peak, 334, 339, 343, 344, 345
Polariton(s), 333
exciton, 219
Polaron(s), 45, 307, 308, 310, 312, 313, 317
effect(s), 308, 310, 312, 315, 317
mass, 312, 314, 315

Quantum Hall (QH)
effect (QHE), 7, 58, 59, 60, 85, 86, 96, 97, 99, 102, 103, 104, 105, 297
regium, 96, 102, 297, 304
resistance, 299, 300, 301
Quantum well(s) (QW), 11, 12, 13, 14, 15, 16, 17, 18, 29, 32, 33, 35, 36, 43, 52, 54, 60, 78, 82, 98, 99, 107, 141, 147, 159, 160, 169, 171, 172, 180, 181, 195, 199, 200, 202, 205, 207, 208, 209, 210, 211, 217, 219, 220, 221, 222, 226, 227, 228, 229, 230, 231, 232, 233, 235, 237, 240, 241, 243, 245, 246, 247, 248, 249, 250, 253, 255, 256, 257, 258, 259, 260, 261, 262, 263, 264, 265, 271, 272, 275, 277, 278, 281, 285, 286, 287, 288, 290, 291, 292, 293, 294, 320, 333, 345
heterostructure, 29, 33
states, 194, 255, 256, 259, 260, 263
superlattices, 29
Quantum wire(s), 265, 266, 267, 268

Raman
intensity, 247, 249, 259, 345
measurements, 217, 246, 249, 333, 334
peak(s), 250, 333, 334, 337
results, 342
scattering, 37, 39, 245, 246, 247, 248, 249, 253, 333, 334, 335

Raman (continued)
spectra, 39, 247, 248, 249, 250, 252, 253
Random GaAs/AlAs superlattices, 29, 39
Resonant tunneling, 25, 43, 54, 56, 255, 256, 257, 259, 260, 262, 265, 266, 268

Screening, 86, 92, 94, 95, 102, 128, 185, 248, 307, 308, 310, 312, 315, 317
Secondary ion mass spectrometry (SIMS), 20, 22, 23, 46, 47
Semiconductor
heterostructure(s), 307
interface(s), 63, 71
superlattice(s), 37, 319, 320, 346
Shallow acceptor, 11
Shallow donor, 11, 32, 179, 182
Shallow donor level(s), 132, 176, 179
Shallow donor states, 128, 132
Shallow hydrogenic impurities, 147
Shallow impurity (impurities), 11, 18, 135, 159, 175, 178, 179, 195, 272, 280
Shallow impurity level(s), 178
Shallow impurity states, 147, 330
Shallow level(s), 19, 147, 175, 176, 177, 178, 179, 180
Shubnikov-de Haas
amplitude, 299, 304
effect, 6, 96, 123, 297
experiments, 249
measurements, 121, 123, 131, 249, 250, 252, 253, 303, 304
oscillations, 4, 54, 57, 58, 123, 124, 249, 297, 298, 300, 303, 304
resistance 299, 300, 301, 303
structure, 123, 126
Si
deep level, 180
donor(s), 1, 3, 85, 86, 88, 92, 96, 102, 132, 152, 301, 302, 333
donor state(s), 150, 152
/Ge(Si_n/Ge_n), 152, 153, 154, 157, 320, 322, 324, 325, 326, 327, 328, 330
/Ge($(Si)_n/(Ge)_n$) heterostructure(s), 320
/Ge($(Si)_n/(Ge)_n$)(SiGe) superlattices, 152, 153, 154, 157, 319, 320, 321, 322, 323, 330
impurity, 149, 150, 151
impurity band, 149, 150
Spacer, 25, 26, 33, 85, 86, 90, 91, 92, 98, 102, 104, 135, 136, 222, 223, 245, 257, 258, 259
Spin density fluctuations, 248, 249, 252, 253
Stability of superlattices, 324
Staggered band alignment, 152, 154, 157
Strain, 67, 272, 277, 278, 279, 280, 281, 282, 320, 325, 327

Strain(continued)
  energy, 323
  field, 176
Strained layer(s), 272, 279, 281, 320
Strained-layer superlattices (SLS), 271, 273, 275, 281, 282
Strained quantum well(s), 271, 272, 278, 279, 282
Strained SiGe, 319, 321, 324
Strained Si/Ge superlattice(s), 153, 323, 330
Strained superlattice(s), 271, 272, 278, 319, 323
Strained superlattice semiconductor, 152
Structural characterization, 63, 66, 69, 74
Subband(s), 2, 3, 4, 6, 13, 35, 54, 58, 59, 160, 161, 162, 163, 167, 171, 207, 214, 216, 217, 219, 223, 224, 225, 226, 227, 228, 248, 249, 250, 251, 255, 285, 286, 292, 293, 294, 304, 309, 314
  energies (energy), 3, 222, 290, 315
  levels, 6
  masses, 222
  states, 1, 15, 222, 225, 226
  structure 195
Superlattice(s) (SL), 11, 16, 29, 30, 31, 32, 33, 34, 35, 36, 37, 39, 43, 45, 46, 47, 48, 50, 54, 55, 57, 60, 77, 78, 82, 83, 84, 107, 108, 109, 110, 112, 113, 114, 115, 116, 117, 118, 119, 135, 140, 141, 142, 143, 144, 147, 148, 149, 150, 151, 159, 160, 164, 168, 175, 180, 181, 182, 183, 185, 195, 206, 213, 219, 262, 285, 319, 320, 321, 322, 323, 324, 326, 327, 330, 333, 334, 335, 337, 340, 341, 345, 346
  crystal, 33
  semiconductors, 147, 195, 217

Theory of impurity states, 195
Thue-Morse
  sequence, 30, 37, 38
  superlattices, 29, 30, 31, 37, 38, 39
Transfer of electron, 96, 97, 222
Transition-metal (TM) impurity level(s), 189, 190
Transmission electron microscopy (TEM), 43, 48, 50, 56, 64
  high resolution (HRTEM), 63, 64, 66, 67, 68, 71
Tunneling, 3, 4, 54, 57, 60, 96, 98, 107, 108, 138, 170, 171, 225, 255, 256, 257, 262, 263, 265
2D Electron Gas (2DEG), 33, 54, 60, 85, 86, 87, 96, 97, 98, 135, 137, 138, 140, 141, 142, 144, 245, 246, 297, 298, 300, 301, 304, 307, 308, 310, 312, 317

Ultrathin-layer ($(GaAs)_m(AlAs)_n$) superlattices (UTLS) 29, 31, 32, 33, 34, 35, 36, 37

Zone folding 33, 39, 321, 326, 327